OCR GCSE Mathematics A

Foundation Student's Book

Series Editor: Brian Seager

- **Howard Baxter**
- **Mike Handbury**
- **Jean Matthews**
- **Colin White**
- **Ruth Crookes**
- **John Jeskins**
- **Brian Seager**

DYNAMIC LEARNING

HODDER EDUCATION
AN HACHETTE UK COMPANY

The Publishers would like to thank the following for permission to reproduce copyright material:
Photo credits p 146 © Stuart Walker / Alamy; p 227 © Daniel Bosworth / Britain On View
/ VisitBritain; p 336 © Don Hammond / Design Pics Inc. / Rex Features; p 475 © Simon Stacpoole /
Offside / Rex Features; p 594 © Yoshikazu Tsuno / AFP / Getty Images; p 629 © free photo /
Fotolia.com; p 642 © Dr Juerg Alean / Science Photo Library

Acknowledgements
Every effort has been made to trace all copyright holders, but if any have been inadvertently overlooked
the Publishers will be pleased to make the necessary arrangements at the first opportunity.

Although every effort has been made to ensure that website addresses are correct at time of going to press,
Hodder Education cannot be held responsible for the content of any website mentioned in this book. It is
sometimes possible to find a relocated web page by typing in the address of the home page for a website
in the URL window of your browser.

Hachette UK's policy is to use papers that are natural, renewable and recyclable products and made from
wood grown in sustainable forests. The logging and manufacturing processes are expected to conform to
the environmental regulations of the country of origin.

Orders: please contact Bookpoint Ltd, 130 Milton Park, Abingdon, Oxon OX14 4SB.
Telephone: (44) 01235 827720. Fax: (44) 01235 400454. Lines are open 9.00 – 5.00, Monday to Saturday,
with a 24-hour message answering service. Visit our website at www.hoddereducation.co.uk

© Howard Baxter, Ruth Crookes, Mike Handbury, John Jeskins, Jean Matthews, Mark Patmore,
Brian Seager, Eddie Wilde, Colin White 2010
First published in 2010 by
Hodder Education
An Hachette UK Company
338 Euston Road
London NW1 3BH

Impression number   5 4 3 2 1
Year   2015 2014 2013 2012 2011 2010

Cover illustration by Oxford Design & Illustrators
Illustrations by Tech-Set Ltd., Gateshead
Typeset in 12/13 Bembo by Tech-Set Ltd., Gateshead, Tyne & Wear
Printed in Italy

A catalogue record for this title is available from the British Library

ISBN: 978 1444 112 801

Introduction

About this book

This book covers the complete specification for the Foundation Tier of GCSE Mathematics. It has been written especially for students following OCR's 2010 Specification GCSE Mathematics A (J562). All three units are covered in the book.

There are three important elements in the new specifications: **problem solving**, elements of **functional mathematics** and **quality of written communication**. All of these are covered in this book.

- Each chapter is presented in a way which will help you to understand the mathematics, with straightforward explanations and worked examples covering every type of problem. These are models of **written communication** for you to follow.
- At the start of each chapter are two lists, one of what you should already know before you begin and the other of the topics you will be learning about in that chapter.
- There are **Check ups** to check understanding of work already covered.
- **Discoveries** encourage you to find out something for yourself, either from an external source such as the internet, or through a guided activity.
- **Challenges** are rather more searching and are designed to make you think mathematically and enhance your **problem-solving** skills.
- There are plenty of exercises to work through to practise your skills. Many of the questions require **problem solving**.

  These are indicated by this icon.
- Many exercises contain questions which involve elements of **functional** mathematics.
- Most of the questions in Unit B are designed to be done without a calculator so that you can practise for this non-calculator paper.
- Look out for the **Tips**. These give advice on how to improve examination performance, direct from the experienced examiners who have written this book.
- At the end of each chapter there is a short **summary** of what you have learned.
- Finally, there is a **Mixed exercise** to help you revise all the topics covered in that chapter.

Other components in the series

- **A Homework Book**
  This contains parallel exercises to those in this book to give you more practice.
- **Dynamic Learning online**
  This offers interactive online assessment and access to audio-visual worked examples to help you to understand fully key concepts. Access to online versions of the textbooks and homework books is also available.

Top ten tips

Here are some general tips from the examiners who wrote this book to help you to do well in your tests and examinations.

Practise
1 **taking time** to work through each question carefully.
2 answering questions **without** a calculator.
3 answering questions which require **explanations**.
4 your **problem-solving** skills.
5 **accurate** drawing and construction.
6 answering questions which **need a calculator**, trying to use it efficiently.
7 **checking answers**, especially for reasonable size and degree of accuracy.
8 making your work **concise** and well laid out.
9 checking that you have **answered the question**.
10 **rounding** numbers, but only at the appropriate stage.

Unit A Contents

Integers

> ▶ **This chapter is about**

- Adding, subtracting, multiplying and dividing integers
- Multiples and factors
- Rounding numbers to the nearest 10, 100, 1000, …
- Squares and cubes
- Other powers
- Square roots
- Negative numbers

> ▶ **You should already know**

- An integer is a whole number, for example 7, 18 or 253
- How to do simple additions, subtractions, multiplications and divisions
- Your multiplication tables up to the 10 times table

Arithmetic check

When you are doing calculations, write the numbers in columns: units under units, tens under tens, and so on.

Example 1.1

Work out these.

(a) $46 + 32$ **(b)** $78 - 32$ **(c)** $38 + 126$ **(d)** $164 - 38$

Solution

(a)
$$\begin{array}{r} 4\ 6 \\ +\ 3\ 2 \\ \hline 7\ 8 \end{array}$$
Simply add the digits in each column.
$6 + 2 = 8$ and $4 + 3 = 7$.

(b)
$$\begin{array}{r} 7\ 8 \\ -\ 3\ 2 \\ \hline 4\ 6 \end{array}$$
Simply subtract the digits in each column.
$8 - 2 = 6$ and $7 - 3 = 4$.

(c)
$$\begin{array}{r} 3\ 8 \\ +1\ 2\ 6 \\ \hline 1\ 6\ 4 \\ {\scriptstyle 1} \end{array}$$
$8 + 6 = 14$.
You write 4 in the units column and a small 1 at the bottom to show you are carrying 1 'ten' over from the units column to the tens column.
$3 + 2 = 5; 5 + 1$ carried over $= 6$.

(d)
$$\begin{array}{r} 1\ {}^5\!\!\not{6}\ {}^1\!4 \\ -\ \ 3\ 8 \\ \hline 1\ 2\ 6 \end{array}$$
You cannot take 8 from 4 so you change the 6 tens into 5 tens and 10 units.
$14 - 8 = 6$ and $5 - 3 = 2$.

● Challenge 1.1

(a) Look again at Example 1.1.
   (i) What other calculation can be made using the three numbers 32, 46 and 78?
   (ii) What other calculation can be made using the three numbers 38, 126 and 164?

(b) $56 + 79 = 135$. Write down two other calculations that can be made using these numbers.

(c) Write down three calculations that can be made using the numbers 78, 83 and 161.

Example 1.2

Work out these.

(a) 32×3 (b) $96 \div 3$

(c) 18×7 (d) $126 \div 7$

Solution

(a)
$$\begin{array}{r} 3\ 2 \\ \times \quad 3 \\ \hline 9\ 6 \end{array}$$
You multiply first the units and then the tens by 3.
$2 \times 3 = 6$ and $3 \times 3 = 9$.

(b)
$$\begin{array}{r} 3\ 2 \\ 3\overline{)9\ 6} \end{array}$$
You divide first the tens and then the units by 3.
$9 \div 3 = 3$ and $6 \div 3 = 2$.

(c)
$$\begin{array}{r} 1\ 8 \\ \times \quad 7 \\ \hline 1\ 2\ 6 \\ {\scriptstyle 5} \end{array}$$
$8 \times 7 = 56$.
You write the 6 in the units column and carry 5 tens over.
$1 \times 7 = 7; 7 + 5$ carried over $= 12$.

(d)
$$\begin{array}{r} 1\ 8 \\ 7\overline{)12^5 6} \end{array}$$
7 into 1 does not go so you look at the next digit.
7 into 12 is 1 remainder 5.
Look at the remainder together with the next digit.
7 into 56 is 8.

TIP

$21 \div 7$, $7\overline{)21}$ and $\frac{21}{7}$ all mean 21 divided by 7.

Challenge 1.2

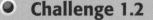

(a) Look again at Example 1.2.
   (i) What other calculation can be made using the three numbers 3, 32 and 96?
   (ii) What other calculation can be made using the three numbers 7, 18 and 126?

(b) $65 \times 6 = 390$. Write down two other calculations that can be made using these numbers.

(c) Write down three calculations that can be made using the numbers 43, 20 and 860.

Exercise 1.1

1 Work out these.
   (a) $46 + 53$ (b) $54 + 37$ (c) $78 + 46$
   (d) $158 + 23$ (e) $136 + 282$ (f) $264 + 189$

2 Work out these.
   (a) $96 - 55$ (b) $64 - 27$ (c) $75 - 28$
   (d) $147 - 53$ (e) $236 - 129$ (f) $562 - 286$

3 Work out these.
   (a) 23×3 (b) 19×4 (c) 36×5
   (d) 68×7 (e) 123×6 (f) 262×4

4 Work out these.
   (a) $84 \div 4$ (b) $72 \div 3$ (c) $75 \div 5$
   (d) $91 \div 7$ (e) $144 \div 6$ (f) $184 \div 4$

5 Jamie bought a CD for £14, a pair of trainers for £38 and a ticket for a football match for £17. What was the total cost?

6 Jatindar was given £80 for her birthday. She bought some clothes for £53. How much did she have left?

7 Emma bought six packets of biscuits at 46p each. What was the total cost in pence? How much is the total in £s?

8 A school has £182 to spend on books. The books they want to buy cost £7 each. How many books can they buy?

Multiples

The numbers in the five times table are $5, 10, 15, 20, 25, \ldots$.

$5, 10, 15, 20, 25, \ldots$ are called **multiples** of 5.

You should know your five times table up to '12 fives are 60' but the multiples of five do not stop at 60. They go on $65, 70, 75, \ldots$. In fact there is no end to the list of multiples.

Example 1.3

(a) List the multiples of 2 that are less than 35.
(b) List the multiples of 6 that are less than 100.
(c) List the multiples of 9 that are less than 100.

Solution

You list the 2, the 6 and the 9 times tables and carry on until you get to 35 or 100, as instructed.

(a) $2, 4, 6, 8, 10, 12, 14, 16, 18, 20, 22, 24, 26, 28, 30, 32, 34$
(b) $6, 12, 18, 24, 30, 36, 42, 48, 54, 60, 66, 72, 78, 84, 90, 96$
(c) $9, 18, 27, 36, 45, 54, 63, 72, 81, 90, 99$

The multiples of two are also called the **even numbers**.
Notice that they all end in 0, 2, 4, 6 or 8.
So 1398 is an even number because it ends in 8.

All the other integers, $1, 3, 5, 7, 9, 11, 13, 15, 17, 19, 21, 23, \ldots$ are called **odd numbers.**
Notice that they all end in 1, 3, 5, 7 or 9.
So 6847 is an odd number because it ends in 7.

Look again at Example 1.3.

Notice that 18, 36, 54, 72 and 90 are in the list of multiples for both 6 and 9.

18, 36, 54, 72 and 90 are called **common multiples** of 6 and 9 because they are in, or common to, both lists.

Factors

◉ Discovery 1.1

There are 70 sweets in a jar.

(a) Can the sweets be shared equally between three people?

(b) Find all the numbers of people the sweets can be shared between. How many do they each receive?

A number that will divide into a number exactly is called a **factor** of the number.

For example,   2 is a factor of 8,
7 is a factor of 21,
10 is a factor of 100,
1 is a factor of 6,
9 is a factor of 9.

Notice that 1 is a factor of every number and every number is a factor of itself.

🔍 Check up 1.1

Find all the factors of these numbers.

(a) 12   (b) 25   (c) 48   (d) 100

The factors of 30 are 1, 2, 3, 5, 6, 10, 15 and 30.
The factors of 50 are 1, 2, 5, 10, 25 and 50.

Notice that 1, 2, 5 and 10 are in both lists of factors.

1, 2, 5 and 10 are called the **common factors** of 30 and 50 because they are common to both lists.

TIP

Once you have gone beyond half the number, there will be no new factors except the number itself.

◉ Challenge 1.3

One light flashes every 25 seconds.

Another light flashes every 30 seconds. At a certain time they flash together.

How many seconds will it be before they flash together again?

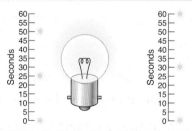

Rounding numbers

The Earth is approximately a sphere.
It is not exact however and the radius varies from
about 6356 km at the North Pole to 6378 km
at the Equator.
The average radius is 6367 km.

Since it varies so much, an
approximate value for the radius will
do for most calculations.

6400 km is the usual approximation.
It is accurate to the nearest 100 km.

For very large numbers it is usual to approximate to the nearest
hundred, thousand, ten thousand, etc.

The distance from the Earth to the Sun varies but it is usually given as
93 million miles, that is 93 000 000, to the nearest million.

This headline appeared in a newspaper.

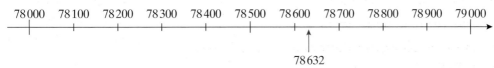

Local Man Wins £79 000!

This does not necessarily mean that the man won exactly £79 000.

The actual prize may have been £78 632 but the headline makes more impact if it is rounded
to the nearest thousand.

Counting in thousands, 78 632 is between 78 000 and 79 000.

```
78 000   78 100   78 200   78 300   78 400   78 500   78 600   78 700   78 800   78 900   79 000
  |--------|--------|--------|--------|--------|--------|---↑----|--------|--------|--------|---→
                                                      78 632
```

It is nearer to 79 000.
So 78 632 to the nearest thousand is 79 000.

Here is a quick method of rounding to the nearest thousand.

Step 1:   Put a ring round the thousands digit. For example, 7⑧632.

Step 2:   Look at the next digit to the right.
          If it is less than 5, leave the thousands digit as it is.
          If it is 5 or more, add 1 to the thousands digit.

Step 3:   Replace the remaining digits by zeros. For the example above, 79 000.

A similar method can be used to round to the nearest 100, 10 000, etc.

Example 1.4

(a) Round 45 240 to the nearest 100.

(b) Round 458 000 to the nearest 10 000.

(c) Round 6375 to the nearest 10.

Solution

(a) 45②40   2 is the 100s digit.
    45 200   4 is less than 5.

(b) 4⑤8 000   5 is the 10 000s digit.
    460 000   8 is greater than 5.

(c) 63⑦5   7 is the tens digit.
    6380   5 is 5 or more.

◉ Discovery 1.2

Find some examples of numbers in newspapers or magazines that have probably been rounded.

For each one, decide whether the number has been rounded to the nearest 100, 1000, million, etc.

Exercise 1.2

1 List the following.

   (a) The multiples of 6 less than 100   (b) The multiples of 8 less than 100

2 Use your answers to question 1 to list the common multiples of 6 and 8 less than 100.

3 Look at these numbers.   2, 6, 15, 18, 30, 33

   (a) Which have 2 as a factor?

   (b) Which have 3 as a factor?

   (c) Which have 5 as a factor?

4 List the following.

   (a) The multiples of 12 less than 100   (b) The multiples of 15 less than 100

5 Use your answers to question 4 to find a common multiple of 12 and 15 less than 100.

Exercise continues …

6 List the following.

(a) The factors of 18   (b) The factors of 24

7 Use your answers to question 6 to list the common factors of 18 and 24.

8 List the following.

(a) The factors of 40   (b) The factors of 36

9 Use your answers to question 8 to list the common factors of 40 and 36.

10 Round these numbers to the nearest 1000.

(a) 23 400   (b) 196 700   (c) 7800   (d) 147 534   (e) 5 732 498

11 Round these numbers to the nearest 100.

(a) 7669   (b) 17 640   (c) 789   (d) 654 349   (e) 4980

12 Here are some newspaper headlines.
Round the numbers so that they have more impact.

(a) 67 846 watch United!   (b) Lottery winner scoops £5 213 198!

(c) Waiting lists down by 7863!   (d) Chairman gets £684 572 bonus!

Multiplication and division

Multiplying by 10, 100, 1000, …

Here are two entries in the 10 times table.

$5 \times 10 = 50$ $12 \times 10 = 120$

You can see that to multiply by 10 you move the units into the tens column, the tens into the hundreds column and so on. You put a zero in the units column.

So, for example,

$25 \times 10 = 250,$ $564 \times 10 = 5640,$ $120 \times 10 = 1200.$

In the same way

$4 \times 100 = 400,$ $6 \times 100 = 600.$

You can see that to multiply by 100 you move the units into the hundreds column, the tens into the thousands column and so on. You put zeros in the units and the tens columns.

In the same way, to multiply by 1000 you move the digits three places to the left and add three zeros.

Example 1.5

Write down the answers to these.

(a) 56×10 **(b)** 47×100

(c) 156×1000 **(d)** 420×100

(e) $65 \times 10\,000$

Solution

(a) 560   **(b)** 4700

(c) 156 000   **(d)** 42 000

(e) 650 000

Dividing by 10, 100, 1000, …

Since dividing is the reverse of multiplying, to divide by 10 you move the digits one place to the right and take a zero off.

To divide by 100 you move the digits two places to the right and take two zeros off.

Example 1.6

Write down the answers to these.

(a) $580 \div 10$

(b) $1400 \div 100$

(c) $362\,000 \div 1000$

(d) $60\,000 \div 100$

Solution

(a) $58\cancel{0} \rightarrow 58$

(b) $14\cancel{00} \rightarrow 14$

(c) $362\,\cancel{000} \rightarrow 362$

(d) $600\cancel{00} \rightarrow 600$

Multiplying by multiples of 10, 100, 1000, …

You can multiply by 30 by first multiplying by 3 and then by 10.

You can multiply by 500 by first multiplying by 5 and then by 100.

Example 1.7

(a) 300×40 **(b)** 42×30 **(c)** 54×40 **(d)** 27×500

Solution

(a) First, you multiply 300×4:

```
    3 0 0
×       4
  1 2 0 0
```

Then you need to multiply by 10.
When you are working with integers (whole numbers), you can do this by just adding 0 to your answer.

$$1200 \times 10 = 12\,000$$

(b)
```
    4 2
×    3
  1 2 6
```
$126 \times 10 = 1260$

(c)
```
    5 4
×    4
  2 1 6
    1
```
$216 \times 10 = 2160$

(d)
```
    2 7
×    5
  1 3 5
    3
```
$135 \times 100 = 13\,500$

More difficult multiplications

You need to be able do questions like 53×38 or 258×63 without a calculator.

There are several methods to do this. Two are shown here but your teacher may show you more. Choose a method you are happy with and stick with it.

Method 1

```
      5 3
×     3 8
  1 5 9 0    (53 × 30)
    4 2 4    (53 × 8)
       2
  2 0 1 4    Add
    1 1
```

```
      2 5 8
  ×     6 3
  1 5₃4₄8 0    (258 × 60)
      7₁7₂4    (258 × 3)
  1 6 2 5 4
      1  1
```

TIP

63 × 258 would give the same answer as 258 × 63 but it is usually easier to have the smaller number on the bottom.

This is the traditional method, called 'long multiplication'. The second method uses a grid.

Method 2

53 × 38

×	50	3
30	1500	90
8	400	24

```
  1 5 0 0
    4 0 0
      9 0
+     2 4
  2 0 1 4
    1 1
```

258 × 63

×	200	50	8
60	12 000	3000	480
3	600	150	24

```
  1 2 0 0 0
    3 0 0 0
      4 8 0
      6 0 0
      1 5 0
+       2 4
  1 6 2 5 4
      1 1
```

Exercise 1.3

1 Work out these.

 (a) 52 × 10
 (b) 63 × 100
 (c) 54 × 1000
 (d) 361 × 100
 (e) 56 × 10 000
 (f) 60 × 100
 (g) 549 × 1000
 (h) 8100 × 100
 (i) 530 × 1000
 (j) 47 × 10 000
 (k) 923 × 100 000
 (l) 62 × 1 000 000

2 Work out these.

 (a) 530 ÷ 10
 (b) 14 000 ÷ 100
 (c) 532 000 ÷ 1000
 (d) 64 000 ÷ 100
 (e) 6 400 000 ÷ 1000
 (f) 536 000 ÷ 10
 (g) 675 400 ÷ 100
 (h) 7 300 000 ÷ 100
 (i) 58 000 000 ÷ 10 000

Exercise continues …

3 Work out these.
   (a) 30 × 50   (b) 70 × 80   (c) 70 × 200   (d) 200 × 300
   (e) 800 × 30   (f) 50 × 40   (g) 600 × 3000   (h) 600 × 500
   (i) 800 × 7000   (j) 4000 × 3000 (k) 70 000 × 40   (l) 9000 × 8000

4 Work out these.
   (a) 64 × 30   (b) 72 × 60   (c) 234 × 30   (d) 56 × 200
   (e) 63 × 400   (f) 78 × 300   (g) 432 × 600   (h) 58 × 4000

5 Work out these.
   (a) 54 × 32   (b) 38 × 62   (c) 57 × 82   (d) 98 × 18
   (e) 66 × 29   (f) 84 × 74   (g) 123 × 27   (h) 264 × 35
   (i) 483 × 72   (j) 691 × 43   (k) 542 × 81   (l) 88 × 236

6 (a) How many pence are there in £632?
  (b) Change 5600 pence into pounds.

7 1 kilometre is 1000 metres.
  How many metres is 47 kilometres?

8 Trainers cost £40 per pair.
  What do six pairs cost?

9 Gary walks 400 metres to school and 400 metres back.
  How far does he walk in 195 school days?

10 28 people attended Rajvee's
   birthday party.
   She gave them each a packet of
   sweets which cost 34p each.
   What was the total cost in pence?

Squares and cubes

You read 5^2 as '5 squared'.

5^2 means $5 \times 5 = 25$.

All the **squares** from 1^2 to 10^2 are in your tables so you should know them.

TIP

It is a very common mistake to think that 5^2 means 5×2.

For harder squares you can use your calculator.

Look for the square function on your calculator. The key may look like $\boxed{x^2}$ or $\boxed{\blacksquare^2}$

Example 1.8

Find 18^2.

Solution

There are two ways of doing this on a calculator.

Method 1

Work out $18 \times 18 = 324$.

TIP

Calculators vary. Make sure you are familiar with the functions on your calculator.

Method 2

Use the square function on your calculator.
Enter 18 and then press $\boxed{x^2}$ and $\boxed{=}$. The display should read 324.

You read 2^3 as '2 cubed'.

   2^3 means $2 \times 2 \times 2 = 8$.

You should be able to work out 2^3, 3^3, 10^3 and possibly even 4^3 and 5^3 in your head but for other cubes you may need your calculator.

TIP

It is a very common mistake to think that 2^3 means 2×3.

Some calculators do not have a 'cube' button, $\boxed{x^3}$, so it is probably best to use the $\boxed{\times}$ button twice.

Example 1.9

Work out 17^3.

Solution

$17 \times 17 \times 17 = 4913$

Other powers

Squares and cubes are examples of **powers**. Another way of saying 2^2 is '2 to the power 2' and of saying 2^3 is '2 to the power 3'.

Squares and cubes are the only powers which have special names.

You read 5^4 as '5 to the power 4'.

5^4 means $\underbrace{5 \times 5 \times 5 \times 5}_{\text{four fives multiplied together}} = 625$

At this stage you will not need to find powers of most numbers other than squares or cubes.

The powers of ten, however, form a sequence which you already know.

Discovery 1.3

$10^2 = 10 \times 10 = 100$
$10^3 = 10 \times 10 \times 10 = 1000$

Work out $10^4 = 10 \times 10 \times 10 \times 10 =$
Work out 10^5.

What do you notice about the power of 10 and the number of zeros?

Write down the value of   (a) 10^6.   (b) 10^8.

Square roots

Ashraf thought of a number and then multiplied it by itself.

The answer was 36.

What number did Ashraf start with? $? \times ? = 36$

Using your tables you should realise that Ashraf started with 6 because $6^2 = 36$.

Discovery 1.4

Work with a friend.

Take turns to think of a number and multiply it by itself. Tell your friend your answer. The other person must find what number you started with.

Continue until you cannot find any more.

What you have been doing in Discovery 1.4 is finding **square roots**. That is the reverse of finding squares.

'Square root' is written $\sqrt{}$ so $\sqrt{36} = 6$.

For harder square roots you will need your calculator. Look for the $\boxed{\sqrt{}}$ button on your calculator.

Example 1.10

Find $\sqrt{289}$.

Solution

Press $\boxed{\sqrt{}}$ and then $\boxed{2}\,\boxed{8}\,\boxed{9}$ and then $\boxed{=}$.

The display should read 17.

Check that $17 \times 17 = 289$.

Exercise 1.4

1 Work out these without your calculator.
  (a) 7^2 (b) 9^2 (c) 11^2 (d) 12^2 (e) 30^2
  (f) 50^2 (g) 60^2 (h) 200^2 (i) 400^2 (j) 800^2

2 Use your calculator to work out these.
  (a) 14^2 (b) 22^2 (c) 31^2 (d) 47^2 (e) 89^2
  (f) 56^2 (g) 34^2 (h) 180^2 (i) 263^2 (j) 745^2

3 Use your calculator to work out these.
  (a) 6^3 (b) 9^3 (c) 11^3 (d) 14^3
  (e) 25^3 (f) 37^3 (g) 43^3 (h) 147^3

4 Use your calculator to work out these.
  (a) $\sqrt{225}$ (b) $\sqrt{196}$ (c) $\sqrt{361}$ (d) $\sqrt{529}$
  (e) $\sqrt{1521}$ (f) $\sqrt{7569}$ (g) $\sqrt{4624}$ (h) $\sqrt{2916}$

5 Work out these without your calculator.
  (a) $\sqrt{400}$ (b) $\sqrt{900}$ (c) $\sqrt{2500}$ (d) $\sqrt{6400}$ (e) $\sqrt{40\,000}$

Negative numbers

Numbers less than zero are called **negative numbers**.

Example 1.11

The temperature at 4 p.m. is 3°C. By midnight it has fallen 8 degrees.
What is the temperature at midnight?

Solution

Moving down 8 degrees goes to 5 below zero.
The answer is written -5°C and you say 'negative 5' or 'minus 5'.

A number line is very useful when adding and subtracting with negative numbers.

Notice that the further left, the smaller the number. For example, -2 is smaller than 1.

Example 1.12

Use a number line to work out these calculations.

(a) $-2 + 4$ **(b)** $5 - 7$

Solution

(a) Start at -2 and move 4 to the right.

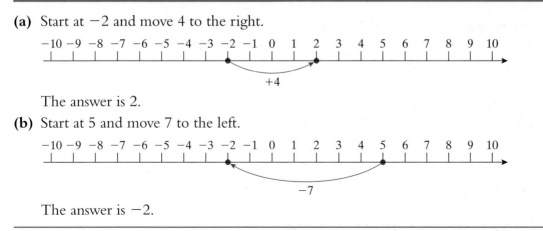

The answer is 2.

(b) Start at 5 and move 7 to the left.

The answer is -2.

You can also do additions and subtractions involving negative numbers without a number line.
- When a smaller number is subtracted from a bigger number, the answer is positive.
- When a bigger number is subtracted from a smaller number, the answer is negative.
- When you subtract a negative number from another negative number, you add the numbers and make the answer negative.
- Adding a negative number is the same as subtracting a positive number.
- Subtracting a negative number is the same as adding a positive number.

Example 1.13

Work out these.

(a) $4 - 6$ **(b)** $-2 - 6$ **(c)** $-4 + (-3)$ **(d)** $-2 - (-5)$

Solution

(a) $4 - 6 = -2$ $6 - 4 = 2$ and the number being subtracted is bigger.

(b) $-2 - 6 = -8$ $2 + 6 = 8$ and both numbers are negative.

(c) $-4 + (-3) = -7$ Adding a negative number is the same as subtracting a positive number: $-4 + (-3) = -4 - 3$.

(d) $-2 - (-5) = 3$ Subtracting a negative number is the same as adding a positive number: $-2 - (-5) = -2 + 5$.
The order does not matter when you add: $-2 + 5 = 5 + (-2)$.
Adding a negative number is the same as subtracting a positive number: $5 + (-2) = 5 - 2 = 3$.

Exercise 1.5

1   Work out these.
    (a) -4 add 7   (b) -7 add 4   (c)   9 subtract 12

2   The temperature is $-6°C$. Find the new temperature after
    (a)   a rise of 5°C.   (b)   a rise of 10°C.   (c)   a fall of 2°C.

3   Find the difference in temperature between
    (a)   5°C and 21°C.   (b) $-5°C$ and 21°C.   (c) $-18°C$ and $-4°C$.

4   Arrange these numbers in order, smallest first.
    (a) $1, -3, 7, -8$ (b) $0, -4, 5, -6$ (c) $1, 2, -3, -4, 5, -6$

5   An office building has 20 floors and three levels of underground car park.
In the lift, the ground floor button is labelled 0.
What should be the label on the button for the lowest car park level?

6 Work out these.

(a) $4 - 5$

(b) $6 - 2$

(c) $7 - 5$

(d) $3 - 6$

(e) $6 - 4$

(f) $5 + 3$

(g) $-2 + 3$

(h) $-4 + 3$

(i) $4 + 1$

(j) $-2 + 1$

(k) $2 - 3$

(l) $-5 + 6$

(m) $-5 + 2$

(n) $-6 + 5$

(o) $-3 - 2$

(p) $-3 + 4$

(q) $4 + (-2)$

(r) $-6 - (-3)$

(s) $2 + (-1)$

(t) $-1 - (-4)$

What you have learned

- An integer is a whole number
- A multiple is a result of multiplying one integer by another integer; a common multiple of two integers is a multiple of both of the integers
- An integer that will divide into a number exactly is called a factor of the number; a common factor of two numbers is an integer that will divide exactly into both numbers
- To round numbers to the nearest 10, you look at the digit in the units place; if it is less than 5 you leave the tens digit as it is; if it is 5 or more you add one to the tens digit; you replace the remaining digits with zeros
- You use similar methods to round to the nearest 100, 1000, …
- To multiply an integer by 10 you move the digits one place to the left and add a zero in the units place; to multiply by 100 you move the digits two places to the left and add zeros in the units and tens place; to multiply by 1000 you move the digits three places to the left and add zeros in the units, tens and hundreds place
- To divide by 10 you move the digits one place to the right; to divide by 100 you move the digits two places to the right; to divide by 1000 you move the digits three places to the right
- A square is a number multiplied by itself; you can write '5 squared' as 5^2 and it means 5×5
- You can write '2 cubed' as 2^3 and it means $2 \times 2 \times 2$
- You can write '3 to the power 4' as 3^4 and it means $3 \times 3 \times 3 \times 3$
- You can write 'the square root of 36' as $\sqrt{36}$; finding a square root is the reverse of finding a square so $\sqrt{36} = 6$
- You can find squares, cubes, other powers and square roots using your calculator
- Numbers less than zero are negative
- Adding a negative number is the same as subtracting a positive number
- Subtracting a negative number is the same as adding a positive number

Mixed exercise 1

1 Work out these.

(a) $59 + 73$ (b) $62 - 18$ (c) $456 - 187$

(d) 58×6 (e) 254×4 (f) $441 \div 7$

2 Jane bought six pens at 38p each and a notebook for 43p.
Find the total cost in pence.

3 (a) List the multiples of 20 less than 125.

(b) List the multiples of 12 less than 125.

(c) List the common multiples of 12 and 20 less than 125.

4 (a) List the factors of 12.   (b) List the factors of 18.

(c) List the factors of 30.   (d) List the common factors of 12, 18 and 30.

5 Round

(a) 5632 to the nearest 100.   (b) 17 849 to the nearest 1000.

(c) 273 490 to the nearest 1000.   (d) 273 490 to the nearest 100.

(e) 5 836 492 to the nearest million.   (f) 3498 to the nearest 10.

6 Work out these.

(a) 93×100 (b) 630×100 (c) 572×1000

(d) $7800 \div 100$ (e) $6\,300\,000 \div 1000$ (f) 50×80

(g) 70×300 (h) 47×30 (i) 58×600

(j) 28×5000 (k) 456×70 (l) 732×400

7 Work out these. Show your working.

(a) 63×28 (b) 83×57 (c) 256×38

8 In a sponsored walk 186 people each walked 45 km.
What was the total distance they walked? Show your working.

9 Work out these without your calculator.

(a) 8^2 (b) 40^2 (c) 500^2 (d) 10^5 (e) 20^3

10 Use your calculator to work out these.

(a) 29^2 (b) 12^3 (c) 53^2 (d) $\sqrt{484}$ (e) $\sqrt{5184}$

11 Copy and complete the table.

Temperature (°C)	5		−4	
10° warmer		7		
5° colder				−15

12 Work out these.

(a) $-4 - 2$ (b) $-2 + 4$ (c) $2 - 6$ (d) $5 + (-3)$ (e) $3 - (-2)$

Algebra 1

- Using letters to represent numbers
- Writing simple expressions
- Collecting like terms

- How to add, subtract, multiply and divide simple numbers

Using letters to represent numbers

John has 4 marbles and Leroy has 6.
You find how many marbles they have altogether by adding.

Total number of marbles $= 4 + 6 = 10$.

If you do not know how many marbles one of the boys has, you can use a letter to stand for the unknown number.

For example, John has n marbles and Leroy has 6.
Then

Total number of marbles $= n + 6$.

$n + 6$ is an example of an **expression**. An expression can include numbers and letters but no equals sign.

$n - 6$, $x + 3$, $n + m$ and other similar expressions cannot be written more simply.

🔍 Check up 2.1

(a) Julie walked 2 miles in the morning and 3 miles in the afternoon. How far did she walk in total?
(b) Penny walked 2 miles in the morning and x miles in the afternoon.
Write an expression for the total distance she walked.

Example 2.1

(a) There are five girls and four boys at a bus stop.
How many children are there at the bus stop?

(b) There are *p* girls and six boys at a bus stop.
Write an expression for the number of children at the bus stop.

(c) There are *p* girls and *q* boys at a bus stop.
Write an expression for the number of children at the bus stop.

Solution

(a) $5 + 4 = 9$ You simply add the number of girls and the number of boys.

(b) $p + 6$ The number of girls is represented by *p*.
You add the number of boys, 6. You cannot write this more simply.

(c) $p + q$ Now you have letters representing both the number of girls and the number of boys but they are different letters so you cannot write the expression more simply.

> **TIP**
>
> $6 + p$ is the same as $p + 6$.
> $p + q$ is the same as $q + p$.

Example 2.2

(a) This line is made of two pieces.
What is the length of the line?

2   5

(b) This line is made up of two pieces.
Write an expression for the length of the line.

x   5

(c) Write an expression for the length of this line.

x   *y*

Solution

(a) $2 + 5 = 7$ You simply add the two lengths.

(b) $x + 5$ One of the lengths is now represented by a letter. You add the two lengths but you cannot simplify the expression further.

(c) $x + y$ Both lengths are now represented by letters. You add the two lengths but you cannot simplify the expression further.

> **TIP**
>
> You can check your expression is correct by using a number for each letter.
> For example, in part **(c)** of Example 2.2, if $x = 3$ and $y = 2$ you can see that the length of the line would be 5. This is the same as the value of the expression $x + y = 2 + 3 = 5$, so you can see that the answer is correct.

Example 2.3

(a) Tina has 3 red crayons, 2 blue crayons and 4 green crayons.
How many crayons does she have altogether?

(b) Sam has x red crayons, y blue crayons and 2 green crayons.
Write an expression for the number of crayons he has altogether.

(c) Rhian has x red crayons, 4 blue crayons and x green crayons.
Write an expression for the number of crayons she has altogether.

Solution

(a) $3 + 2 + 4 = 9$

(b) $x + y + 2$

(c) $x + 4 + x = 2x + 4$
$x + x$ is $2 \times x$.
You can write this more simply as $2x$.

Rhian has x red crayons and x green crayons. It is the number of crayons that is important here, not the colour.

Example 2.4

A chocolate biscuit costs 15p.

(a) How much do 6 biscuits cost?

(b) Write an expression for the cost, in pence, of b biscuits.

Solution

(a) $6 \times 15 = 90$p   You multiply 15p by 6 to find the cost of 6 biscuits.

(b) $b \times 15 = 15b$ You multiply 15p by b to find the cost of b biscuits.

TIP

Do not write 15b p as this is confusing. You could write 15b pence.

🔍 Check up 2.2

(a) Sally has x sweets. James also has x sweets.
How many sweets do they have altogether?

(b) Dan has y sweets. Sarah has three times as many sweets as Dan.
How many sweets does Sarah have?

Exercise 2.1

1 (a) Tom has 4 blue pens and 2 red pens.
        How many pens does he have altogether?

   (b) Sarah has *x* blue pens and 2 red pens.
        Write an expression for the number of pens she has altogether.

   (c) Robert has *b* blue pens and *r* red pens.
        Write an expression for the number of pens he has altogether.

2 (a) Mrs Khan buys 2 pints of milk on Monday and 3 pints on Tuesday.
        How many pints does she buy altogether?

   (b) Mrs Lundy buys 2 pints of milk on Monday and *p* pints on Tuesday.
        Write an expression for the number of pints of milk she buys altogether.

   (c) Mr Mansfield buys *q* pints of milk on Monday and *r* pints on Tuesday.
        Write an expression for the number of pints of milk he buys altogether.

3 (a) What is the length of this line?   2   3

   (b) Write an expression for the length   *x*   3
        of this line.

   (c) Write an expression for the length   2   *x*
        of this line.

4 (a) What is the length of this line?   4   9

   (b) Write an expression for the length   *p*   9
        of this line.

   (c) Write an expression for the length   4   *q*
        of this line.

5 (a) Write an expression for the length   *x*   5
        of this line.

   (b) Write an expression for the length   *x*   3   5
        of this line.

   (c) Write an expression for the length   *x*   2   *y*
        of this line.

   (d) Write an expression for the length   *x*   4   *x*
        of this line.

Exercise continues …

6 (a) David is 4 years old and Sam is x years old.
Write an expression for the sum of their ages.

(b) Patrick is x years old and Mary is y years old.
Write an expression for the sum of their ages.

(c) Gamal and Arabella are both x years old.
Write an expression for the sum of their ages.

7 Simone is 6 years older than Paula.

(a) How old was Simone when Paula was 4 years old?

(b) How old was Simone when Paula was 8 years old?

(c) Write an expression for Simone's age when Paula was x years old.

8 Crisps cost 25p a packet.

(a) How much do 3 packets cost?

(b) How much do 6 packets cost?

(c) Write an expression for the cost, in pence, of x packets.

9 A bag of flour costs x pence.
Write an expression for the cost of

(a) 2 bags.   (b) 5 bags.   (c) 7 bags.

10 Amanda, Pat and Annabel were given some sweets.
Pat received twice as many as Amanda.
Annabel received six more than Amanda.
Amanda received m sweets.
Write an expression for the number of sweets received by

(a) Pat.   (b) Annabel.

Writing simple expressions

So far, all the questions in this chapter have involved adding or multiplying. We will now look at expressions that use subtracting and dividing.

Example 2.5

In Class 3A there are x students.
Write an expression for the number of students present when

(a) three students are absent.

(b) y students are absent.

Solution

(a) $x - 3$ To find the number of students present you subtract the number of students absent, 3, from the number of students in the class, x. You cannot write this more simply.

(b) $x - y$ The calculation is similar but the number of students absent is also represented by a letter.

Example 2.6

Write expressions for the length of the red part of these lines.

(a)

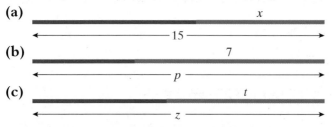

(b)

(c)

Solution

(a) $15 - x$ To find the length of the red part of the line you subtract the length of the blue part, x, from the total length, 15.

(b) $p - 7$ To find the length of the red part of the line you subtract the length of the blue part, 7, from the total length, p.

(c) $z - t$ To find the length of the red part of the line you subtract the length of the blue part, t, from the total length, z.

Example 2.7

Tony bought four oranges.

(a) How much did one orange cost if he paid 60p?

(b) Write an expression for the cost of one orange if he paid x pence.

Solution

(a) $60 \div 4 = 15p$ To find the cost of one orange you divide the total cost, 60p, by the number of oranges, 4.

(b) $x \div 4$ To find the cost of one orange you divide the total cost, x, by the number of oranges, 4.

Check up 2.3

(a) Pam has x books. Selina has five fewer.
Write an expression for the number of books that Selina has.

(b) Betty paid 85p for five plums.
How much did one plum cost?

(c) David paid x pence for eight apples.
How much did one apple cost?

Exercise 2.2

1   A dancing school has seven fewer boys than girls.

   (a) How many boys are there if there are
        (i)   15 girls?   (ii) 10 girls?

   (b) Write an expression for the number of boys if there are g girls.

2   Beth has three fewer pens than Grace.

   (a) How many pens does Beth have if Grace has
        (i)   five pens?   (ii) ten pens?

   (b) Write an expression for the number of pens Beth has if Grace has p pens.

3   Write expressions for the length of the red part of these lines.

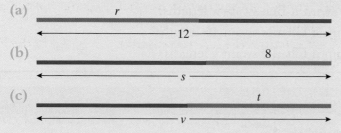

4   The width of a rectangle is 3 cm less than the length.

   (a) What is the width of the rectangle if the length is
        (i)   9 cm?   (ii) 14 cm?

   (b) Write an expression for the width of the rectangle if the length is p cm.

5   Linford is 5 cm shorter than Earl.

   (a) How tall is Linford if Earl's height is
        (i)   160 cm?   (ii) 189 cm?

   (b) Write an expression for Linford's height if Earl's height is h cm.

Exercise continues ...

6 There were x people on a bus when it arrived at a bus stop. Some got off.
Write an expression for the number of people left on the bus if the number getting off was

(a) 14.   (b) 21.   (c) p.

7 (a) Simon bought six baking potatoes for 72p.
How much did one potato cost?

(b) Susan bought six baking potatoes for b pence.
Write an expression for the cost of one potato.

8 There are y children at Swallow Vale Nursery.
Write an expression for the number of boys if the number of girls is

(a) 10.   (b) 15.   (c) g.

9 There are a fewer apples than bananas on a fruit stall.
Write an expression for the number of apples when there are

(a) 25 bananas.   (b) 15 bananas.   (c) b bananas.

10 Five tins of beans cost b pence.
Write an expression for the cost of one tin.

● Challenge 2.1

Ben and Charlie went swimming. Charlie swam four lengths more than Ben.

(a) How many lengths did Charlie swim if Ben swam 15 lengths?
(b) How many lengths did Charlie swim if Ben swam x lengths?
(c) How many lengths did Ben swim if Charlie swam 16 lengths?
(d) How many lengths did Ben swim if Charlie swam y lengths?

● Challenge 2.2

This square is made from four matchsticks.

This pattern of two squares is made from seven matchsticks.

(a) Draw the pattern with three squares.
How many matchsticks would you need?
(b) Draw the patterns with four, five and six squares.
Can you see any pattern in the number of matchsticks?
(c) How many matchsticks would you need for the pattern with ten squares?
(d) Write an expression for the number of matchsticks needed for the pattern with s squares.

Collecting like terms

When you write algebra, you do not need to write the $\times$ sign.

You write $1 \times a$ as a.

You write $2 \times a$ as $2a$, $3 \times a$ as $3a$, ...

If the answer is $0a$ you write it as 0, for example, $2a - 2a = 0$.

Example 2.8

Write, as simply as possible, an expression for the length of this line.

$$r \qquad r \qquad r \qquad r$$

Solution

$r + r + r + r = 4 \times r$ There are four pieces each of length r.
$\qquad\qquad\quad = 4r$ You do not need to write the $\times$ sign.

The expression $r + r + r + r$ has been simplified to $4r$.

Simplify means to write as simply as possible.

r and $4r$ are examples of **terms**. Terms using the same letter or combination of letters are called **like** terms. You can simplify like terms by adding or subtracting.

Example 2.9

Simplify these.

(a) $3a + 4a$ **(b)** $5a - a$ **(c)** $2b + 3b$ **(d)** $3a - 3a$

Solution

(a) $3a + 4a = 7a$ $3a$ and $4a$ are like terms.
$\qquad\qquad\qquad\quad$ 3 lots of a plus 4 lots of a is 7 lots of a in total.

(b) $5a - a = 4a$ 5 lots of a minus 1 lot of a is 4 lots of a.

(c) $2b + 3b = 5b$ $2b$ and $3b$ are like terms.
$\qquad\qquad\qquad\quad$ 2 lots of b plus 3 lots of b is 5 lots of b in total.

(d) $3a - 3a = 0$ 3 lots of a minus 3 lots of a is zero lots of a.
$\qquad\qquad\qquad\quad$ You write $0a$ as 0.

TIP

When adding or subtracting a it is best to think of it as $1a$. For example,
$a + 4a = 1a + 4a$
$\qquad\quad = 5a$ and
$4a - a = 4a - 1a$
$\qquad\quad = 3a$.
A common error is to write $4a - a = 4$. Thinking of a as $1a$ should help to avoid this.

Check up 2.4

Simplify these expressions.

(a) $x + x + x + x + x$

(b) $3 \times a$

(c) $6p - 2p$

(d) $2c + 3c$

(e) $3a - 2a + 4a$

Exercise 2.3

Simplify these expressions.

1 $p + p + p + p + p$	2 $a + a + a + a + a + a + a$	3 $5 \times x$
4 $4 \times c$	5 $4p + 3p$	6 $b + 2b + 3b$
7 $p \times 3$	8 $s + 2s + s$	9 $4a - 2a$
10 $8c - 3c$	11 $5x - 2x + 4x$	12 $2m - m + 3m$
13 $2a + 3a + 5a - 2a$	14 $2c + 3c - c - 2c$	15 $4b - 3b + b - 2b$
16 $2p - p$	17 $4b - 2b$	18 $4x + 5x$
19 $a + 2a - 3a + 4a + 5a$	20 $3 \times a + 2 \times a$	

Like and unlike terms

Terms using different letters are called **unlike** terms. You cannot simplify expressions such as $x + y$ any further because unlike terms cannot be added or subtracted.

You have already seen that $2 \times a$ can be written more simply as $2a$. Similarly, you write $a \times b$ as ab and $3 \times a \times b \times c$ as $3abc$.

You write $a \times a = a^2$.

> **TIP**
>
> You cannot add or subtract unlike terms. For example, $2p + 5q$ cannot be simplified.
>
> You can multiply unlike terms. For example, $2p \times 5q = 10pq$.

Example 2.10

Simplify these expressions where possible.

(a) $3a + 4b$ **(b)** $a + b + 3a - 3b$ **(c)** $2 \times a \times c$
(d) $b \times b$ **(e)** $2a \times 3b$

Solution

(a) $3a + 4b$ $3a + 4b$ are unlike terms; they cannot be added.

(b) $4a - 2b$ You simplify the expression by collecting all the a terms into a single term and collecting all the b terms into a single term:
$a + 3a = 4a$ and $b - 3b = -2b$.

(c) $2ac$ You can multiply unlike terms. You write the product without the $\times$ sign.

(d) b^2 When you multiply like terms together you write them as powers.

(e) $6ab$ $2a \times 3b = 2 \times 3 \times a \times b = 6ab$

> **TIP**
>
> Do not confuse $2a$ and a^2. $2a$ is $2 \times a$; a^2 is $a \times a$.

Check up 2.5

Simplify these expressions where possible.

(a) $x + 3y + 3x - y$ **(b)** $x \times y \times 7$ **(c)** $4x + 4y$
(d) $x \times x$ **(e)** $3y + 7x - y - 2x - 2y$

Exercise 2.4

Simplify these expressions where possible.

1 $2a + 3b - a$

2 $3x - 2x + 3y$

3 $4 \times a \times b$

4 $3p + q$

5 $3a + 2b + 3a + 4b$

6 $2 \times a + 3 \times c$

7 $2 \times p \times 4 \times q$

8 $4x + y + 3y + 2x$

9 $3 \times p \times p$

10 $5a + 2b + 1 + 2b + a + 3$

11 $3ab + 2bc$

12 $a \times a$

13 $4a + 2c - 3a + c$

14 $4 \times a \times b + 2a \times a$

15 $4s + 2s - 3s$

16 $3a + 2 - a + 2$

17 $a \times a \times b$

18 $3x + y + 2y - 2x$

19 $4a + 2b + 6a - 4b$

20 $3 \times a \times a \times b \times b$

Challenge 2.3

Simplify these expressions where possible.

(a) $3a \times b + 2a$

(b) $a \times a + 3a$

(c) $3a + 3b + 3a \times 3b$

(d) $2a - 3a \times a + 4a^2$

(e) $4ab + 3ba - 2ab$

Challenge 2.4

The length of a rectangle is 3 cm longer than its width.
The width of the rectangle is x cm.

Write down, as simply as possible, an expression for

(a) the length of the rectangle.

(b) the perimeter of the rectangle.

x

Challenge 2.5

David is y years old. Pat is 4 years older than David and Simon is 6 years younger than David.

(a) Write an expression for each of their ages.

(b) Write, as simply as possible, an expression for the sum of their ages.

Challenge 2.6

Okera bought five cans of coke at $2a$ pence each and 3 bars of chocolate at a pence each.

Find and simplify the amount he spent altogether.

What you have learned

- Letters can be used to stand for numbers whose value you do not know
- An expression is a combination of letters and numbers, without an equals sign
- Like terms use exactly the same letters but can use different numbers, for example a and $5a$ are like terms, ab and $4ab$ are like terms but a and ab are not like terms
- You can simplify an expression by collecting like terms

Mixed exercise 2

1 At Fred's Fishery, a fish cost f pence more than a packet of chips.
Write an expression for the cost of a fish if a packet of chips costs
   (a) 35p. (b) 60p. (c) c pence.

2 (a) Write an expression for the length of this line.
   (b) Write an expression for the length of the red part of this line.
   (c) Write an expression for the length of the blue part of this line.

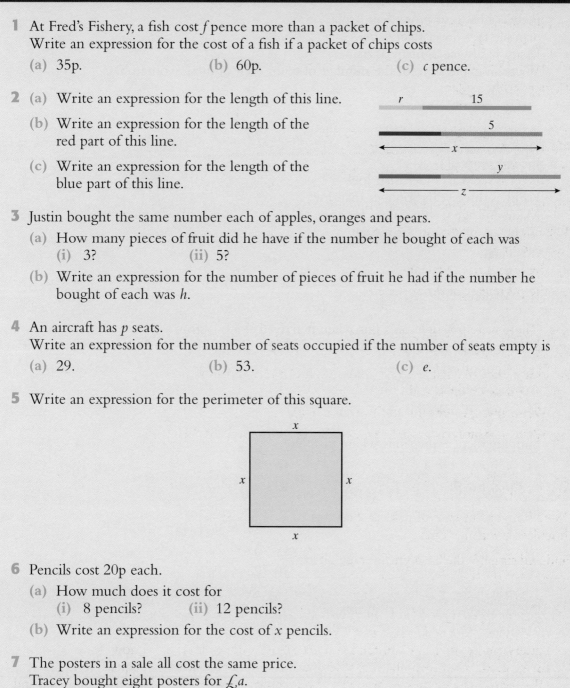

3 Justin bought the same number each of apples, oranges and pears.
   (a) How many pieces of fruit did he have if the number he bought of each was
   (i) 3? (ii) 5?
   (b) Write an expression for the number of pieces of fruit he had if the number he bought of each was h.

4 An aircraft has p seats.
Write an expression for the number of seats occupied if the number of seats empty is
   (a) 29. (b) 53. (c) e.

5 Write an expression for the perimeter of this square.

6 Pencils cost 20p each.
   (a) How much does it cost for
   (i) 8 pencils? (ii) 12 pencils?
   (b) Write an expression for the cost of x pencils.

7 The posters in a sale all cost the same price.
Tracey bought eight posters for £a.
Write an expression for the cost of one poster.

Mixed exercise 2 continues …

8 Asma has t coins in her purse.
Rebecca has four more than Asma.
Sian has two fewer than Asma.
Jessica has twice as many as Asma.
Write an expression for the number of coins each of these women has.

(a) Rebecca

(b) Sian

(c) Jessica

9 Joe is x years old.
Peter is six years older than Joe.
Kathy is three years younger than Joe.
Maureen's age is three times Joe's age.
Write an expression for the age of

(a) Peter.

(b) Kathy.

(c) Maureen.

10 There were p people on a bus when it arrived at a bus stop.
Write an expression for the number of people on the bus when it left the bus stop if

(a) 4 got off and 6 got on.

(b) 3 got off and q got on.

(c) r got off and s got on.

11 Simplify these expressions where possible.

(a) $a + a + a + a$

(b) $2 \times a$

(c) $a \times b \times c$

(d) $y \times y$

(e) $c + c + c - c$

(f) $4a - 2b + 3b$

(g) $2 \times y + 3 \times s$

(h) $2a + 3b + 4a - 2b$

(i) $2 \times p \times p$

(j) $3x + 2y$

(k) $x + 3y + 2x - 3y$

(l) $2a \times 3b$

(m) $5ab + 3ac - 2ab$

(n) $5x + 3y - 2x - y$

(o) $3a + 2b + 3c + b + 3a - 3c$

(p) $4 + 2x - 3 + 2y + 2 + 3x$

(q) $3 \times a \times b \times b$

(r) $1 + a + 2 - b + 3 + c$

(s) $2p \times 4q$

(t) $3a - 2b + 3 - a + 4 + 5b$

Data collection

▶ This chapter is about

- Deciding what type of data to collect and how to collect it
- Discrete and continuous data
- Tally charts, frequency tables and two-way tables
- Vertical line graphs and bar charts

▶ You should already know

- How to draw simple graphs

Collecting data

Different types of data

Most things can either be counted or measured in some way.

Data that are the result of objects being counted are called **discrete data** and data that are the result of measurement are called **continuous data**.

Example 3.1

Which of the following are discrete data and which are continuous?

| Height | Number of children | Amount of money |
| | Shoe size | Time |

Solution

Number of children, amount of money and shoe size are discrete data. They are all counted.

Shoe size can be tricky. You would measure your feet but your shoe size is a number.

Height and time are continuous data.
They are both measured.

Collection of data

Usually the results of collecting data are shown in a table or diagram of some kind. When you collect data it is sensible to first think about how you will show the results as this may influence how you collect the data.

You could quite easily collect data from every student in your class on the number of children in their family. One way to collect this data would be to ask each person individually and make a list like this.

Class 10G

1	2	1	1	2	3	2	1	2	1	1	2	4	2	1
5	2	3	1	1	4	10	3	2	5	1	2	1	1	2

Using the same method to collect data for all the students in your year could get very messy and one way to make the collection of data easier is to use a data collection sheet. Designing a table like the one on the left below can make collecting the data easy and quick.

The data for Class 10G is shown in the table on the right below. A vertical line, called a **tally**, is drawn for each response with a slanting line for the fifth to collect the tallies in bundles. This makes the number of tallies easier to count.

Frequency is another word for number of times and the completed table is called a frequency table. A frequency table shows the number of times a certain response was received.

Number of children	Tally
1	
2	
3	
4	
5	
6	
7	
8	
9	
10	

Number of children	Tally	Total (Frequency)
1	ⅧⅧ‖	12
2	ⅧⅧ	10
3	‖|	3
4	‖	2
5	‖	2
6		0
7		0
8		0
9		0
10			1

Before you design a data collection sheet it is useful to know what the answers might be but this is not always possible. For example, what would happen if you were using the table above and someone said 13?

One way to deal with this problem is to have an extra line at the end of the table to record all other responses, as in the table on the right.

Adding the 'More than 5' line allows all possible responses to be recorded and gets rid of some of the lines where there are no responses. It could just as easily have been 'More than 6' with zero shown as the frequency for 6. It is all a matter of what you think the answers to your question might be.

Number of children	Tally	Frequency
1	ЖЖ ЖЖ ||	12
2	ЖЖ ЖЖ	10
3					3
4				2
5				2
More than 5			1

Exercise 3.1

1 Which of the following are discrete data and which are continuous data?
   Weight   Distance   Number of pets   Temperature   Money   Age in years

2 List at least three more examples each of continuous data and discrete data.

3 Design a data collection sheet for each of the following.
   (a) Month of birth   (b) Day of birth   (c) Shoe size
   (d) Family car colours   (e) Favourite soft drink

4 Collect the data for your class and produce a frequency table for each of the items in question 3.

● Challenge 3.1

Collect the data for your year group and produce a frequency table for the following.
(a) Type of pet owned   (b) Favourite type of music
(c) Favourite fruit   (d) Favourite vegetable

Data display

It is possible to take the data from a frequency table and draw a diagram to show the results. Two different ways of doing this for discrete data are shown on the next page. Notice that there is a gap between each of the lines or bars. When you draw a bar chart for continuous data there is no gap between the bars, as you will see in Chapter 16.

You do not draw vertical line graphs for continuous data.
These diagrams both show the results for number of children in a family.

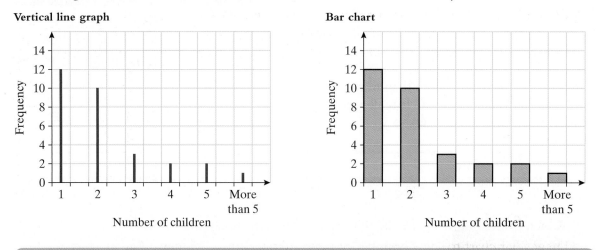

Vertical line graph

Frequency / Number of children

Bar chart

Frequency / Number of children

Example 3.2

Draw a vertical line graph and a bar chart to show these data.

Number of pets	Frequency
0	7
1	9
2	5
3	2
4	2
5	1
More than 5	4

Solution

You draw a line or a bar for each class. The height of the line or bar shows the frequency of the class.

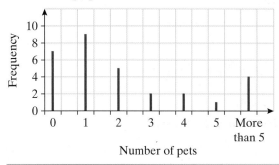

Vertical line graph

Frequency / Number of pets

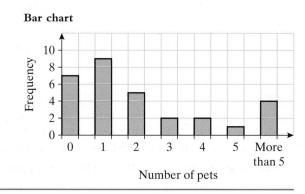

Bar chart

Frequency / Number of pets

Exercise 3.2

1 Draw a vertical line graph to show each of these sets of data.

(a)

Type of coin	Frequency
1p	6
2p	8
5p	12
10p	7
20p	9
50p	4
£1	2
£2	1

(b)

Number of cars	Frequency
0	3
1	9
2	11
3	4
4	2
More than 4	1

(c)

Number of bedrooms	Frequency
1	1
2	8
3	10
4	7
5	3
More than 5	1

2 Draw a bar chart to show each of these sets of data.

(a)

Type of pet	Frequency
Bird	3
Cat	9
Dog	7
Fish	6
Horse	1
Rabbit	3
Other	1

(b)

Number of brothers	Frequency
0	7
1	11
2	9
3	1
More than 3	2

(c)

Eye colour	Frequency
Blue	5
Brown	7
Green	10
Grey	7
Other	1

(d)

Number of fillings	Frequency
0	15
1	12
2	8
3	5
More than 3	2

(e)

Favourite crisps	Frequency
Plain	55
Beef	31
Chicken	12
Cheese and onion	15
Salt and vinegar	23
Other	14

Challenge 3.2

Choose a question of your own that involves discrete data.

Design a data collection sheet and collect the data for your year group.

Draw an appropriate graph to show the results.

Two-way tables

Sometimes the data collected involve more than one factor. Look at this example.

Example 3.3

Peter has collected data about cars in a car park. For each car he has recorded the colour and where it was made.

He can show both of these factors in a two-way table.

He has only completed some of the entries. Complete the table.

	Made in Europe	Made in Asia	Made in the USA	Total
Red	15	4	2	
Not red	83			154
Total		73		

Solution

	Made in Europe	Made in Asia	Made in the USA	Total
Red	15	4	2	21
Not red	83	69	2	154
Total	98	73	4	175

The total number of cars made in Asia is 73, so there are $73 - 4 = 69$ cars made in Asia that are not red.

The total number of cars that are not red is 154, so there are $154 - 83 - 69 = 2$ that are not red that are made in the USA.

All the totals can now be completed by adding across the rows or down the columns.

TIP

A useful check is to calculate the grand total (the number in the bottom right corner of the table) twice. The number you get by adding down the last column should be the same as the number you get by adding across the bottom row.

Exercise 3.3

1 Here is a two-way table showing the results of a car survey.

(a) Copy and complete the table.

	Japanese	Not Japanese	Total
Red	35	65	
Not red	72	438	
Total			

(b) How many cars were surveyed?

(c) How many Japanese cars were in the survey?

(d) How many of the Japanese cars were not red?

(e) How many red cars were in the survey?

2 A drugs company compared a new type of drug for hay fever with an existing drug. The two-way table shows the results of the trial.

(a) Copy and complete the table.

	Existing drug	New drug	Total
Symptoms eased	700	550	
No change in symptoms	350	250	
Total			

(b) How many people took part in the trial?

(c) How many people using the new drug had their symptoms eased?

3 A group of students voted on what they wanted to do for an activity day. Copy and complete the table.

	Riding	Sport	Total
Boys		18	
Girls	15		
Total		25	48

4 A group of students were surveyed about which sports they play.

(a) Copy and complete the table.

	Hockey	Not hockey	Total
Badminton	33		
Not badminton			39
Total	57		85

(b) How many students do not play either hockey or badminton?

Exercise continues …

5 At the indoor athletics championships, the USA, Germany and China won most medals.

(a) Copy and complete the table.

	Gold	Silver	Bronze	Total
USA	31		10	
Germany	18	16		43
China		9	11	42
Total		43		

(b) Which country won the most gold medals?

(c) Which country won the most bronze medals?

Grouping data

It is possible to create frequency tables and to draw bar charts to show large amounts of data but, when there are many different items, it often makes sense to arrange them in groups.

This can be done for discrete or continuous data. The next example uses discrete data.

Example 3.4

The data shows the numbers of apples produced by 100 trees.

43	56	89	64	74	52	48	55	63	74
52	75	59	46	77	55	80	93	63	58
63	57	81	57	58	59	51	63	67	62
81	62	68	68	59	61	39	78	46	49
57	66	57	79	48	72	47	54	70	34
49	54	37	67	83	67	78	47	59	84
53	59	79	53	69	53	67	66	83	89
77	70	42	48	72	64	56	52	73	71
38	84	62	32	78	77	41	64	58	44
48	90	57	50	49	60	36	72	48	68

Create a frequency table using tallies and draw a bar chart to show these data.

Solution

When you group data, you should make the groups the same size.
Here the groups are in intervals of 10.

Number of apples	Tally	Frequency
30 to 39	ЖΙ	6
40 to 49	Ж Ж ЖΙ	16
50 to 59	Ж Ж Ж Ж Ж ΙΙ	27
60 to 69	Ж Ж Ж Ж ΙΙ	22
70 to 79	Ж Ж Ж ΙΙΙ	18
80 to 89	Ж ΙΙΙΙ	9
90 to 99	ΙΙ	2

The height of the bars shows the frequency of each group.

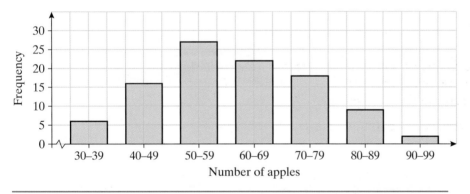

Exercise 3.4

1 Draw a frequency table using tallies for each of these sets of data.

(a) Number of birds in a garden
Use groups of 1 to 5, 6 to 10, 11 to 15, 16 to 20, … .

4	26	11	24	3	7	8	12	23	14
22	15	5	6	7	8	11	3	8	28
3	7	5	7	17	9	1	13	7	12
1	2	8	18	13	12	3	6	6	9
17	15	11	9	8	17	7	14	16	4
9	7	9	7	3	7	8	7	9	13

Exercise continues …

(b) Number of stamps collected per day
Use groups of 1 to 10, 11 to 20, 21 to 30, … .

41	64	12	34	17	32	18	27	37	14
23	25	15	43	3	24	33	13	28	21
13	37	45	27	18	39	31	23	6	19
14	2	28	19	35	41	33	46	27	16
7	35	24	19	9	23	37	24	10	17
26	27	39	26	23	37	28	17	8	13

(c) Number of cars in a car park
Use groups of 1 to 20, 21 to 40, 41 to 60, 61 to 80, … .

41	84	42	34	67	37	88	37	67	74
63	65	55	43	53	44	63	53	48	61
33	67	47	27	48	59	51	73	36	59
14	52	28	17	35	51	43	86	47	46
7	35	14	39	49	63	37	74	50	37
56	27	39	56	53	77	68	57	38	43

2 Draw a bar chart to show each of these sets of data.

(a)

Grapes per bunch	Frequency
30 to 49	35
50 to 69	49
70 to 89	27
90 to 109	18
110 or more	0

(b)

Leaves per branch	Frequency
1 to 25	12
26 to 50	23
51 to 75	31
76 to 100	17
101 or more	9

(c)

Greenfly per plant	Frequency
0 to 19	15
20 to 39	32
40 to 59	45
60 to 79	41
80 to 99	24
100 or more	43

(d)

Number of ants	Frequency
Less than 30	15
30 to 49	37
50 to 69	46
70 to 89	29
90 or more	13

Exercise continues …

(e)	Eggs per day	Frequency
	0 to 9	21
	10 to 19	65
	20 to 29	88
	30 to 39	59
	40 or more	17

(f)	Apples per box	Frequency
	140 to 149	1
	150 to 159	13
	160 to 169	38
	170 to 179	49
	180 to 189	16
	190 to 199	6
	200 or more	2

What you have learned

- Data that are measured are continuous data; data that are counted are discrete data
- Vertical line graphs and bar charts are ways of displaying discrete data
- Two-way tables can be used to record data when you are interested in more than one aspect of the data
- Large amounts of data can be grouped; this can be done for both discrete and continuous data

Mixed exercise 3

1 Which of the following are discrete data and which are continuous data?

Hair colour   Vehicle type   Height   Number of sisters

Crowd size   Length   Score on a dice

2 Design a data collection sheet for each of the following.

(a) Hair colour   (b) Favourite variety of apple

(c) Colour of house door   (d) Handspan

Mixed exercise 3 continues …

3 Draw a vertical line graph for each of these sets of data.

(a)

Score when rolling a dice	Frequency
1	7
2	10
3	6
4	12
5	9
6	6

(b)

Number of goals per game	Frequency
0	4
1	11
2	15
3	7
4	2
More than 4	1

(c)

Hair colour	Frequency
Brown	26
Black	8
Blonde	21
Red	7
Grey	3
Other	5

(d)

Games machine owned	Frequency
Playstation	45
X-box	17
Nintendo Wii	9
DS	26
Other	3

4 Draw a bar chart for each of these sets of data.

(a)

Type of car	Frequency
Ford	31
Vauxhall	28
Toyota	7
Nissan	6
Volkswagen	15
Volvo	3
Other	10

(b)

Number of pets	Frequency
0	15
1	12
2	7
3	2
More than 3	4

(c)

Type of phone	Frequency
Motorola	12
Samsung	7
Nokia	19
Ericsson	8
Siemens	1
Other	4

Mixed exercise 3 continues …

5 A group of students voted on where they wanted to go for a day out at the end of the term. The two-way table shows some of their votes.
Copy and complete the table.

	Alton Towers	Legoland	Thorpe Park	Total
Girls	31	25		
Boys		11		75
Total	74		48	

6 Draw a frequency table using tallies for each of these sets of data.

(a) Number of CDs owned
Use groups of 1 to 10, 11 to 20, 21 to 30, 31 to 40, … .

43	25	51	24	43	23	52	62	23	24
27	14	15	16	37	34	10	34	38	38
31	7	55	8	26	45	41	13	27	15
67	13	48	18	13	56	39	53	3	39
16	35	31	50	18	67	57	44	18	46

(b) Number of flowers per plant
Use groups of 1 to 20, 21 to 40, 41 to 60, 61 to 80, … .

41	74	43	37	57	27	58	26	64	68
63	45	50	45	43	54	43	43	38	51
33	57	41	28	47	51	31	54	46	39
24	62	29	16	25	55	46	66	42	40
37	35	19	34	39	43	37	73	55	32
55	28	36	36	56	77	48	59	28	49

7 Draw a bar chart for each of these sets of data.

(a)

Flowers on a plant	Frequency
1 to 5	8
6 to 10	19
11 to 15	14
16 to 20	6
21 to 25	13

(b)

Passengers on a bus	Frequency
0 to 9	13
10 to 19	31
20 to 29	39
30 to 39	37
40 to 49	22
50 to 59	8

(c)

Mice in a nest	Frequency
1 to 5	4
6 to 10	15
11 to 15	26
16 to 20	27
21 to 25	18

Chapter 4

Decimals

▶ This chapter is about

- Place value in decimals
- Converting familiar fractions into decimals
- Adding and subtracting decimals
- Multiplying and dividing decimals by 10, 100, 1000, …
- Multiplying decimals by integers
- Multiplying simple decimals

▶ You should already know

- The meaning of place value in integers
- How to add and subtract integers
- How to use fraction notation
- How to multiply and divide integers by 10, 100, 1000, …
- How to multiply integers

Place value

Look at a ruler.

Some rulers are marked in millimetres, like this.
The arrow points to 38 mm.
This is between 30 mm and 40 mm and is $\frac{8}{10}$ of the way from 30 mm towards 40 mm.

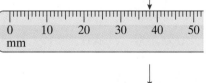

Some rulers are marked in centimetres, like this.
The arrow points to 3.8 cm.
This is between 3 cm and 4 cm and is $\frac{8}{10}$ of the way from 3 cm towards 4 cm.

◉ Discovery 4.1

Find the position of 3.8 cm or 38 mm on your ruler.
Then find the position of 7.4 cm or 74 mm on your ruler.
Work in pairs and describe its position in a similar way as is done above.
Choose some more measurements that are on your ruler.
Describe their positions on the ruler in a similar way.

Decimals are a way of describing numbers which are not integers.
We use them when measuring lengths, as in Discovery 4.1.

We also use them in connection with money. For instance, £0.42 means 42p, or $\frac{42}{100}$ of a pound.

Discovery 4.2

Think of other situations where you know that decimals are used. How many can your group find?

This table shows place values, including decimals.
It shows some of the numbers you have met in this chapter so far.
It can be used with other numbers you meet.

Th	H	T	U	.	$\frac{1}{10}$	$\frac{1}{100}$	$\frac{1}{1000}$
		7	4				
			3	.	8		
			0	.	4	2	

Example 4.1

What is the place value of the digit 4 in these numbers?
(a) 74 000   **(b)** 643.2   **(c)** 8.415   **(d)** 0.04

Solution

Use the place value table.

Ten Th	Th	H	T	U	.	$\frac{1}{10}$	$\frac{1}{100}$	$\frac{1}{1000}$
7	4	0	0	0				
		6	4	3	.	2		
				8	.	4	1	5
				0	.	0	4	

(a) 4 thousands   **(b)** 4 tens   **(c)** 4 tenths   **(d)** 4 hundredths

Changing fractions to decimals

Using place value, you already know that $0.7 = \frac{7}{10}$ and $0.17 = \frac{17}{100}$.

Working backwards, you can use place value to change tenths and hundredths into decimals.

For example $\frac{29}{100} = 0.29$.

Look at the shading on this diagram.

You can see that $\frac{1}{2} = \frac{5}{10} = 0.5$.

The shading in this diagram shows that $\frac{1}{5} = \frac{2}{10} = 0.2$.

Similarly, the unshaded area $= \frac{4}{5} = 0.8$.

This grid has 100 squares.

25 of the squares are shaded.

This is $\frac{1}{4}$ of the grid.

So $\frac{1}{4} = \frac{25}{100} = 0.25$.

75 of the squares are unshaded.

This is $\frac{3}{4}$ of the grid.

So $\frac{3}{4} = \frac{75}{100} = 0.75$.

You can check these results on your calculator by dividing 3 by 4 to get 0.75, and so on.

Exercise 4.1

1 Write in words the place value of the digit 4 in each of these numbers.

   (a) 40   (b) 0.4   (c) 40 000   (d) 8.74   (e) 0.014

2 Write these numbers as decimals.

   (a) $\frac{3}{10}$ (b) $4\frac{7}{10}$ (c) $\frac{9}{100}$ (d) $52\frac{79}{100}$ (e) $\frac{21}{1000}$

3 Write these decimals as fractions or mixed numbers (whole numbers and fractions) in their lowest terms.

   (a) 0.6   (b) 4.3   (c) 14.1   (d) 0.75   (e) 9.03

4 Use the fact that there are 1000 grams in a kilogram to write these weights in kilograms.

   (a) 468 g   (b) 1645 g   (c) 72 g   (d) 6 g   (e) 2450 g

5 Write these lengths in metres.

   (a) 12 cm   (b) 874 mm   (c) 21.8 cm   (d) 56 mm   (e) 138 cm

6 Draw a grid to show that $\frac{3}{5} = \frac{6}{10} = 0.6$.

7 Write these fractions as decimals.

   (a) $\frac{1}{10}$ (b) $\frac{1}{2}$ (c) $\frac{2}{5}$ (d) $\frac{1}{4}$ (e) $\frac{3}{100}$

Multiplying and dividing decimals by 10, 100, 1000, ...

You learned how to multiply and divide integers by 10, 100, 1000, ... in Chapter 1.

You can use place value tables to help you multiply and divide decimals by 10, 100, 1000, ...

Example 4.2

Work out these.

(a) 35.6×10 (b) 35.6×100 (c) $435.2 \div 10$

(d) $435.2 \div 100$ (e) $79.2 \div 100$ (f) $79.2 \div 1000$

Solution

(a)
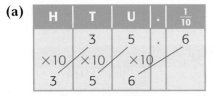

H	T	U	.	$\frac{1}{10}$
	3	5	.	6
×10	×10	×10		
3	5	6		

To multiply by 10 you move each digit one place to the left.

$35.6 \times 10 = 356$

(b)

Th	H	T	U	.	$\frac{1}{10}$
		3	5	.	6
	×100	×100	×100		
3	5	6	0		

To multiply by 100 you move each digit two places to the left.

Notice that you have to put a zero in the units column otherwise the number would read 356 rather than 3560.

$35.6 \times 100 = 3560$

(c)

H	T	U	.	$\frac{1}{10}$	$\frac{1}{100}$
4	3	5	.	2	
÷10	÷10	÷10		÷10	
	4	3	.	5	2

To divide by 10 you move each digit one place to the right.

$435.2 \div 10 = 43.52$

(d)

H	T	U	.	$\frac{1}{10}$	$\frac{1}{100}$	$\frac{1}{1000}$
4	3	5	.	2		
	÷100	÷100		÷100	÷100	
		4	.	3	5	2

To divide by 100 you move each digit two places to the right.

$435.2 \div 100 = 4.352$

(e)

T	U	.	$\frac{1}{10}$	$\frac{1}{100}$	$\frac{1}{1000}$
7	9	.	2		
	÷100		÷100	÷100	
	0	.	7	9	2

To divide by 100 you move each digit two places to the right.

You usually write a zero when there are no digits in front of the decimal point. This makes the number easier to read.

$$79.2 \div 100 = 0.792$$

(f)

T	U	.	$\frac{1}{10}$	$\frac{1}{100}$	$\frac{1}{1000}$	$\frac{1}{10\,000}$
7	9	.	2			
	÷1000		÷1000	÷1000		
	0	.	0	7	9	2

To divide by 1000 you move each digit three places to the right.

You have to write a zero in the tenths column to show that there are no tenths in the number.

$$79.2 \div 1000 = 0.0792$$

Exercise 4.2

1 Multiply each of these numbers by 10.

(a) 56 (b) 8 (c) 7.9 (d) 8.26

(e) 7.34 (f) 15.89 (g) 8.3 (h) 0.6

(i) 5.9 (j) 0.08 (k) 9.62 (l) 27.48

(m) 7 (n) 35

> **TIP**
>
> You don't have to draw a place value table. You can move the digits in your head.

2 Multiply each of these numbers by 100.

(a) 56 (b) 8 (c) 7.9 (d) 8.26

(e) 7.34 (f) 15.89 (g) 8.3 (h) 0.6

(i) 5.9 (j) 0.08 (k) 9.62 (l) 27.48

(m) 9 (n) 27

3 Multiply each of these numbers by 1000.

(a) 56 (b) 8 (c) 7.9 (d) 8.26

(e) 7.34 (f) 15.89 (g) 8.3 (h) 0.6

(i) 5.9 (j) 0.08 (k) 9.62 (l) 27.48

(m) 6 (n) 43

Exercise continues …

4 Divide each of these numbers by 10.

(a) 1.8   (b) 2.7   (c) 8.3   (d) 18   (e) 27

(f) 483   (g) 0.06   (h) 7.28   (i) 28.39   (j) 2.4

(k) 6.1   (l) 7.8   (m) 51   (n) 43   (o) 682

(p) 8.34   (q) 5.07   (r) 42.93

5 Divide each of these numbers by 100.

(a) 1.8   (b) 2.7   (c) 8.3   (d) 18   (e) 27

(f) 483   (g) 0.06   (h) 7.28   (i) 28.39   (j) 2.4

(k) 6.1   (l) 7.8   (m) 51   (n) 43   (o) 682

(p) 8.34   (q) 5.07   (r) 42.93

6 Divide each of these numbers by 1000.

(a) 1.8   (b) 2.7   (c) 8.3   (d) 18   (e) 27

(f) 483   (g) 0.06   (h) 7.28   (i) 28.39   (j) 2.4

(k) 6.1   (l) 7.8   (m) 51   (n) 43   (o) 682

(p) 8.34   (q) 5.07   (r) 42.93

7 Work out these.

(a) $15.6 \times 10\,000$ (b) $0.07 \times 10\,000$ (c) $17.28 \times 100\,000$

(d) $48.3 \times 1\,000\,000$ (e) $4.32 \times 100\,000$ (f) $1.03 \times 10\,000$

(g) $37.48 \times 10\,000$ (h) $84.2 \times 1\,000\,000$

8 Work out these.

(a) $956 \div 10\,000$ (b) $7863 \div 10\,000$ (c) $236.4 \div 100\,000$

(d) $2364 \div 1\,000\,000$ (e) $975.3 \div 10\,000$ (f) $269.43 \div 10\,000$

(g) $6925 \div 100\,000$ (h) $692.5 \div 100\,000$

9 In 100 years earnings have multiplied on average by 1000.
If someone earned £0.30 per week in 1900, what would they earn in 2000?

10 The value of some shares in a company increased by 100 times from when they were bought to when they were sold.
The shares cost £0.16 each when bought.
How much were they worth when sold?

11 On a model of a jumbo jet, all real lengths are divided by 100.
The length of a jumbo jet is 70 m.
What is the length of the model?

Exercise continues ...

12 On a road map, all real distances are divided by 100 000 to
get the distance on the map.

(a) The distance from Longcroft to Winter Bay is 8000 m.
How far is it on the map?

(b) On the map the distance from Gradbridge to
Lowton Waters is 0.3 m.
How far is it actually?

Adding and subtracting decimals

◉ Discovery 4.3

Use your ruler to draw a line which is 8.5 cm long.
Mark a point on the line which is 4.7 cm from one end.
Measure the distance from this point to the other end of your line.

Do an appropriate addition or subtraction to find out if your
measurement is accurate.
Work in pairs. Repeat the instructions above using different
measurements.
One person finds the distance by drawing, the other by doing a
subtraction.

For a line 75 mm long and a point 52 mm from one end, the distance
from the point to the other end is found by doing this subtraction.

$$\begin{array}{r} 75 \\ -52 \\ \hline 23 \text{ mm} \end{array}$$

Working in centimetres, for a line 7.5 cm long and a point 5.2 cm
from one end, the distance from the point to the other end is found by
doing this subtraction.

$$\begin{array}{r} 7.5 \\ -5.2 \\ \hline 2.3 \text{ cm} \end{array}$$

When you use column methods of adding or subtracting, make sure
you line up all the decimal points under each other. Then add or
subtract as you would with integers.

Multiplying a decimal by an integer

Compare these two multiplications for finding the cost of three CDs at £4.95 each.

Working in pence

You know that £4.95 = 495p.

$$\begin{array}{r} 495 \\ \times\ 3 \\ \hline 1485p \end{array} = £14.85$$

Multiply first the units, then the tens and then the hundreds by 3.
Write the units under the units, the tens under the tens, and so on.
Convert your answer from pence to pounds by dividing by 100.

Working in pounds

$$\begin{array}{r} 4.95 \\ \times\ 3 \\ \hline £14.85 \end{array}$$

To multiply a decimal by an integer, put the decimal points under each other.
Make sure you line up your work carefully.
Put the first digit you work out under the last decimal place.

In each case, the digits are the same.

> **TIP**
>
> A quick estimate can help you check that your answer is sensible. Here the cost will be less than 3 × £5 = £15.

Example 4.3

Clare bought four melons at £1.45 each.
How much change did she get from £10?

Solution

First, find the cost of the melons.

$$\begin{array}{r} 1.45 \\ \times\ 4 \\ \hline 5.80 \end{array}$$

Then subtract from £10 to find the change.

$$\begin{array}{r} 10.00 \\ -5.80 \\ \hline 4.20 \end{array}$$

Answer: £4.20

> **TIP**
>
> Using a calculator to solve this problem would give the answer 4.2.
>
> Remember that, when working with money, you must write the answer as £4.20.

Challenge 4.1

What strategies could you use to solve the problem in Example 4.3 mentally?

Example 4.4

A piece of wood is 2.3 m long. 75 cm is cut off.
How much remains?

Solution

First, make the units the same.   75 cm = 0.75 m

Then do the subtraction.
$$\begin{array}{r} 2.3\,0 \\ -0.7\,5 \\ \hline 1.5\,5 \end{array}$$

Answer: 1.55 m

You could also solve this problem by working in centimetres and then changing your answer into metres.

Multiplying a decimal by a decimal

Discovery 4.4

(a) Work out the answers to 120×4, 12×4 and 1.2×4.

(b) Work out the answers to 216×7, 21.6×7 and 2.16×7.

(c) Compare the answers in each part. What do you notice?

Look again at the calculations in Discovery 4.4 and your answers.
For each set of answers, the digits are the same, but the place values of the digits are different.

This helps you in finding the answer to a calculation such as 0.2×0.3.

Look at these calculations.

$2 \times 3 = 6$

$0.2 \times 3 = 0.6$ Multiplying 2 tenths by 3 means the answer is 6 tenths.

$2 \times 0.3 = 0.3 \times 2 = 0.6$ Multiplying 2 by 3 tenths means the answer is 6 tenths.

These help you to see that

$0.2 \times 0.3 = 0.06$ Multiplying 2 tenths by 3 tenths gives 6 hundredths.

These are steps you take to multiply decimals.

1 Carry out the multiplication ignoring the decimal points.
The digits in the answer will be the same as the digits in the final answer.

2 Count the total number of decimal places in the two numbers to be multiplied.

3 Put the decimal point in the answer you got in step 1 so that the final answer has the same number of decimal places as you found in step 2.

This also works when you multiply an integer by a decimal.

Example 4.5

Work out 0.8×0.7.

Solution

1 First do $8 \times 7 = 56$.
2 The total number of decimal places in 0.8 and 0.7 $= 1 + 1 = 2$.
3 The answer is 0.56.

TIP

Notice that when you multiply by a number between 0 and 1, such as 0.7, you decrease the original number (0.8 to 0.56).

Exercise 4.3

1 Work out these.

(a) $\begin{array}{r} 6.72 \\ +7.19 \end{array}$

(b) $\begin{array}{r} 18.95 \\ +23.14 \end{array}$

(c) $\begin{array}{r} 27.54 \\ +83.61 \end{array}$

(d) $\begin{array}{r} 5.91 \\ +8.72 \end{array}$

(e) $\begin{array}{r} 16.74 \\ +43.97 \end{array}$

(f) $\begin{array}{r} 33.51 \\ +79.86 \end{array}$

2 Work out these.

(a) $\begin{array}{r} 16.78 \\ -7.13 \end{array}$

(b) $\begin{array}{r} 28.75 \\ -13.84 \end{array}$

(c) $\begin{array}{r} 128.36 \\ -73.52 \end{array}$

(d) $\begin{array}{r} 13.49 \\ -5.18 \end{array}$

(e) $\begin{array}{r} 47.51 \\ -26.74 \end{array}$

(f) $\begin{array}{r} 439.87 \\ -218.03 \end{array}$

3 Work out these.
(a) £6.84 + 37p + £9.41
(b) £16.83 + 94p + £6.81 + 32p
(c) £61.84 + 76p + £9.72 + £41.32 + 83p
(d) £3.89 + 73p + 68p + £91.80

Exercise continues …

4 Find the cost of five CDs at £11.58 each.

5 In the long jump, Jim jumps 13.42 m and Dai jumps 15.18 m.
Find the difference between the lengths of their jumps.

6 The times for the first and last places in a 200–metre race were 24.42 seconds and 27.38 seconds.
Find the difference between these times.

7 Kate buys three of these packs of meat.
(a) What is the total weight?
(b) What is the total cost?

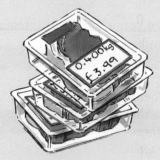

8 Find the cost of 5 kg of new potatoes at £1.18 per kilogram.

9 Work out these. Give your answers in the larger unit.
(a) 6.1 m + 92 cm + 9.3 m   (b) 3.2 m + 28 cm + 6.74 m + 93 cm
(c) 7.2 m − 165 cm   (d) 8.5 m − 62 cm
(e) 7.6 cm − 8 mm   (f) 8.5 cm − 12 mm

10 Work out these. Where applicable, give your answers in the larger unit.
(a) 300 g + 1.4 kg + 72 g + 2.8 kg   (b) 3.9 kg + 760 g − 2.7 kg
(c) 2.4 kg − 786 g   (d) 2 litres − 525 ml
(e) 4 × 0.468 litres   (f) $\frac{1}{2}$ litre + 200 ml

11 Pali buys two shirts at £8.95 each and a pair of trousers at £17.99.
How much change does he get from £50?

12 Gemma buys two cucumbers at 68p each and three cauliflowers at £1.25 each.
How much change does she get from £10?

13 Work out these.
(a) 5 × 0.7   (b) 0.3 × 6   (c) 4 × 0.6
(d) 0.7 × 9   (e) 0.3 × 0.1   (f) 0.9 × 0.6
(g) 50 × 0.3   (h) 0.6 × 70   (i) 0.4 × 0.2
(j) 0.5 × 0.3   (k) $(0.5)^2$ (l) $(0.1)^2$

What you have learned

- The place value of a digit in a number tells you the value of that digit; the place value table has columns getting greater in value as you move left
- You can use place value to convert tenths, hundredths, etc. to fractions, for example $0.1 = \frac{1}{10}, 0.03 = \frac{3}{100}, 0.13 = \frac{13}{100}$ and $2.13 = 2\frac{13}{100}$
- The decimal equivalents of some other common fractions besides tenths and hundredths

Fraction	$\frac{1}{2}$	$\frac{1}{4}$	$\frac{3}{4}$	$\frac{1}{5}$	$\frac{2}{5}$	$\frac{3}{5}$	$\frac{4}{5}$
Decimal	0.5	0.25	0.75	0.2	0.4	0.6	0.8

- To multiply a number by 10 you move the digits one place to the left; to multiply a number by 100 you move the digits two places to the left; to multiply a number by 1000 you move the digits three places to the left
- To divide a number by 10 you move the digits one place to the right; to divide a number by 100 you move the digits two places to the right; to divide a number by 1000 you move the digits three places to the right
- To add and subtract decimals, write the digits in columns with the decimal points lining up
- To multiply a decimal by an integer, write the decimal point of the answer under the decimal point of the number in the question; work from the right and put the first digit you work out in the last decimal place
- To multiply a decimal by a decimal, carry out the multiplication ignoring the decimal points then place the decimal point so that there are the same number of decimal places in the answer as there are in total in the two numbers being multiplied

Mixed exercise 4

1. What is the place value of the digit 6 in these numbers?
   (a) 6000
   (b) 4.6
   (c) 8462
   (d) 9.46
   (e) 176.09

2. Write these numbers as decimals.
   (a) $\frac{9}{10}$
   (b) $2\frac{1}{10}$
   (c) $\frac{7}{100}$
   (d) $16\frac{23}{100}$
   (e) $\frac{19}{1000}$

3. Write these fractions as decimals.
   (a) $\frac{3}{10}$
   (b) $\frac{1}{5}$
   (c) $\frac{3}{4}$
   (d) $\frac{7}{10}$
   (e) $\frac{3}{5}$

4. Write these lengths in centimetres.
   (a) 2.36 m
   (b) 83 mm
   (c) 0.57 m
   (d) 5.8 m
   (e) 470 mm

Exercise continues …

5 Work out these.

    (a) 5.3 × 10   (b) 6.4 × 100   (c) 6.79 × 1000   (d) 53.92 × 10 000

6 Work out these.

    (a) 5.6 ÷ 10   (b) 8.9 ÷ 100   (c) 23.8 ÷ 100   (d) 72.89 ÷ 1000

7 On a plan all distances are divided by 1000.
A building is actually 48.6 m long.
What will the length of the building be on the plan?

8 Work out these.

    (a)   6.82   (b)   26.92   (c)   27.36
        +2.49   +18.54   +91.48

    (d)   9.16   (e)   13.84   (f)   38.53
        +7.72   +37.67   +89.76

9 Work out these.

    (a) 21.74   (b) 36.86   (c) 130.46
      −8.13   −12.78   −83.92

    (d) 12.59   (e) 35.57   (f) 409.15
      −7.16   −28.74   −213.08

10 Find the cost of seven CDs at £8.59 each.

11 Two pieces of wood are put end to end. Their lengths are 2.5 m and 60 cm.
Find, in metres, the total length of the wood.

12 Maisy buys two magazines at £1.69 each and a bunch of flowers at £3.70.
How much change does she get from £10?

13 Work out these.

    (a) 3 × 0.4   (b) 0.5 × 0.1   (c) 0.7 × 0.8   (d) 1.2 × 0.4

Chapter 5

Formulae

▶ **This chapter is about**

- Formulae written in words
- Formulae written using letters
- Substituting numbers into a formula

▶ **You should already know**

- How to add, subtract, multiply and divide whole numbers and decimals
- How to use letters to represent numbers

Formulae written in words

A **formula** is a rule used to work something out. You can write a formula in words or algebraically using letters.

Here are two examples showing how you use a formula written in words.

Example 5.1

To work out John's weekly wage, multiply the number of hours he works by £6.

How much does John earn when he works

(a) 10 hours? **(b)** 40 hours? **(c)** 25 hours?

Solution

(a) $10 \times 6 = £60$ **(b)** $40 \times 6 = £240$ **(c)** $25 \times 6 = £150$

Example 5.2

To find the perimeter of a rectangle, add the length and the width and multiply the total by two.

Work out the perimeter of the rectangles with these dimensions.

(a) Length 5 cm and width 4 cm

(b) Length 19 m and width 8 m

(c) Length 3.2 cm and width 6.1 cm

Solution

(a) $5 + 4 = 9$ You add the length and the width together first.
$9 \times 2 = 18 \,\text{cm}$ Then you multiply the total by 2.

(b) $19 + 8 = 27$ $27 \times 2 = 54 \,\text{m}$

(c) $3.2 + 6.1 = 9.3$ $9.3 \times 2 = 18.6 \,\text{cm}$

> **TIP**
>
> Notice that the units are not included in the calculation. They can get in the way. However, you *must* include them in your answer.

Exercise 5.1

1. The cost of a carpet is found by multiplying the area of a room by £5.
   Work out the cost of a carpet for rooms with these areas.
   (a) $9 \,\text{m}^2$ (b) $20 \,\text{m}^2$ (c) $12 \,\text{m}^2$ (d) $15 \,\text{m}^2$

2. To find Bob's age, subtract 14 from Janice's age.
   How old is Bob when Janice is
   (a) 17?   (b) 46?   (c) 30?   (d) 51?

3. Tanya is doing a sponsored walk for charity.
   She will receive £4 for each mile that she walks.
   How much money will she raise if she walks
   (a) 6 miles?   (b) 15 miles?   (c) $3\frac{1}{2}$ miles?   (d) 23 miles?

4. A group of people win some money in a quiz.
   To work out the amount each person receives, divide the amount of money by the number of people.
   How much does each person receive when
   (a) 4 people win £100?   (b) 5 people win £60?
   (c) 3 people win £840?   (d) 2 people win £37?

5. The cost of hiring a cement mixer is £50 plus £10 per hour.
   How much will it cost to hire the mixer for
   (a) 4 hours?   (b) 10 hours?   (c) $5\frac{1}{2}$ hours?   (d) 24 hours?

6. The area of a triangle is found by multiplying the base by the height and dividing the answer by two.
   Work out the areas of these triangles.
   (a) Base 3 cm and height 6 cm   (b) Base 10 cm and height 15 cm
   (c) Base $2\frac{1}{2}$ m and height 4 m   (d) Base 8.2 mm and height 3 mm

Exercise continues ...

7 To find the speed, divide the distance travelled by the time taken.
   Work out these.
   (a) The speed of a car, in miles per hour, which travels 80 miles in 2 hours.
   (b) The speed of a train, in kilometres per hour, which travels 240 km in 3 hours.
   (c) The speed of a runner, in metres per second, who runs 200 m in 25 seconds.
   (d) The speed of an aircraft, in miles per hour, which travels 920 miles in 4 hours.

8 To find the cost of going to the cinema, multiply the number of adults by £6 and the
   number of children by £2.50. Then add the two answers together.
   Work out the cost for
   (a) 2 adults and 2 children.
   (b) 3 adults and 4 children.
   (c) 8 adults and 1 child.
   (d) 5 adults and 3 children.

9 The volume of a box is found by multiplying the length by the width by the height.
   Work out the volume of boxes with the following dimensions.
   (a) Length 3 cm, width 4 cm and height 5 cm
   (b) Length 2 cm, width 7 cm and height 2 cm
   (c) Length 5 cm, width 5 cm and height 5 cm
   (d) Length 15 mm, width 20 mm and height 35 mm

10 To find the amount of tax someone pays, divide their wage by 5.
   Work out the tax payable when a person earns
   (a) £300.   (b) £2000.   (c) £50.   (d) £240.

Some rules of algebra

- You do not need to write the $\times$ sign:
       $4 \times t$ is written $4t$.
- When being multiplied, the number is always written in front of the
  letter:
       $p \times 6 - 30$ is written $6p - 30$.
- You always start a formula with the single letter you are finding:
       $2 \times l + 2 \times w = P$ is written $P = 2l + 2w$.
- When there is a division in a formula it is always written as a
  fraction:
       $y = k \div 6$ is written $y = \dfrac{k}{6}$.

Check up 5.1

Write each of these formulae in the correct algebraic way.

(a) $I = 7 \times V$ **(b)** $p = s \times 4$ **(c)** $m \times a = F$

(d) $10 - x = y$ **(e)** $r = d \div 2$ **(f)** $t \times 10 = v$

(g) $z \div y = w$ **(h)** $t = 30 \times n + 50$ **(i)** $A = w \times 6 \times h$

(j) $m = k \times 5 \div 8$ **(k)** $u - t \times 10 = v$

Formulae written using letters

You can write a word formula as a formula using letters. It is useful to use a letter which tells you something about what it represents.

Example 5.3

To work out John's weekly wage, multiply the number of hours he works by £6.

Write a formula using letters to work out John's weekly wage, in £.

Solution

Use w to represent John's wage in £ and h to represent the number of hours he works.

$w = h \times 6$ which is written as $w = 6h$

> **TIP**
>
> Notice that none of the units are included in the formula.

Example 5.4

To find the time needed to cook a piece of meat, allow 30 minutes for each kilogram and then add on an extra 20 minutes.

Write a formula to work out the time, in minutes, needed to cook a piece of meat.

Solution

Use t to represent the time in minutes and k to represent the weight of the meat in kilograms.

$t = 30 \times k + 20$ which is written as $t = 30k + 20$

Exercise 5.2

In questions 1 to 10, write down a formula for the situation using the letters in **bold**.

1 The **c**ost of a carpet is found by multiplying the **a**rea of a room by £5.

2 To find **B**ob's age, subtract 14 from **J**anice's age.

3 Tanya is doing a sponsored walk for charity.
   She will **r**eceive £4 for each **m**ile that she walks.

4 A group of people win some money in a quiz.
   To work out the **a**mount each person receives, divide the amount of **m**oney by the **n**umber of people.

5 The **c**ost of hiring a cement mixer is £50 plus £10 per **h**our.

6 The **a**rea of a triangle is found by multiplying the **b**ase by the **h**eight and dividing the answer by two.

7 To find the **s**peed, divide the **d**istance travelled by the **t**ime taken.

8 To find the **t**otal cost of going to the cinema, multiply the number of **a**dults by £6 and the number of **c**hildren by £2.50. Then add the two answers together.

9 The **v**olume of a box is found by multiplying the **l**ength by the **w**idth by the **h**eight.

10 To find the amount of **t**ax someone pays, multiply their **w**age by 0.2.

In questions 11 to 15, write down a formula for the situation using appropriate letters and say what each letter stands for.

11 The total amount saved when Tim saves £4.50 a week.

12 The number of 3 m strips of paper that can be cut from a roll of paper.

13 The sale price of a computer when a discount is taken from the normal price.

14 A taxi company works out a fare by dividing the distance covered by 5 and then adding 3.

15 The total cost of a school outing when the coach costs £100 to hire and the entrance fee is £7 for each student.

Challenge 5.1

A family of five visit a safari park.
They use the formula $F = 10a + 6s + 4c$
to work out the total entry fee.

Safari Park
Adults   £10
Students £6
Children £4

(a) What do the letters F, a, s and c stand for?

(b) The total entry fee was £34.
Find the possible combinations of people that
could be in the family.
Check with a partner to see how many different answers they found.

Substituting numbers into a formula

If you are told the values of the letters in an expression or formula you
can find the value of that expression or formula. You **substitute**, or
replace, the letters with the values you have been given.

Example 5.5

Find the value of these expressions when $p = 2$, $q = 3$ and $r = 4$.

(a) $p + q$ **(b)** $r - p$ **(c)** $5p$ **(d)** $r + 2q$

(e) $4p + 6q$ **(f)** $5pq$ **(g)** pqr **(h)** $\dfrac{qr}{6}$

(i) q^2 **(j)** $r^2 - 3p^2$

Solution

(a) $p + q = 2 + 3$
          $= 5$

(b) $r - p = 4 - 2$
          $= 2$

(c) $5p = 5 \times p$
       $= 5 \times 2$
       $= 10$

(d) $r + 2q = r + 2 \times q$
            $= 4 + 2 \times 3$
            $= 4 + 6 = 10$

(e) $4p + 6q = 4 \times 2 + 6 \times 3$
               $= 8 + 18$
               $= 26$

(f) $5pq = 5 \times p \times q$
        $= 5 \times 2 \times 3$
        $= 30$

(g) $pqr = p \times q \times r$

       $= 2 \times 3 \times 4$

       $= 24$

(h) $\dfrac{qr}{6} = \dfrac{q \times r}{6}$

      $= \dfrac{3 \times 4}{6}$

      $= 2$

(i) $q^2 = q \times q$
     $= 3 \times 3$
     $= 9$

(j) $r^2 - 3p^2 = r \times r - 3 \times p \times p$
             $= 4 \times 4 - 3 \times 2 \times 2$
             $= 16 - 12 = 4$

Exercise 5.3

1 Find the value of these expressions when $a = 5$, $b = 4$ and $c = 2$.

(a) $a + b$ (b) $b + c$ (c) $a - c$

(d) $a + b + c$ (e) $2a$ (f) $3b$

(g) $5c$ (h) $3a + b$ (i) $3c - b$

(j) $a + 6c$ (k) $4a + 2b$ (l) $2b + 3c$

(m) $a - 2c$ (n) $8c - 2b$ (o) bc

(p) $4ac$ (q) abc (r) $ac + bc$

(s) $ab - bc - ca$ (t) a^2 (u) $\dfrac{a + b}{3}$

(v) $\dfrac{ab}{c}$ (w) $b^2 + c^2$ (x) $3c^2$

(y) $a^2 b$ (z) c^3

2 Find the value of these expressions when $t = 3$.

(a) $t + 2$ (b) $t - 4$ (c) $5t$

(d) $4t - 7$ (e) $2 + 3t$ (f) $10 - 2t$

(g) t^2 (h) $10t^2$ (i) $t^2 + 2t$

(j) t^3

3 Use the formula $A = 5k + 4$ to find A when

(a) $k = 3$.   (b) $k = 7$.   (c) $k = 0$.

(d) $k = \frac{1}{2}$.   (e) $k = 2.1$.

4 Use the formula $y = mx + c$ to find y when

(a) $m = 3$, $x = 2$, $c = 4$.   (b) $m = \frac{1}{2}$, $x = 8$, $c = 6$.

(c) $m = 10$, $x = 15$, $c = 82$.

5 Use the formula $D = \dfrac{m}{v}$ to find D when

(a) $m = 24$, $v = 6$.   (b) $m = 150$, $v = 25$.

(c) $m = 17$, $v = 2$.   (d) $m = 4$, $v = \frac{1}{2}$.

● Challenge 5.2

Find the value of these expressions when $x = -2$, $y = 3$ and $z = -4$.

(a) $x + y$ (b) $y - z$ (c) $3y + z$

(d) $5z + 2x$ (e) yz (f) xz

● **Challenge 5.3**

A plumber charges £20 per hour plus a call-out fee of £30.
The total cost, £C, for a job taking h hours is given by this formula.

$$C = 20h + 30$$

(a) Find the total cost of a job which takes
    (i)   2 hours.
    (ii) 4 hours.
    (iii) 8 hours.
    (iv) $6\frac{1}{2}$ hours.

(b) How many hours does a job last if the total cost is
    (i)   90?
    (ii) £130?
    (iii) £200?

● **Challenge 5.4**

Formula codes

Give each letter of the alphabet a number, in order, from 1 to 26.

$$a = 1 \qquad b = 2 \qquad c = 3 \qquad d = 4 \qquad e = 5 \qquad \dots$$
$$x = 24 \qquad y = 25 \qquad z = 26$$

(a) Substitute the value of the letters into each of the formulae below.
Find the letters for each of the answers and they will spell a word.
The first one has been done for you.

$a + c = 1 + 3 = 4 \rightarrow d$
$t - s$
$2g$
$\dfrac{u}{c}$
$2p - 3i$
$dt + b - h^2$

(b) Make some messages of your own.
Exchange your messages with a friend and decipher them.

(c) Try a different numbering of the letters of the alphabet such as
$a = 26, b = 25, c = 24, \dots$ and write some messages using these.
Exchange your messages with a friend and see if you can 'crack'
the new code.

● Challenge 5.5

(a) Write down a formula in x for the total, T, of the angles on this straight line.

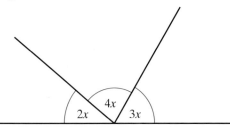

(b) The angles on a straight line add up to 180°.
    (i)   Use this fact to write down an equation in x.
    (ii) Solve the equation to find x.
    (iii) Work out the size of each of the three angles.

● Challenge 5.6

(a) Write down a formula in x for the perimeter, P, of this triangle. Write your answer as simply as possible.

(b) The perimeter of the triangle is 22 cm.
    (i)   Write down an equation in x.
    (ii) Solve the equation to find x.
    (iii) Work out the lengths of the three sides of the triangle.

$3x + 1$

$2x - 1$

$x + 4$

What you have learned

- A formula is a rule used to work something out; you can write a formula in words or algebraically using letters
- How to use a formula written in words
- Some rules for writing expressions using algebra
- How to write a formula using letters
- How to substitute numbers into a formula

Mixed exercise 5

1 The approximate distance around a circular pond is found by multiplying the diameter of the pond by three.
Work out the distance around circular ponds with these diameters.

(a) 5 m   (b) 14 m   (c) 25 m   (d) 3.5 m

2 To change a temperature in degrees Celsius to a temperature in degrees Fahrenheit, multiply the temperature in degrees Celsius by 1.8 and then add on 32.
Work out the Fahrenheit equivalents to the following.

(a) 10 °C   (b) 0 °C   (c) 30 °C   (d) 12 °C

3 The time taken for a journey can be found by dividing the distance covered by the speed.
Work out the time it takes

(a) a train to cover a distance of 200 miles at a speed of 100 mph.

(b) a car to cover a distance of 180 miles at a speed of 40 mph.

(c) a plane to cover a distance of 3500 miles at a speed of 250 mph.

(d) a runner to cover a distance of 400 m at a speed of 5 metres per second.

4 Write down a formula using letters for each of the situations in Questions **1** to **3** and say what each letter stands for.

5 Find the value of these expressions when $x = 2$, $y = 3$ and $z = 5$.

(a) $x + 7$ (b) $6 - y$ (c) $6z$ (d) $9y$

(e) yz (f) $4xy$ (g) $8z - y$ (h) $4x + 6y$

(i) $8z - 12x$ (j) $x + 4y - 2z$ (k) $\dfrac{xyz}{6}$ (l) $\dfrac{5x + 9y + 8z}{x + y + z}$

(m) z^2 (n) $y^2 + x^2$ (o) x^3 (p) $5z^2 - 2y^2$

Chapter 6

Equations 1

▶ This chapter is about

- Solving equations

▶ You should already know

- That you use a letter to stand for the number of objects not for the objects themselves
- How to write expressions and formulae
- How to simplify an expression by collecting like terms

One-step equations

An **equation** contains an equals sign linking an algebraic expression and a number or two algebraic expressions.

To solve an equation you must always do each operation to both sides of the equation. This is shown in the next example.

Example 6.1

Solve the following equations.

(a) $3d = 12$ **(b)** $m - 5 = 9$

Solution

(a) $3d = 12$ The d has been multiplied by 3.
To find d you must divide by 3.

$3d \div 3 = 12 \div 3$ To keep the sides the same you must also divide the 12 by 3.

$d = 4$

(b) $m - 5 = 9$ 5 has been subtracted from m.
To find m you must add 5.

$m - 5 + 5 = 9 + 5$ To keep the sides the same you must also add 5 to the 9.

$m = 14$

Exercise 6.1

Solve these equations.

TIP

Always do the same operation to the whole of both sides.

1 $3a = 15$ 2 $4p = 20$ 3 $2x = 16$

4 $4m = 8$ 5 $a + 1 = 8$ 6 $x + 3 = 6$

7 $n - 3 = 9$ 8 $h + 6 = 9$ 9 $r + 5 = 16$

10 $n - 3 = 1$ 11 $x - 17 = 11$ 12 $m - 12 = 1$

13 $b - 12 = 4$ 14 $p - 22 = 14$ 15 $x + 6 = 10$

16 $x + 8 = 3$ 17 $x + 6 = 2$ 18 $x + 2 = -6$

19 $x - 2 = -6$ 20 $x + 4 = -10$

Two-step equations

Sometimes more than one step is needed to solve an equation.
At each stage you must do the same to both sides of the equation.
This is shown in the next example.

Example 6.2

Solve the following equations.

(a) $12 = 14 - x$

(b) $3x - 1 = 8$

Solution

(a)

$$12 = 14 - x \qquad \text{x has been subtracted from 14.}$$
$$12 + x = 14 - x + x \qquad \text{First add x to each side.}$$
$$12 + x = 14$$
$$12 + x - 12 = 14 - 12 \qquad \text{Now subtract 12 from each side.}$$
$$x = 2$$

(b)

$$3x - 1 = 8$$
$$3x - 1 + 1 = 8 + 1 \qquad \text{Add 1 to each side.}$$
$$3x = 9$$
$$3x \div 3 = 9 \div 3 \qquad \text{Divide each side by 3.}$$
$$x = 3$$

Exercise 6.2

Solve these equations.

1 $11 = 17 - x$	2 $1 = 12 - m$	3 $4 = 12 - b$	4 $14 = 22 - p$
5 $5x + 2 = 17$	6 $4x - 11 = 5$	7 $2x - 5 = 9$	8 $3x - 7 = 8$
9 $3x + 7 = 13$	10 $5x - 8 = 12$	11 $4x - 12 = 8$	12 $5x - 6 = 39$
13 $2x - 6 = 22$	14 $6x - 7 = 41$	15 $4x - 3 = 29$	16 $5x + 10 = 5$

● Challenge 6.1

(a) Imagine that you have just solved an equation.
The operations were add 2 and divide by 5.
The solution was $x = 4$.
Can you find the equation you solved?

(b) Work in pairs. Take turns to write down other operations and
solutions and find the equations.

Equations involving division

When you see a fraction in an equation, a division has taken place.

To solve the equation you multiply both sides of the equation by the
denominator of the fraction.

Example 6.3

Solve the equation $\frac{x}{7} = 4$.

Solution

$\frac{x}{7} = 4$ Remember: $\frac{x}{7}$ means $x \div 7$.

$\frac{x}{7} \times 7 = 4 \times 7$ Multiply each side by 7.

$x = 28$

Exercise 6.3

Solve these equations.

1 $\dfrac{x}{3} = 2$ 2 $\dfrac{p}{2} = 6$ 3 $\dfrac{p}{5} = 5$ 4 $\dfrac{x}{2} = 9$ 5 $\dfrac{a}{6} = 1$

6 $\dfrac{m}{4} = 12$ 7 $\dfrac{t}{2} = 2$ 8 $\dfrac{b}{8} = 16$ 9 $\dfrac{d}{3} = 6$ 10 $\dfrac{y}{10} = 100$

11 $\dfrac{x}{3} = 18$ 12 $\dfrac{x}{2} = 9$ 13 $\dfrac{x}{4} = 1$ 14 $\dfrac{x}{7} = 3$ 15 $\dfrac{x}{6} = 12$

16 $12 = \dfrac{x}{2}$ 17 $20 = \dfrac{x}{4}$ 18 $3 = \dfrac{x}{10}$ 19 $-2 = \dfrac{x}{5}$ 20 $-4 = \dfrac{x}{4}$

Word problems

With word problems you need to work out the equation to be solved.

Example 6.4

A packet of sweets is divided equally between five children.
Each child receives four sweets.
How many sweets were there in the packet?

Solution

Use x to represent the number of sweets in the packet.
The equation to solve is therefore

$\dfrac{x}{5} = 4$ since when the sweets are divided between five
                        children they each receive four sweets.

$\dfrac{x}{5} \times 5 = 4 \times 5$ Multiply both sides by 5.

$x = 20$ There were 20 sweets in the packet.

Exercise 6.4

1   The angles on a straight line add up to 180°.
    Write down and solve an equation for each of these diagrams.

_____ 121° a _____   _____ b 88° _____   _____ c 31° _____   _____ 40° d 42° _____

Exercise continues …

2 Sam is two years older than his brother. His brother is 16.
Write down an equation which uses x to represent Sam's age.
Solve your equation to find Sam's age.

3 Eight children take the same amount of money, £x, to school for a trip.
The total amount of money collected is £72.
Write down an equation and solve it to find x.

4 5 added to twice a number gives an answer of 13.
Use x to represent the number.
Write down and solve an equation to find x.

5 6 subtracted from three times a number gives an answer of 18.
Use x to represent the number.
Write down and solve an equation to find x.

6 Lauren thinks of a number.
She multiplies it by three and then subtracts 5. The answer is 10.
Use x to represent Lauren's number.
Write down and solve an equation to find Lauren's number.

Challenge 6.2

(a) Can you think of a word problem to give the equation $4x - 3 = 7$?
Work in pairs and give your problem to your partner to find the
equation.
Is it the same equation?

(b) Think of your own equation.
Turn it into a word problem and give it to your partner to find
the equation.

What you have learned

- An equation contains an equals sign linking an algebraic expression and a number or two algebraic expressions
- To solve an equation you must always do the same operations to both sides of the equation

Mixed exercise 6

Solve the equations in questions **1** to **20**.

1 $5x = 20$ **2** $4x = 24$ **3** $5x = 35$ **4** $2x = 12$

5 $x - 19 = 4$ **6** $x + 10 = -16$ **7** $x - 12 = 2$ **8** $6 + x = 5$

9 $4 = 11 - x$ **10** $5x + 6 = 31$ **11** $3m - 9 = 0$ **12** $4p + 4 = 12$

13 $7y - 6 = 50$ **14** $3x + 2 = 14$ **15** $2x - 2 = 8$ **16** $3x - 6 = 12$

17 $8x - 1 = 15$ **18** $2x + 4 = 36$ **19** $\dfrac{x}{5} = 15$ **20** $\dfrac{x}{2} = -8$

21 Jack thinks of a number.
He multiplies it by 10, then he adds 5. The answer is 95.
Use x to represent Jack's number.
Write down and solve an equation to find Jack's number.

Coordinates

▶ This chapter is about

- Reading coordinates
- Plotting coordinates
- Horizontal and vertical lines

▶ You should already know

- How to substitute positive and negative numbers into simple formulae

Coordinates

On this grid the lines at the bottom and on the left are called **axes**.

The bottom line is called the **x-axis**.
The left-hand line is called the **y-axis**.

TIP

Notice that the lines are numbered, not the spaces.

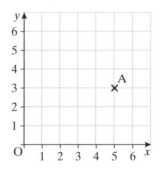

Coordinates are a pair of numbers which fix the position of a point.

The coordinates of point A are (5, 3).

5 is the distance across the grid. It is called the **x-coordinate**.
3 is the distance up the grid. It is called the **y-coordinate**.
You write them in a bracket.

The point with coordinates (0, 0) is called the **origin**.
The letter O is often used to mark the origin.

TIP

The 'across' coordinate always comes first.
One way to remember this is to think of an aircraft.
It always goes along the runway before it goes up.
across →   before up ↑

Check up 7.1

Here is a map of a village on a grid.

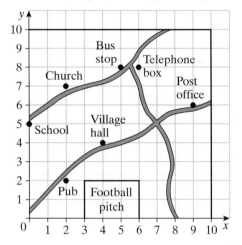

(a) Write down the coordinates of the village hall, the school and the crossroads.

(b) What is at the point $(9, 6)$?

(c) Which two things have the same y-coordinate?

(d) Which two things have the same x-coordinate?

(e) What are the coordinates of the corners of the football pitch?

Exercise 7.1

1 Write down the coordinates of the points A, B, C, D, E, F, G, H, I and J.

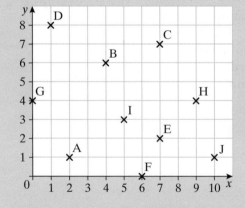

2 On a grid, draw x- and y-axes from 0 to 8.
Plot and label these points.

A(5, 2)   B(7, 6)   C(3, 4)   D(7, 0)   E(0, 2)

● Challenge 7.1

In map reading, six–figure map references are used. These are a form of coordinates.

For each coordinate you need to estimate how many tenths of a unit the point is to the right of or above the number printed on the grid.

So, for the inn on the map the 'across' coordinate is between 23 and 24. It is about 4 tenths of the way across the square so the across coordinate is given as 234.

The 'up' coordinate is between 76 and 77.
It is about 7 tenths of the way up the square so the 'up' coordinate is given as 767.

The full six–figure map reference for the inn is 234 767.
Notice that you do not use brackets or a comma but otherwise you use exactly the same rule as for coordinates: 'across before up'.

The third and sixth figures are only estimates.

(a) What is at 243 776?

(b) Write down the six–figure references for these.

    (i)   The station   (ii) The church
    (iii) The post office   (iv) The golf club
    (v)   The museum

Points in all four quadrants

By using both positive and negative numbers, you can fix points anywhere in two dimensions.

This grid shows

- the *x*-axis going across the page and labelled from −5 to +5.
- the *y*-axis going up the page and labelled from −5 to +5.

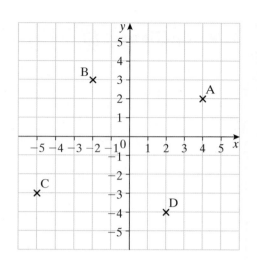

The axes divide the grid into four parts called **quadrants**. Each quadrant contains one of the four points, A, B, C and D.

A is the point $(4, 2)$.
It is 4 to the right on the *x*-axis and 2 up on the *y*-axis.

B is the point $(−2, 3)$.
It is 2 to the left on the *x*-axis and 3 up on the *y*-axis.

C is the point $(−5, −3)$.
It is 5 to the left on the *x*-axis and 3 down on the *y*-axis.

D is the point $(2, −4)$.
It is 2 to the right on the *x*-axis and 4 down on the *y*-axis.

Example 7.1

Write down the coordinates of the points A, B, C and D.

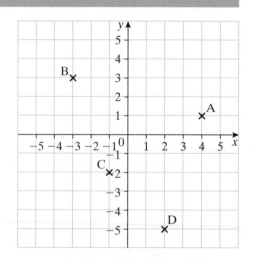

Solution

To get to point A you go 4 across and 1 up, so the coordinates of point A are $(4, 1)$.

Similarly, the coordinates of the other points are
B$(−3, 3)$ C$(−1, −2)$ D$(2, −5)$.

Example 7.2

Draw x- and y-axes from -5 to $+5$.

Plot and label the points $A(5, 2)$, $B(-2, 4)$, $C(-4, -3)$ and $D(0, -2)$.

Solution

To plot point A you go 5 across and 2 up.
You plot the other points in a similar way.

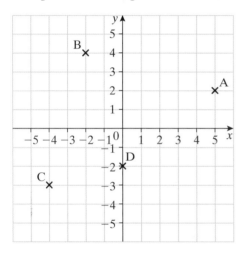

TIP

The best way of plotting points is to mark a cross with a sharp pencil. This is accurate and easily seen. To use a dot large enough to be seen would make it inaccurate.

● Challenge 7.2

- Draw x- and y-axes from -7 to 17.
- Plot each pair of points and join them with straight lines.
- Find the coordinates of four other points on the line.

(a) $(-5, 1)$ and $(5, 1)$ (b) $(7, 5)$ and $(7, -1)$

(c) $(-3, -6)$ and $(1, 2)$ (d) $(0, 3)$ and $(2, 0)$

Can you find some points with fractional coordinates?
Can you find some more points by extending the lines?

Exercise 7.2

1 Write down the coordinates of the
   points A, B, C, D, E, F, G, H, I and J.

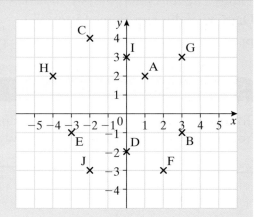

In questions 2 to 4 you will need to draw x- and y-axes from -5 to $+5$.

2 Plot and label the points A(4, 3), B(4, -2), C(-1, -2) and D(-1, 3).
   Join the points to make shape ABCD.

3 Plot and label the points A(3, 4), B(3, -2) and C(-5, 1).
   Join the points to make shape ABC.

4 Plot and label the points A(3, 3), B(3, -5), C(-2, -3) and D(-2, 2).
   Join the points to make shape ABCD.

Lines parallel to the axes

The equation of a line is a relationship in x or in y or
between x and y which is true for every point on the line.

Look at the points A, B, C, D, E and F on this grid.

They are all on a line parallel to the y-axis.

The coordinates of the points are (2, 4), (2, 3), (2, 1), (2, 0),
(2, -2) and (2, -3) respectively.

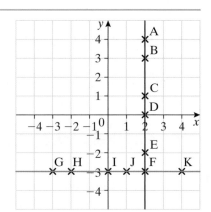

Notice that the x-coordinate of all the points is 2.
This is true for every point on the line.

The equation of the line is $x = 2$.

Now look at the points G, H, I, J, F
and K.

The coordinates of the points are
(-3, -3), (-2, -3), (0, -3), (1, -3),
(2, -3) and (4, -3).

TIP

It is very common to get lines parallel to the axes mixed up.
The equation of a line across the page is $y = $ a number.
The equation of a line up the page is $x = $ a number.

Notice that the y-coordinate of all the points is -3.
This is true for every point on the line.

The equation of the line is $y = -3$.

Check up 7.2

What is the equation of
(a) the x-axis?   **(b)** the y-axis?

Challenge 7.3

Other lines

Look at the red line on this grid.
Write down the coordinates of
six points on the red line.
What do you notice about the
coordinates?
What do you think the equation
of the red line is?

Now look at the coordinates of some
of the points on the blue line.
What do you think the equation of
the blue line is?

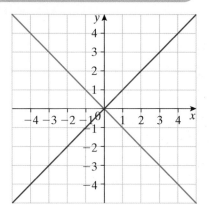

Exercise 7.3

1 Write down the equation of each of the lines **(a)**, **(b)**, **(c)** and **(d)**.

What you have learned

- The x-axis goes across the page and the y-axis goes up the page
- The point where the axes cross has coordinates $(0, 0)$ and is called the origin
- How to plot coordinates in all four quadrants
- A line across the page has the equation $y = a$
- A line up the page has the equation $x = b$

Mixed exercise 7

1 Write down the coordinates of the points A, B, C, D, E, F and G.

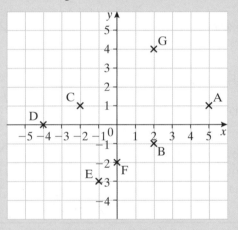

For each of questions **2** to **5**, draw a separate grid with both axes labelled from -5 to $+5$.

2 Plot and label these points.

A$(5, 3)$ B$(-4, 2)$ C$(4, -1)$ D$(0, -3)$ E$(-2, 0)$ F$(-5, -4)$ G$(2.5, 1)$

3 Plot and label the points A$(-5, 0)$, B$(3, 0)$, C$(4, 4)$ and D$(-4, 4)$.
   Join the points to make shape ABCD.

4 Plot and label the points A$(-4, -3)$, B$(-4, 4)$, C$(2, 3)$ and D$(2, -1)$.
   Join the points to make shape ABCD.

5 Draw the straight lines with these equations.
   Remember to label your lines.

   (a) $x = 3$ (b) $y = 5$ (c) $x = -2$ (d) $y = -3$

Statistical calculations 1

▶ This chapter is about

- Finding the mean, median, mode and range of sets of data
- Finding the modal class of data which is grouped

▶ You should already know

- How to order numbers
- How to divide, using a calculator if necessary

Median

The **median** is the middle value of an ordered set of data.

If there is an even number of values the median is halfway between the two middle values.

Example 8.1

(a) Here are the results of a survey on the price of a CD in five shops.

£7.50   £9.00   £12.50   £10.00   £9.00

What is the median price?

(b) Another shop is surveyed. The price of the CD there is £11.00. What is the median price now?

Solution

(a) First you put the prices in order. You would usually start with the smallest.

£7.50   £9.00   £9.00   £10.00   £12.50

The middle price is £9.00 so this is the median.

(b) You add the new price in the correct place in the ordered list.

£7.50   £9.00   £9.00   £10.00   £11.00   £12.50

There is now an even number of values.
You add the middle two values together then divide by 2.
The median price is now (£9.00 + £10.00) ÷ 2 = £9.50.

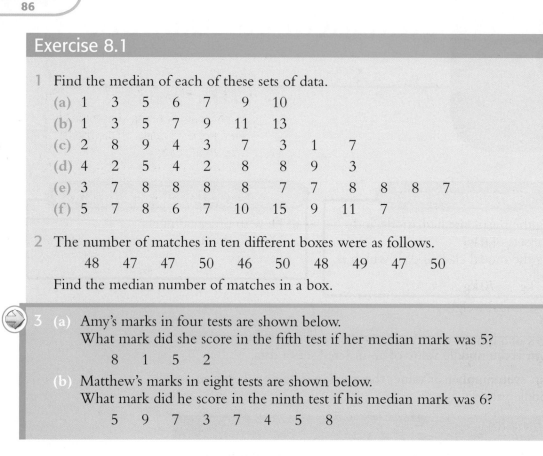

Exercise 8.1

1 Find the median of each of these sets of data.

   (a) 1   3   5   6   7   9   10

   (b) 1   3   5   7   9   11   13

   (c) 2   8   9   4   3   7   3   1   7

   (d) 4   2   5   4   2   8   8   9   3

   (e) 3   7   8   8   8   8   7   7   8   8   8   7

   (f) 5   7   8   6   7   10   15   9   11   7

2 The number of matches in ten different boxes were as follows.

    48   47   47   50   46   50   48   49   47   50

   Find the median number of matches in a box.

3 (a) Amy's marks in four tests are shown below.
        What mark did she score in the fifth test if her median mark was 5?

        8   1   5   2

   (b) Matthew's marks in eight tests are shown below.
        What mark did he score in the ninth test if his median mark was 6?

        5   9   7   3   7   4   5   8

Mode

The **mode** is the most common value in a set of data.

There can be more than one mode.

Example 8.2

Find the mode of these numbers.

   1   5   2   4   8   3   1

Solution

It helps to rewrite the numbers in order.

   1   1   2   3   4   5   8

The mode is 1.

Exercise 8.2

1 The marks scored in a test were as follows.

20	16	18	17
16	18	14	13
18	18	15	18
19	9	12	13

Find the modal number of marks scored.

TIP

Another way of saying 'the mode of the number of marks' is 'the modal number of marks'.

2 Here is a list of the weights of people in a 'Keep Fit' class.

73 kg	58 kg	61 kg
43 kg	81 kg	53 kg
73 kg	70 kg	62 kg

Find the modal weight.

3 The ages of a group of people are as follows.

19	23	53
19	16	26
77	19	27

Find the modal age.

4 The number of matches in ten different boxes were as follows.

48	47	47	50	46
50	49	49	47	50

Find the modal number of matches in a box.

● Challenge 8.1

(a) Write down a set of five marks with a median of 3 and a mode of 2.

(b) Write down a set of six marks with a median of 1 and a mode of 0.

Mean

The **mean** of a set of data is found by adding the values together and dividing the total by the number of values used. Remember that a value of zero still counts as a result.

Example 8.3

Find the mean of this set of data.

| 5 | 7 | 8 | 6 | 7 | 10 | 15 | 9 | 11 | 7 |

Solution

$$\text{Mean} = \frac{\text{Sum of the values}}{\text{Number of values}}$$

$$= \frac{5 + 7 + 8 + 6 + 7 + 10 + 15 + 9 + 11 + 7}{10}$$

$$= \frac{85}{10} = 8.5$$

The mean is 8.5.

Exercise 8.3

1 Find the mean of the following sets of data.
   (a) 3   12   4   6   8   5   4
   (b) 7   21   2   17   3   13   7   4   9   7   9
   (c) 12   1   10   1   9   3   4   9   7   9

2 In a maths test the marks for the boys were as follows.
       6   3   9   8   2   2
   Here are the marks for the girls in the same test.
       9   7   8   7   5
   (a) Find the mean mark for the boys.
   (b) Find the mean mark for the girls.
   (c) Find the mean mark for the whole class.

3 Here are the times, in minutes, for a bus journey.
       15   7   9   12   9   19   6   11   9   14
   Find the mean time for the journey.

Range

The **range** of a set of data is the difference between the highest value and the lowest value. The range gives an idea of how spread out the data are.

Example 8.4

Here are the times, in minutes, for a bus journey.

15   7   9   12   9   19   6   11   9   14

Find the range.

Solution

Range = Highest value − Lowest value
= 19 − 6
= 13 minutes

Exercise 8.4

1   Find the range of each of these sets of data.
   (a) 15   17   12   29   21   18   31   22
   (b) 2.7   3.8   3.9   5.0   4.5   1.8   2.3   4.7
   (c) 313   550   711   365   165   439   921   264

2   Five people work in a shop. Their weekly wages are as follows.
       £157   £185   £189   £177   £171
   (a) Find the range of these wages.
   A new employee starts who earns £249.
   (b) What is the new range of the wages?

● Challenge 8.2

(a) Find a set of five numbers with mean 6, median 5 and range 4.
(b) Find a set of ten numbers with mean 5, median 6 and range 7.

Modal class

You often group data when there are a lot of different values or when
the data is continuous. This makes the data easier to understand.
You have to be very clear which values are included in each group.
Inequality symbols are often used to write the groups. In Example 8.5,
$8 \leqslant l < 9$ means all the lengths from 8 cm up to but not including
9 cm. A leaf that measures 9 cm is in the next class, $9 \leqslant l < 10$.

When you group data you can no longer tell the exact value of each data item. In the example below, the lengths of the leaves in the first class could be anywhere from 8 cm up to but not including 9 cm. So it is not possible to find the mode. You can, however, find the modal group or the **modal class**.

Example 8.5

The lengths (l cm) of 100 oak tree leaves are shown in the following table.

Length (l cm)	$8 \leqslant l < 9$	$9 \leqslant l < 10$	$10 \leqslant l < 11$	$11 \leqslant l < 12$	$12 \leqslant l < 13$
Number of leaves	8	18	35	23	16

Find the modal class.

Solution

The highest frequency, 35, is for the class $10 \leqslant l < 11$.

The modal class is $10 \leqslant l < 11$.

> **TIP**
>
> Make sure you write down the class and not the frequency.

Exercise 8.5

1 Sue measures the heights of the students in her class. The table shows her data.

Write down the modal group.

Height in cm (to the nearest cm)	Number of students
111–120	8
121–130	12
131–140	5
141–150	4
151–160	3

2 The table shows the amount of pocket money received by the students in Class 10Y.

Write down the modal class.

Pocket money (£)	Number of students
0–3.99	3
4–7.99	10
8–11.99	12
12–15.99	5

Exercise continues …

3   As part of a survey Masood investigates the number
    of words in sentences in a book.
    He records the number of words in each sentence
    in the first chapter.
    The table shows his data.

    Write down the modal class.

Number of words	Frequency
1–5	16
6–10	27
11–15	29
16–20	12
21–25	10
26–30	6
31–35	3

4   The police measured the speeds of cars along a stretch of road between 0800 and 0830
    one morning.
    The speeds, in miles per hour, are given below.

    40   32   43   47   42   48   51   47   46   45
    38   36   35   39   43   42   39   46   45   41
    42   38   35   33   41   46   36   44   39   40

    Copy and complete the frequency table and write down the modal class.

Speed (mph)	Tally	Frequency
$30 \leqslant s < 35$		
$35 \leqslant s < 40$		
$40 \leqslant s < 45$		
$45 \leqslant s < 50$		
$50 \leqslant s < 55$		

What you have learned

- The median is the middle value or, if there are two middle values, is halfway between the
  two middle values
- The mode is the number or value that occurs most often
- The mean is found by adding up all the values and dividing by the number of values
- The range is the difference between the smallest and largest values
- The modal class is the group or class that has the highest frequency

Mixed exercise 8

1 Tom and Freya go ten-pin bowling.
These are their scores.

Tom	7	8	5	3	7
Freya	10	8	3	1	3

(a) Find the mode, median, mean and range of both Tom and Freya's scores.

(b) Write down two comments on their scores.

2 12 people have their hand span measured.
The results, in millimetres, are shown below.

225　216　188　212　205　198
194　180　194　198　200　194

(a) How many of the group had a hand span greater than 200 mm?

(b) What is the range of the hand spans?

(c) What is the mean hand span?

3 The PE staff of a school measure the time, in seconds, it takes the members of the football team and the hockey team to run 100 metres.

Football team
13　14　15　11　14　12　12　13　11　13　14

Hockey team
12　13　14　11　12　14　15　13　15　14　11

(a) Calculate the mean, median and range for each team.

The PE staff are then timed running over the same distance.
Their times, in seconds, were as follows.

12　11　13　15　11

(b) Calculate the mean, median and range for the PE staff.

(c) Which group do you think is the fastest?

4 (a) Find the range, mean, median and mode of each of these sets of data.

Data set A
1　2　2　3　3　3　4　5　6　7

Data set B
1　2　2　3　3　3　4　5　6　7
1　2　2　3　3　3　4　5　6　7

Data set C
2　4　4　6　6　6　8　10　12　14

(b) Write down anything that you notice.

Mixed exercise 8 continues …

5 (a) Find the mean, median and mode of each of these sets of data.

(i) 1, 2, 3, 3, 4, 5   (ii) 10, 20, 20, 30, 70

(iii) 110, 120, 120, 130, 170   (iv) 7, 10, 13, 16, 19

(b) What do you notice about your answers to parts (ii) and (iii)?

6 In a survey a group of boys and girls wrote down how many hours of television they watched one week.

Boys   17   22   21   23   16   12   15   0   5   13   15   13   14   20

Girls   9   13   15   17   10   12   11   9   8   12   14   15

(a) Find the mean, median, mode and range of these times.

(b) Do the boys watch more television than the girls?

7 Find the median and the mode of each of these data sets.

(a) 4, 3, 15, 9, 7, 6, 11   (b) 60 kg, 12 kg, 48 kg, 36 kg, 24 kg

8 A gardener measures the height, in centimetres, of her sunflower plants.

140   123   131   89   125   123   115   138

Find the median and the mode of these heights.

9 The masses, to the nearest gram, of potatoes bought in bags from a supermarket are as follows.

202   417   301   258   284   290   329   381

315   283   216   329   231   405   350   382

278   394   416   374   367   381   419   381

Copy and complete the frequency table and write down the modal class.

Mass (grams)	Tally	Frequency
$200 \leqslant m < 220$		
$220 \leqslant m < 240$		
$240 \leqslant m < 260$		
$260 \leqslant m < 280$		
$280 \leqslant m < 300$		
$300 \leqslant m < 320$		
$320 \leqslant m < 340$		
$340 \leqslant m < 360$		
$360 \leqslant m < 380$		
$380 \leqslant m < 400$		
$400 \leqslant m < 420$		

Sequences 1

- Sequences of numbers
- Generating sequences using term-to-term rules
- Finding term-to-term rules
- Sequences derived from diagrams
- Generating sequences using position-to-term rules

- What odd and even numbers are
- How to find the difference between two numbers

Sequences

Look at this list of numbers: 1, 2, 3, 4, 5, 6, … .

The pattern continues and you should know that the next number is 7 and the one after that is 8 because these are the **counting numbers**.

Now look at this list of numbers: 2, 4, 6, 8, 10, 12, … .

The next number is 14 and the one after that is 16 because the difference between them is always 2.

These are the **even numbers**.

The sequence 1, 3, 5, 7, 9, 11, … also has a difference of 2 between each of the numbers. These are the **odd numbers**.

Another sequence, 1, 4, 9, 16, 25, 36, … , does not have the same difference between numbers but there is still a pattern. These are the **square numbers**.

Any list of numbers where there is a pattern linking the numbers is called a **sequence**.

Term-to-term rules

In this list of numbers: 3, 8, 13, 18, 23, 28, ... , the next number is 33 and the one after that is 38 because they increase by 5 each time.

The numbers in a sequence are called **terms** and the pattern linking the numbers is called the **term-to-term rule**. Sequences which involve either adding or subtracting the same number each time are called **linear** sequences.

To generate a sequence given the term-to-term rule, you must have a starting number, the first term.

Example 9.1

Find the first four terms of each of these sequences.
(a) First term 5, term-to-term rule Add 3
(b) First term 20, term-to-term rule Subtract 2
(c) First term 1, term-to-term rule Add 10

Solution

(a) The first term is 5.
The second term is $5 + 3 = 8$.
The third term is $8 + 3 = 11$.
The fourth term is $11 + 3 = 14$.
So the first four terms of the sequence are 5, 8, 11, 14.
(b) 20, 18, 16, 14   **(c)** 1, 11, 21, 31

Example 9.2

Write down the next number in each of these sequences and give the term-to-term rule.
(a) 1, 4, 7, 10, 13, 16, ...   **(b)** 2, 7, 12, 17, 22, 27, ...
(c) 3, 7, 11, 15, 19, 23, ...   **(d)** 47, 43, 39, 35, 31, 27, ...

Solution

(a) 19   The numbers are increasing by 3 each time (or $+3$).
(b) 32   The numbers are increasing by 5 each time (or $+5$).
(c) 27   The numbers are increasing by 4 each time (or $+4$).
(d) 23   The numbers are decreasing by 4 each time (or -4).

In the sequences in Example 9.3 some of the numbers are missing. It is possible to work out what they should be by looking at the numbers in the sequence and then finding the term-to-term rule.

Example 9.3

Find the missing numbers in each of these sequences and give the term-to-term rule.

(a) 10, 17, … , … , 38, 45

(b) 35, … , 23, 17, … , 5

(c) … , 15, … , 23, 27, 31

Solution

(a) 24 and 31   The numbers are increasing by 7 each time (or $+7$).

(b) 29 and 11   The numbers are decreasing by 6 each time (or -6).

(c) 11 and 19   The numbers are increasing by 4 each time (or $+4$).

Exercise 9.1

1 Find the first five terms of each of these sequences.

   (a) First term 1, term-to-term rule Add 4

   (b) First term -4, term-to-term rule Add 3

   (c) First term 21, term-to-term rule Subtract 3

   (d) First term 5, term-to-term rule Add 5

   (e) First term 100, term-to-term rule Subtract 40

   (f) First term 1, term-to-term rule Add $\frac{1}{2}$

2 Write down the next two terms in each of these sequences and give the term-to-term rule.

   (a) 1, 5, 9, 13, 17, 21, …   (b) 3, 9, 15, 21, 27, 33, …

   (c) 5, 12, 19, 26, 33, 40, …   (d) 7, 10, 13, 16, 19, 22, …

   (e) 8, 13, 18, 23, 28, 33, …   (f) 23, 32, 41, 50, 59, 68, …

3 Write down the next two terms in each of these sequences and give the term-to-term rule.

   (a) 19, 17, 15, 13, 11, 9, …   (b) 33, 29, 25, 21, 17, 13, …

   (c) 45, 39, 33, 27, 21, 15, …   (d) 28, 24, 20, 16, 12, 8, …

   (e) 23, 19, 15, 11, 7, 3, …   (f) 28, 23, 18, 13, 8, 3, …

Exercise continues …

4 Find the missing numbers in each of these sequences and give the term-to-term rule.
   (a) 1, 8, ... , 22, 29, ...
   (b) 11, ... , 23, 29, ... , 41
   (c) 76, ... , 54, 43, 32, ... , 10
   (d) ... , 57, 48, 39, ... , 21
   (e) 6, 14, ... , ... , 38, 46
   (f) 23, ... , 13, 8, 3, ...

Sequences from diagrams

You can have a series of diagrams that form a sequence.

Example 9.4

Draw the next diagram in each of these sequences.
For each sequence, count the dots in each diagram and find the term-to-term rule.

(a)

(b)

(c)

Solution

(a)   5, 9, 13 and 17   The numbers are increasing by 4 each time (or +4).

(b)   3, 5, 7 and 9   The numbers are increasing by 2 each time (or +2).

(c)   5, 10, 15 and 20   The numbers are increasing by 5 each time (or +5).

Exercise 9.2

1 Draw the next diagram in each of these sequences.
   For each sequence, count the dots in each diagram and find the term-to-term rule.

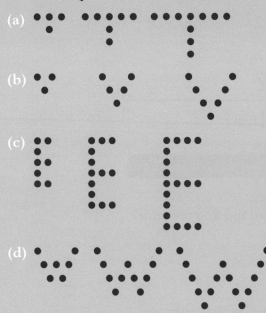

2 Draw the next diagram in each of these sequences.
   For each sequence, count the lines in each diagram and find the term-to-term rule.

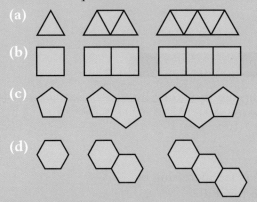

● Challenge 9.1

Draw a sequence of diagrams of your own.

Swap with a partner and find the term-to-term rule of your partner's sequence.

Position-to-term rules

Term-to-term rules are useful if you want to find the next number in a sequence but they are not useful if you want to find a term a long way into a sequence. You can do this more easily if you have a formula which gives the value of a term when you know its position, for example, the 98th term.

Such a formula is called a **position-to-term rule**. It is usually stated for the nth term, for example, nth term $= 3n + 1$.

Example 9.5

For each of these sequences, write down the first four terms and the 98th term.

(a) nth term $= 3n + 2$

(b) nth term $= 2n - 1$

(c) nth term $= 5n + 6$

Solution

(a) For the first term $n = 1$. $3n + 2 = 3 \times 1 + 2 = 5$

For the second term $n = 2$. $3n + 2 = 3 \times 2 + 2 = 8$

For the third term $n = 3$. $3n + 2 = 3 \times 3 + 2 = 11$

For the fourth term $n = 4$. $3n + 2 = 3 \times 4 + 2 = 14$

For the 98th term $n = 98$. $3n + 2 = 3 \times 98 + 2 = 296$

(b) When $n = 1$ $2n - 1 = 2 \times 1 - 1 = 1$

When $n = 2$ $2n - 1 = 2 \times 2 - 1 = 3$

When $n = 3$ $2n - 1 = 2 \times 3 - 1 = 5$

When $n = 4$ $2n - 1 = 2 \times 4 - 1 = 7$

When $n = 98$ $2n - 1 = 2 \times 98 - 1 = 195$

These are the odd numbers.

(c) When $n = 1$ $5n + 6 = 5 \times 1 + 6 = 11$

When $n = 2$ $5n + 6 = 5 \times 2 + 6 = 16$

When $n = 3$ $5n + 6 = 5 \times 3 + 6 = 21$

When $n = 4$ $5n + 6 = 5 \times 4 + 6 = 26$

When $n = 98$ $5n + 6 = 5 \times 98 + 6 = 496$

Exercise 9.3

1 Write down the first five terms of the sequences with these nth terms.

(a) $4n + 7$ (b) $8n + 5$ (c) $7n + 5$

(d) $3n - 1$ (e) $5n + 8$ (f) $9n + 7$

2 Find the 200th term of the sequences with these nth terms.

(a) $3n + 3$ (b) $4n - 1$ (c) $11n - 6$

(d) $6n + 2$ (e) $12n + 13$ (f) $9n - 5$

What you have learned

- A list of numbers with a pattern is called a sequence
- A term-to-term rule is used to move from any term in a sequence to the next term in the sequence
- For a linear sequence, the term-to-term rule is of the form $\pm A$, where A is the difference between one term and the next
- A position-to-term rule is used to find the value of any term in a sequence from its position in the sequence

Mixed exercise 9

1 Find the first five terms of each of these sequences.

(a) First term 1, term-to-term rule Add 5

(b) First term 50, term-to-term rule Subtract 4

(c) First term -6, term-to-term rule Add 2

2 Write down the next term in each of these sequences and give the term-to-term rule.

(a) $2, 6, 10, 14, 18, 22, \ldots$

(b) $3, 11, 19, 27, 35, 43, \ldots$

(c) $4, 9, 14, 19, 24, 29, \ldots$

3 Write down the next two terms in each of these sequences and give the term-to-term rule.

(a) $33, 29, 25, 21, 17, 13, \ldots$

(b) $23, 20, 17, 14, 11, 8, \ldots$

(c) $76, 63, 50, 37, 24, 11, \ldots$

Mixed exercise 9 continues …

4 Find the missing numbers in each of these sequences and give the term-to-term rule.

(a) 7, 12, 17, … , … , 32, …

(b) 25, … , 19, 16, … , 10, …

(c) 4, 15, … , … , 48, 59, …

5 Draw the next pattern in each of these sequences.
For each sequence, count the dots in each pattern and find the term-to-term rule.

(a)

(b)

6 Draw the next pattern in each of these sequences.
For each sequence, count the lines in each pattern and find the term-to-term rule.

(a)

(b)

7 (a) Find the first four terms and the 150th term of the sequence with nth term $4n + 1$.

(b) Find the first four terms and the 78th term of the sequence with nth term $9n - 5$.

(c) Find the first four terms and the 92nd term of the sequence with nth term $8n + 2$.

Chapter 10

Measures

▶ This chapter is about

- Using scales and units
- Changing between metric units
- Changing between metric and imperial units
- Estimating lengths and other measures

▶ You should already know

- The basic units of length, weight, volume and capacity
- How to add and subtract decimals
- How to multiply and divide by 10, 100 and 1000
- How to multiply and divide numbers

Using and reading scales

A scale is marked using equally-spaced divisions. To read the scale you need to decide on what each division represents.

Example 10.1

What are the readings on this scale?

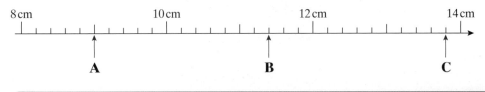

Solution

A is halfway between 8 cm and 10 cm.
A is at 9 cm.

There are five divisions between 9 cm and 10 cm so each small division is 0.2 cm.
B is two divisions past 11 cm.
B is at 11.4 cm.

C is one division before 14 cm.
C is at $14 - 0.2 = 13.8$ cm.

TIP

Check your answers by counting on in steps of 0.2 from the labelled divisions.

Example 10.2

Estimate the reading on this scale.

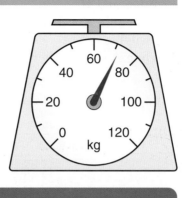

Solution

There is only one mark between 60 kg and 80 kg so that represents 70 kg.
The arrow is pointing to just over 70 kg.
You have to estimate exactly where the arrow is pointing.
It is definitely less than halfway between 70 kg and 80 kg.
About 72 kg is a good estimate.

● Challenge 10.1

Some kitchen scales use weights.

You place whatever you are weighing on one pan.

Then you add weights to the other pan until the two pans balance.

You have these weights.

1 g	2 g	2 g	5 g	10 g
10 g	20 g	50 g	100 g	100 g
200 g	200 g	500 g	1 kg	2 kg

Which weights do you need to weigh the following?

(a) 157 g **(b)** 567 g **(c)** 1.283 kg **(d)** 2091 g **(e)** 2.807 kg

● Discovery 10.1

● Draw a straight line 10 cm long on some strips of paper.
  Mark one end 0 and the other end 10.

- Draw a straight line 20 cm long on some strips of paper.
  Mark one end 0 and the other end 10.
- Show a 10 cm line to some people. Ask them to mark on the line
  where they think 3 and 6 will be.
- Then show them a 20 cm line. Again ask them to mark where they
  think 3 and 6 will be.

Are people better at estimating the position of 3 and 6 on the 10 cm
line or the 20 cm line?

Exercise 10.1

1 How long is this pencil?

2 What is the reading on this scale?

3 How much liquid do you need to add to make 2 litres?

4 There are six identical balls on these scales.
   (a) What is the total weight of the balls?
   (b) How much does one ball weigh?

Exercise continues …

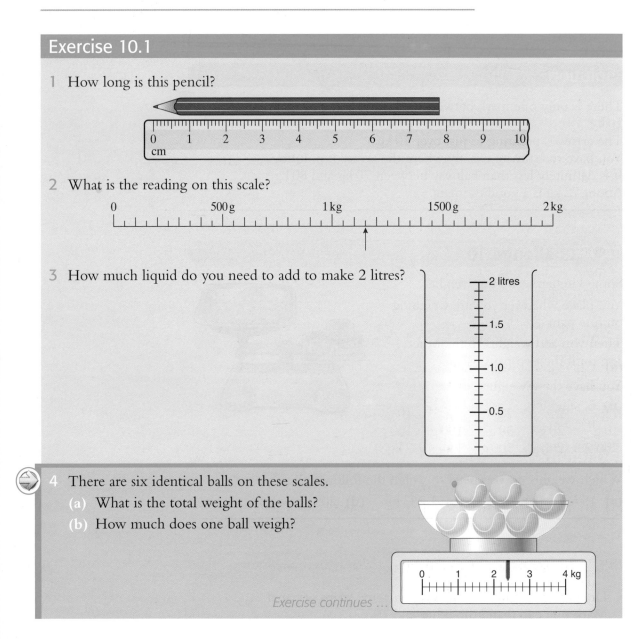

5 (a) What temperatures are shown by arrows A and B?
   (b) What is the difference between the two temperatures?

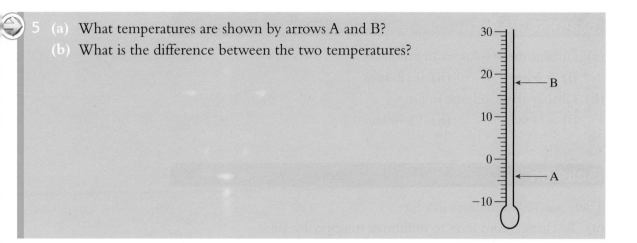

Changing from one unit to another

You need to know and be able to change between the main metric units of measurement.

Length	Mass	Capacity/Volume
1 kilometre = 1000 metres	1 kilogram = 1000 grams	1 litre = 1000 millilitres
1 metre = 100 centimetres	1 tonne = 1000 kilograms	1 litre = 100 centilitres
1 metre = 1000 millimetres		1 centilitre = 10 millilitres
1 centimetre = 10 millimetres		

Example 10.3

(a) Change these weights to grams.
   (i) 3 kg   (ii) 4.26 kg
(b) Change these weights to kilograms.
   (i) 5000 g   (ii) 8624 g

Solution

There are 1000 grams in a kilogram.

(a) To change from kilograms to grams, multiply by 1000.
   (i) $3 \times 1000 = 3000$ g
   (ii) $4.26 \times 1000 = 4260$ g

(b) To change from grams to kilograms, divide by 1000.
   (i) $5000 \div 1000 = 5$ kg
   (ii) $8624 \div 1000 = 8.624$ kg

Example 10.4

(a) Change these volumes to millilitres.
    (i) 5.2 litres   **(ii)** 0.12 litres
(b) Change these volume to litres.
    (i) 724 ml   **(ii)** 13 400 ml

Solution

There are 1000 millilitres in a litre.
(a) To change from litres to millilitres, multiply by 1000.
    (i) $5.2 \times 1000 = 5200$ ml
    (ii) $0.12 \times 1000 = 120$ ml
(b) To change from millilitres to litres, divide by 1000.
    (i) $724 \div 1000 = 0.724$ litres
    (ii) $13\,400 \div 1000 = 13.4$ litres

Example 10.5

Put these lengths in order, smallest first.

    3.25 m   415 cm   302 mm   5012 mm   62.3 cm

Solution

It is usually easiest to change to the smallest unit, in this case millimetres.

    $3.25 \text{ m} = 3.25 \times 1000 \text{ mm} = 3250 \text{ mm}$
    $415 \text{ cm} = 415 \times 10 \text{ mm} = 4150 \text{ mm}$
    $62.3 \text{ cm} = 62.3 \times 10 \text{ mm} = 623 \text{ mm}$

The lengths in millimetres in order are as follows.
    302 mm   623 mm   3250 mm   4150 mm   5012 mm

In your answer you need to give the measurements in the form they are given in the question.

    302 mm   62.3 cm   3.25 m   415 cm   5012 mm

Exercise 10.2

1 Put these volumes in order, smallest first.

      2 litres   1500 ml   1.6 litres   100 ml   0.75 litre

2 Which metric units would you use to measure these lengths?
   (a) The width of a book
   (b) The width of a room
   (c) The width of a car
   (d) The distance between two towns
   (e) The distance round a running track
   (f) The length of a bus
   (g) The length of a finger

3 A tiger is 240 cm long.
   A cheetah is 1.3 m long.
   What is the difference in length between the tiger and the cheetah?

4 Look at this list of measurements.

      72.0   7.2   0.72   0.072

Select an appropriate measurement from
the list to complete this sentence.

The height of the table is

.... metres.

5 Put these weights in order, smallest first.

      2 kg   1500 g   1.6 kg   10 000 g $\frac{3}{4}$ kg

6 Change these volumes to millilitres.
   (a) 14 cl   (b) 2.5 cl   (c) 5 litres   (d) 5.23 litres

7 Harry buys a bag of potatoes weighing 1.5 kg, a bunch of bananas weighing 900 g, a bag
   of sugar weighing 1 kg and a packet of coffee weighing 227 g.
   What is the weight, in kilograms, of his shopping?

Approximate equivalents

Here are some approximate equivalents between imperial and metric units. The ≈ sign means 'approximately equals' or 'is about'.

Length	Weight	Capacity
8 km ≈ 5 miles 1 m ≈ 40 inches 1 inch ≈ 2.5 cm 1 foot (ft) ≈ 30 cm	1 kg ≈ 2 pounds (lb)	4 litres ≈ 7 pints (pt) 1 gallon ≈ $4\frac{1}{2}$ litres

You will be given the conversions to use. Because they are only approximate, the equivalents you are given may be slightly different.

Example 10.6

Change the following measures in imperial units to the approximate metric equivalent.

(a) 15 miles **(b)** 12 pounds **(c)** 5 feet **(d)** 5 pints

Solution

For this question, use the conversions given in the table above.

(a) 8 km ≈ 5 miles

15 miles ≈ 15 × 8 ÷ 5 = 24 km

To change from miles to kilometres, multiply by 8 and then divide by 5.

(b) 1 kg ≈ 2 lb
12 lb ≈ 12 ÷ 2 = 6 kg

To change from pounds to kilograms, divide by 2.

(c) 1 foot ≈ 30 cm
5 feet ≈ 5 × 30 = 150 cm
150 ÷ 100 = 1.5 m

To change from feet to centimetres multiply by 30. Then divide by 100 to get the answer in metres.

(d) 4 litres ≈ 7 pints

5 pints ≈ 5 × 4 ÷ 7 = 2.86 litres

To change from pints to litres, multiply by 4 and then divide by 7.

TIP

When changing between imperial and metric units, decide whether to multiply or divide by considering which is the smaller unit. If the unit you are changing to is smaller, the number of units will be bigger.

⊙ Discovery 10.2

(a) Which *metric* units would you use to weigh the following?
  (i)   A 1p coin   (ii) A bag of potatoes
  (iii) A small chocolate bar (iv) A cow

(b) Which *imperial* units would you use to weigh the objects above?
    You may have to do some research.

🔍 Check up 10.1

Simon has some facts about tennis.
He wants a rough idea of the measurements in metric units.
Use the following conversions.

  1 foot (ft) ≈ 0.3 m   1 ounce (oz) ≈ 25 g   1 inch ≈ 25 mm

A tennis court is 78 feet long and 27 feet wide.
The height of the net must be 3 feet and a tennis ball must weigh 2 oz.
The ball should bounce to a height between 53 inches and 58 inches
when it is dropped from a height of 100 inches on to concrete.

The facts have been rewritten using metric units.
Use the conversions above to fill in the gaps.

A tennis court is ………. metres long and ………. metres wide.
The height of the net must be ………. metres and a tennis ball must
weigh ……….g.

The ball should bounce to a height between ……….cm and ……….cm
when it is dropped from a height of ……….cm on to concrete.

Exercise 10.3

1 1 pound (lb) is about 450 g.
   Write these weights in kilograms and grams.
   (a) 2 lb   (b) 5 lb   (c) 24 lb   (d) $\frac{1}{2}$ lb

2 Here are some conversions.

   miles $\xrightarrow{\times 1.6}$ kilometres   kilometres $\xrightarrow{\div 1.6}$ miles

   kilograms $\xrightarrow{\times 2.2}$ pounds   pounds $\xrightarrow{\div 2.2}$ kilograms

   Use these rules to change
   (a) 30 miles to kilometres.   (b) 10 kg to pounds.   (c) 120 kg to pounds.
   (d) 64 km to miles.   (e) 44 pounds to kilograms.   *Exercise continues…*

3 The distance from Ayton to Beeton is 320 km.
   5 miles is approximately 8 kilometres.
   Use this fact to calculate the approximate distance in miles from Ayton to Beeton.

Estimating measures

To estimate lengths, masses and capacities it is useful to know the length, mass or capacity of some common objects.

Here are some useful examples.
- The height from the floor to a man's waist is about 1 m (= 100 cm).
- The height of a man is about 1.8 m.
- The mass of a bag of sugar is 1 kg.
- Large bottles of lemonade or cola usually hold 2 litres.

You can use other everyday objects to make mental comparisons too. For example, your own weight, the length of your ruler and the capacity of a can of cola. Unless you are told otherwise, you should make estimates using metric units.

Example 10.7

Estimate the following.
(a) The height of a door
(b) The mass of a cup of sugar
(c) The capacity of a glass of lemonade
(d) The length of a car

Solution

(a) 2 m (or 200 cm) Think of a man walking through a door.

(b) 150 g Any answer from about 100 g to 300 g is acceptable.
A cup will hold a lot less than half a bag of sugar.

(c) 300 ml (or 30 cl) Any answer from 200 ml to 500 ml is acceptable.
Glasses vary in size but a glass will hold less than a litre.

(d) 4 m Any answer between about 3 m and 5 m is acceptable.
Compare the length mentally with the height of a man.

Exercise 10.4

1 Estimate the following.
   (a) The height of a single decker bus
   (b) The length of your foot
   (c) The mass of an apple
   (d) The capacity of a bucket
   (e) The mass of an average man
   (f) The length of your arm

2 Estimate the following.
   (a) The height of the fence
   (b) The length of the fence

3 The car is about 4 m long.
   Estimate the length of the lorry.

4 Estimate the height of the lamp post.

Mixed exercise 10

1 Change these lengths to millimetres.

(a) 6 cm (b) 35 cm (c) 4.5 m (d) 62 cm (e) 3.72 cm

2 Change these lengths to metres.

(a) 5 km (b) 4.32 km (c) 46.7 km (d) 1.234 km

3 Change these lengths to kilometres.

(a) 5000 m (b) 6700 m (c) 12 345 m (d) 543.21 m

4 Write these lengths in order of size, smallest first.

2.42 m   1623 mm   284 cm   9.044 m   31.04 cm

5 Which metric units would you use to measure the following?

(a) The length of a swimming pool   (b) The height of a tower

(c) The length of a needle   (d) The distance round your waist

6 Change these masses to grams.

(a) 9 kg (b) 1.129 kg (c) 3.1 kg (d) 0.3 kg

7 Convert these measurements from imperial units to their approximate metric equivalents. Use the table on page 108.

(a) 3 feet (b) 25 miles (c) 1 lb (d) 16 lb

8 Peter drives out of Dover towards London. He sees this sign.
Roughly how far is it from Canterbury to London in kilometres?

> **London 70 miles**
> **Canterbury 15 miles**

9 The car in this picture is about 3 m long. Estimate the length of the bus.

10 Jessica buys a bag of sugar weighing 1 kg, a bag of flour weighing 1.5 kg, a box of breakfast cereal weighing 450 g and two tins of soup weighing 400 g each. What is the weight, in kilograms, of her shopping?

Constructions 1

▶ This chapter is about

- Measuring lengths to the nearest millimetre
- Recognising types of angle
- Measuring angles to the nearest degree
- Drawing triangles and other 2-D shapes using ruler, protractor and compasses
- Using scales on maps and diagrams
- Using bearings

▶ You should already know

- A ruler is marked in centimetres (cm) and millimetres (mm)
- A protractor is marked in degrees

Measuring lengths

Look at this ruler.

The numbers on the scale are the centimetre (cm) marks. These are also the longest marks on the scale. The smallest marks on the scale are the millimetre (mm) marks.

The line measures 4.7 cm. That is 4 cm and 7 mm.

TIP

Notice how the start of the scale is not at the end of the ruler. When you measure lines, always make sure that the start of the scale is at the end of the line.

Exercise 11.1

1 Measure the length of each of these lines in centimetres.

(a) ─────────────────────

(b) ──────────────

(c) ──────────────────────────

Exercise continues …

(d) ——————
(e) ——————————————
(f) ———————————————————
(g) ——————————
(h) ———
(i) ————————————
(j) —————————————————

2 Measure the length of each of these objects.

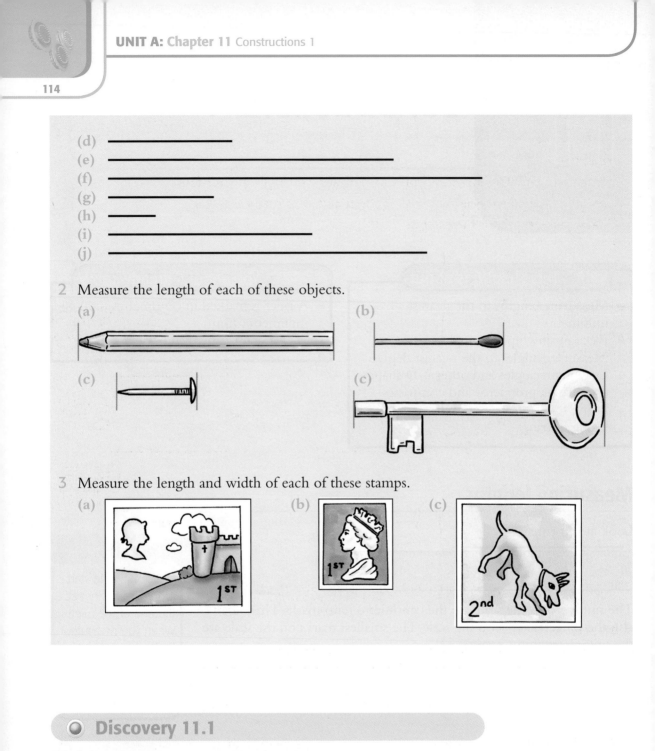

(a)

(b)

(c)

(c)

3 Measure the length and width of each of these stamps.

(a)   (b)   (c)

Discovery 11.1

Work in pairs.

- Measure the length of your middle finger.
  Do it as accurately as you can.

- Now get your partner to measure your middle finger.

Did you both get the same answers? If not, discuss why they are different.

Discovery 11.2

Work in pairs.

- Discuss with your partner how you can measure the length of your foot.
- Use your method to measure your own foot and your partner's.
- Check each other's measurements.

Types of angle

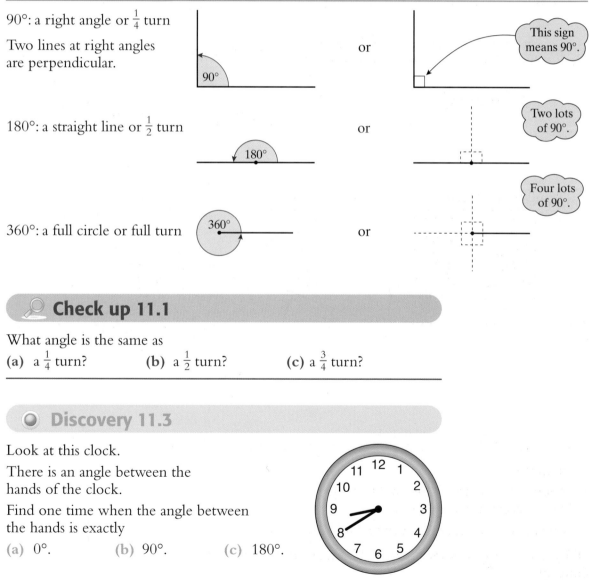

90°: a right angle or $\frac{1}{4}$ turn

Two lines at right angles
are perpendicular.

or

This sign means 90°.

90°

180°: a straight line or $\frac{1}{2}$ turn

or

Two lots of 90°.

180°

360°: a full circle or full turn

or

Four lots of 90°.

360°

Check up 11.1

What angle is the same as
(a) a $\frac{1}{4}$ turn?   **(b)** a $\frac{1}{2}$ turn?   **(c)** a $\frac{3}{4}$ turn?

Discovery 11.3

Look at this clock.

There is an angle between the
hands of the clock.

Find one time when the angle between
the hands is exactly

(a) 0°.   **(b)** 90°.   **(c)** 180°.

○ Discovery 11.4

Look around your classroom. Write down
(a) four places where you can see an angle of 90°.
(b) two places where you can see an angle of 180°.
(c) one place where you can see an angle of 360°.

Angles between 0° and 90° are called acute angles.

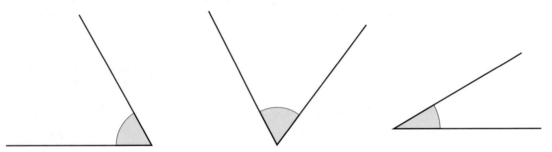

Angles between 90° and 180° are called obtuse angles.

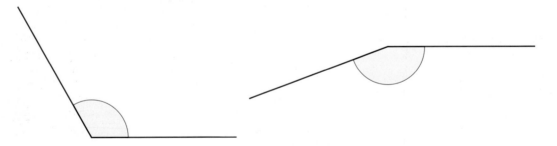

Angles between 180° and 360° are called reflex angles.

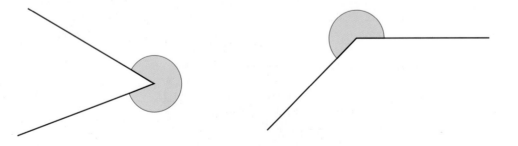

You will need to remember these different types of angle.

Check up 11.2

Here is a picture of the front of a house.

Copy the picture and mark
- all the acute angles with the letter *a*.
- all the right angles with the letter *r*.
- all the obtuse angles with the letter *o*.
- all the reflex angles with the letter *x*.

How many of each type did you find?
Check with your neighbour.
Did they find more?

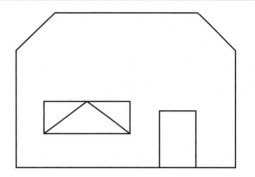

Exercise 11.2

1 Put these angles in order of size, starting with the smallest.

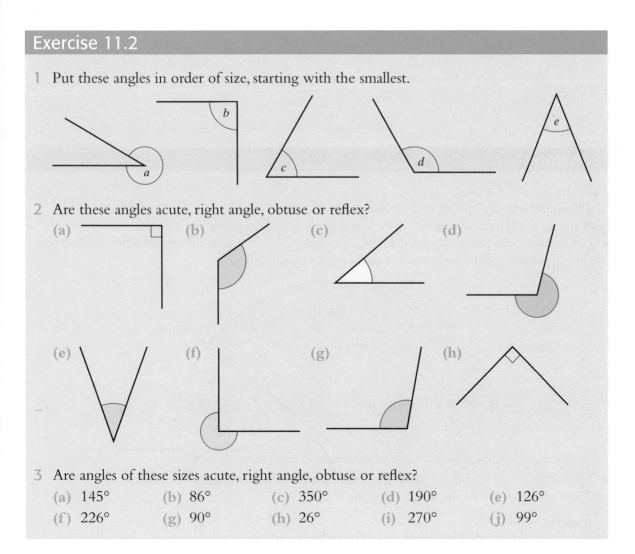

2 Are these angles acute, right angle, obtuse or reflex?

(a)　　　(b)　　　(c)　　　(d)

(e)　　　(f)　　　(g)　　　(h)

3 Are angles of these sizes acute, right angle, obtuse or reflex?

(a) 145°　　(b) 86°　　(c) 350°　　(d) 190°　　(e) 126°

(f) 226°　　(g) 90°　　(h) 26°　　(i) 270°　　(j) 99°

Measuring angles

You can measure angles with a protractor or an angle measurer.

You must be very careful when you use a protractor or angle measurer because they have two scales around the outside. It is important that you use the correct one each time.

Before measuring an angle, it is a good idea to identify which type of angle it is and use this to estimate its size. Knowing roughly what size the angle is should prevent you using the wrong scale on the protractor.

Example 11.1

Measure this angle.

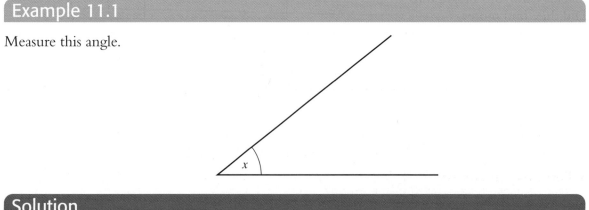

Solution

- First make an estimate.
  The angle is acute and so will be less than 90°.
  A rough estimate is about 40° since it is slightly less than half a right angle.
- Now place your protractor so that the zero line is along one of the arms of the angle. Make sure the centre of the protractor is at the point of the angle.

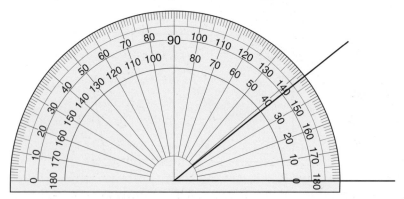

- Start at zero. Go round this scale until you reach the other arm of the angle.
  Then read the size of the angle from the scale.

  Angle $x = 38°$

Example 11.2

Measure angle PQR.

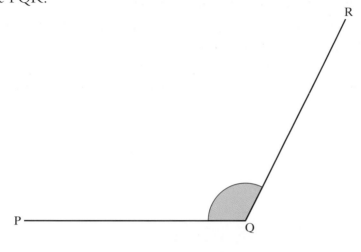

Solution

- First make an estimate.
  The angle is obtuse and so will be between 90° and 180°.
  A rough estimate is about 120°.
- Place your protractor so that the zero line is along one of the arms
  of the angle and the centre is at the point of the angle.

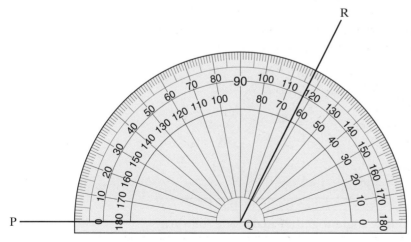

- Start at zero. Go round this scale until you reach the other arm of
  the angle.
  Then read the size of the angle from the scale.

  Angle PQR = 117°

Example 11.3

Measure angle A.

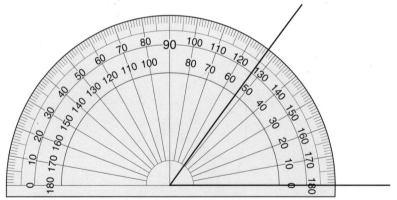

Solution

A reflex angle is between 180° and 360°.
This reflex angle is over $\frac{3}{4}$ of a turn so it is bigger than 270°.
A rough estimate is 300°.

You can measure an angle of this size directly using a 360° angle measurer. However, the scale on a protractor only goes up to 180°. You need to do a calculation as well as measure an angle.

- Measure the acute angle first.
  The acute angle is 53°.

> **TIP**
>
> It's easier with a 360° angle measurer!

The acute angle and the reflex angle together make one full turn.
A full turn is 360°.

- Use the fact that the two angles add up to 360° to calculate the reflex angle.
  Angle A = 360 − 53
         = 307°

Exercise 11.3

- Copy and complete this table.
- Estimate each angle first and then measure it with a protractor.

How good are you at estimating the size of an angle?

Angle	Estimated size	Measured size	Angle	Estimated size	Measured size
a			k		
b			l		
c			m		
d			n		
e			o		
f			p		
g			q		
h			r		
i			s		
j			t		

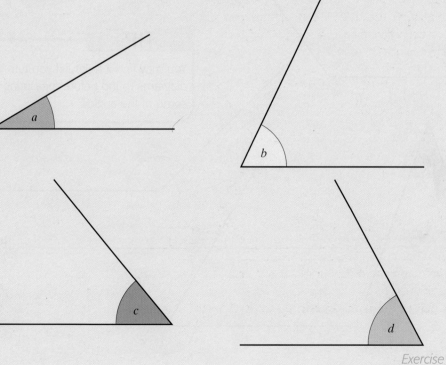

Exercise continues …

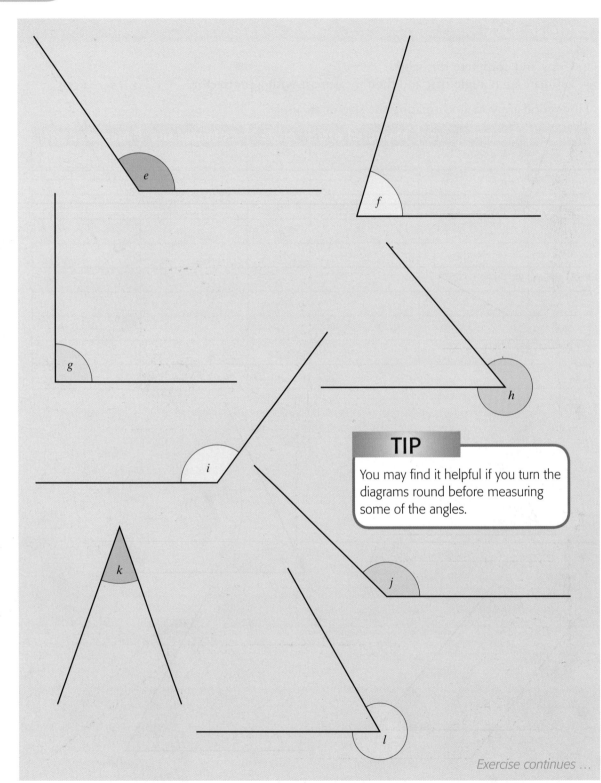

TIP

You may find it helpful if you turn the diagrams round before measuring some of the angles.

Exercise continues ...

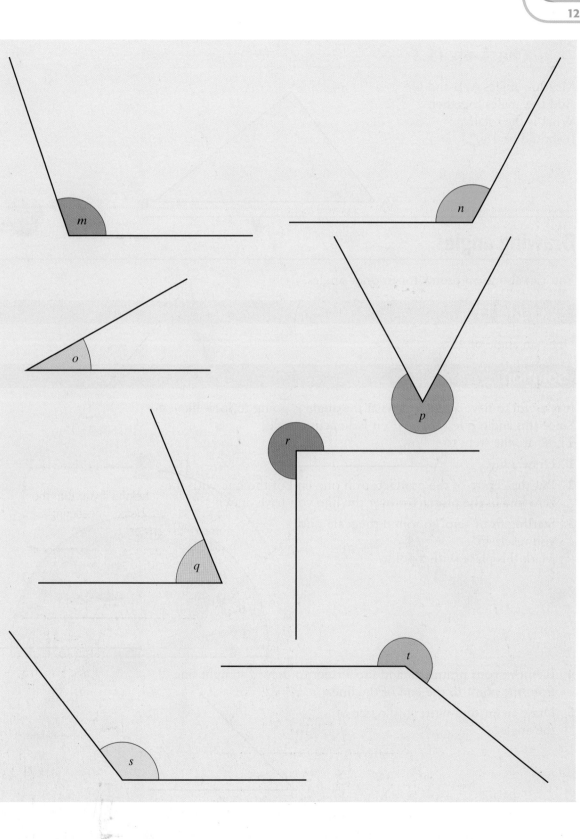

Check up 11.3

Measure angles A, B and C.
Add the angles together.
What is the total?
It should be 180°.

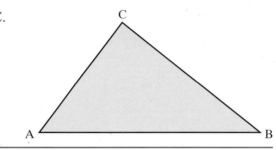

Drawing angles

You can also use a protractor to draw angles.

Example 11.4

Draw an angle of 45°.

Solution

It is useful to have an idea of what the angle is going to look like.
Since this angle is less than 90°, it is an acute angle.
These are the steps to follow.

1 Draw a line.

2 Put the centre of the protractor on one end of the line with the
   zero line of the protractor over the line you have drawn.

3 Starting from zero, go round the scale until
   you reach 45°.
   Mark this place with a point.

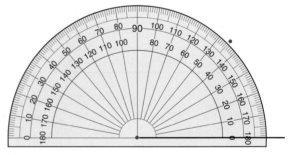

4 Remove your protractor and use a ruler to draw a straight line
   from the point to the end of the line.

5 Draw an arc and write in the size of
   the angle.

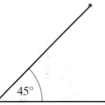

Challenge 11.1

How would you draw an angle of 280°, a reflex angle?

Write a list of instructions like the ones in Example 11.4.

TIP

You may find it useful to look back at Example 11.3.

Exercise 11.4

1 Draw accurately each of these angles.

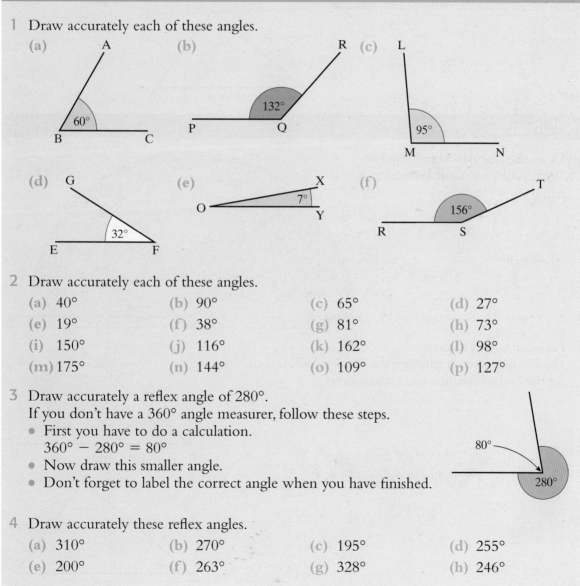

(a) 60° (angle ABC, A top, B bottom left, C bottom right)

(b) 132° (angle PQR, P left, Q bottom, R top right)

(c) 95° (angle LMN, L top, M bottom, N right)

(d) 32° (angle GEF, G top, E bottom left, F bottom right)

(e) 7° (angle XOY, O left, X top right, Y bottom right)

(f) 156° (angle TSR, T top right, S bottom, R left)

2 Draw accurately each of these angles.

    (a) 40°   (b) 90°   (c) 65°   (d) 27°

    (e) 19°   (f) 38°   (g) 81°   (h) 73°

    (i) 150°   (j) 116°   (k) 162°   (l) 98°

    (m) 175°   (n) 144°   (o) 109°   (p) 127°

3 Draw accurately a reflex angle of 280°.
If you don't have a 360° angle measurer, follow these steps.
- First you have to do a calculation.
  360° − 280° = 80°
- Now draw this smaller angle.
- Don't forget to label the correct angle when you have finished.

4 Draw accurately these reflex angles.

    (a) 310°   (b) 270°   (c) 195°   (d) 255°

    (e) 200°   (f) 263°   (g) 328°   (h) 246°

Drawing triangles using a ruler and a protractor only

When you are given two lengths of a triangle and the angle between them, you can draw it using a ruler and a protractor. The method is given in the next example.

Example 11.5

Make an accurate drawing of this triangle.

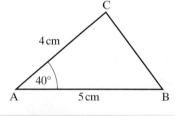

Solution

1 Draw the line AB 5 cm long.
With your protractor centred at A, mark the 40° angle.

2 Remove your protractor.
Draw a line from A through the mark which is 4 cm long.
At the end of this line mark the point C.

3 Finally, draw a straight line to join C to B.

You can also draw a triangle using a ruler and protractor only when you are given two of its angles and the length of the side between them. The method is given in the next example.

Example 11.6

In triangle PQR, PQ = 6 cm, angle QPR = 38° and angle PQR = 70°.

Make an accurate drawing of triangle PQR.

Solution

1   First make a sketch of the triangle.

> **TIP**
>
> If the side is not between the two angles, work out the third angle first.

2   Draw line PQ, 6 cm long. Mark an angle of 38° at P and draw a long line from P through the mark.

3   Mark an angle of 70° at Q. Draw a line from Q to meet the line drawn from P.

4 Label R where the two lines meet.

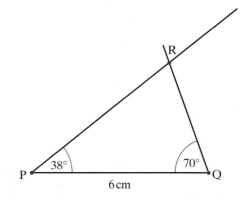

Exercise 11.5

1 Make an accurate full-size drawing of each of these triangles.
For each triangle, measure the unknown length and angles from your drawing.

(a) (b)

(c) (d)

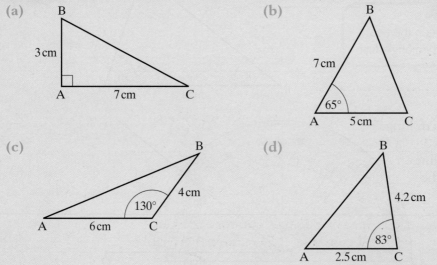

(e) Triangle ABC where AB = 7 cm, angle BAC = 118° and AC = 4 cm.

2 Make an accurate full-size drawing of each of these triangles.
For each triangle, measure the unknown lengths and angle from your drawing.

(a) (b)

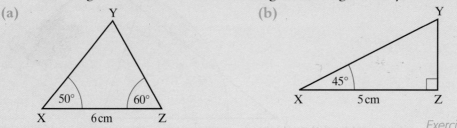

Exercise continues ...

(c)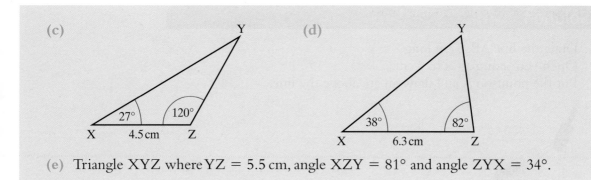

(d)

(e) Triangle XYZ where YZ = 5.5 cm, angle XZY = 81° and angle ZYX = 34°.

Challenge 11.2

Make an accurate drawing of this parallelogram.

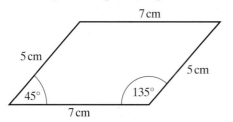

Check your accuracy by seeing if the other two angles are 45° and 135°.

Drawing triangles using compasses

When you are given all the lengths of a triangle but none of its angles, you can draw it using a ruler and compasses. The method is given in the next example.

Example 11.7

Make an accurate drawing of this triangle.

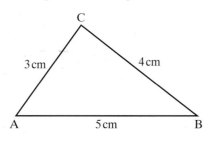

Solution

1   Draw the line AB, 5 cm long.
    Open your compasses to 3 cm.
    Put the point on A and draw an arc above the line.

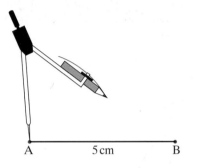

2   Now open your compasses to 4 cm.
    Put the point on B and draw another arc to intersect the first.

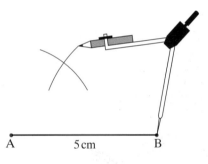

3   The point where the arcs cross is C.
    Join C to A and B using your ruler.

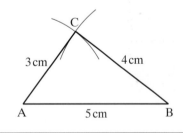

When you are given two lengths of a triangle and one of the angles
that is not between those two lengths, you need to use a ruler,
protractor and compasses to draw the triangle. The method is given in
the next example.

Example 11.8

Make an accurate drawing of triangle PQR where PQ = 6.3 cm, angle RPQ = 60° and
QR = 6 cm.

Solution

1 Draw a sketch of the triangle.

2 Draw the line PQ 6.3 cm long.
Mark an angle of 60° at P and
draw a long line from P
through the mark.

3 Open your compasses to 6 cm.
Put the point of your compasses at Q
and draw an arc to cross the line you
drew from P.

4 The point where the arc crosses the line is R.
Join R to P and Q using your ruler.

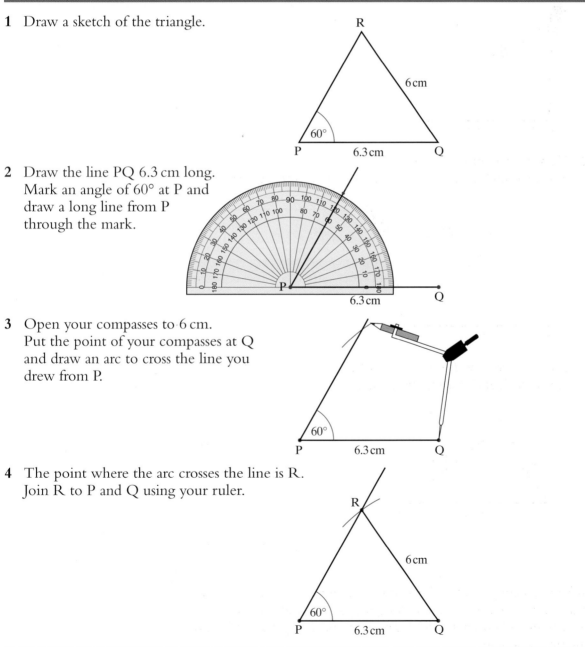

Challenge 11.3

Use the information given in Example 11.8 to draw a different triangle.

Hint: This triangle will have an obtuse angle at R.

Discovery 11.5

Look back at the triangles you have drawn in this chapter.

They can be classified into four groups depending on which measurements were given.

Using S for a given side and A for a given angle, find the different groups.

One group is different. Which is it and why?

Exercise 11.6

1 Make an accurate full-size drawing of each of these triangles. For each triangle, measure all the angles from your drawing.

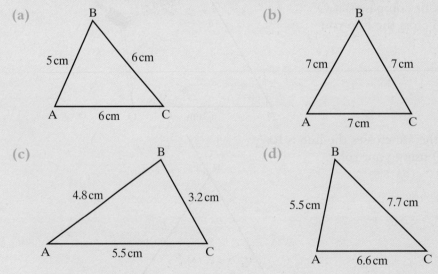

(a) B 5 cm 6 cm A 6 cm C

(b) B 7 cm 7 cm A 7 cm C

(c) B 4.8 cm 3.2 cm A 5.5 cm C

(d) B 5.5 cm 7.7 cm A 6.6 cm C

(e) Triangle ABC where AB = 7 cm, BC = 6 cm and AC = 4 cm.

Exercise continues …

2 Make an accurate full-size drawing of each of these triangles.
For each triangle, measure the unknown length and angles from your drawing.

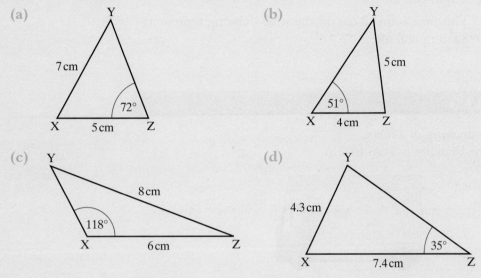

(a)

Y

7 cm

72°

X   5 cm   Z

(b)

Y

5 cm

51°

X   4 cm   Z

(c)   Y

8 cm

118°

X   6 cm   Z

(d)   Y

4.3 cm

35°

X   7.4 cm   Z

(e) Triangle XYZ where YZ = 5.8 cm, angle XZY = 72° and XY = 7 cm.

● **Challenge 11.4**

Make an accurate drawing of each of these shapes.

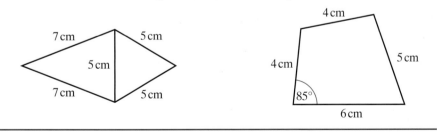

7 cm   5 cm

5 cm

7 cm   5 cm

4 cm

4 cm   5 cm

85°

6 cm

Scale drawings and maps

A scale drawing is exactly the same shape as the original drawing
but is different in size. Large objects are scaled down in size so that,
for example, they can fit on to the page of a book. A map is a scale
drawing of an area of land.

The scale of the drawing can be written like these examples.

1 cm to 2 m   This means that 1 cm on the scale drawing represents 2 m in real life.

2 cm to 5 km   This means that 2 cm on the scale drawing represents 5 km in real life.

Example 11.9

Here is a scale drawing of a lorry.
The scale of the drawing is 1 cm to 2 m.

(a) How long is the lorry?

(b) Will the lorry go safely under a bridge 4 m high?

(c) The lorry driver is 1.8 m tall.
How high will he be on the scale drawing?

Solution

(a) Measure the length of the lorry on the scale drawing.

Length of lorry on the drawing = 4 cm

As 1 cm represents 2 m, multiply the length on the drawing by 2 and change the units.

Length of lorry in real life = 4 × 2 = 8 m

(b) Height of lorry on the drawing = 2.5 cm

Height of lorry in real life = 2.5 × 2 = 5 m

So the lorry will not go under the bridge.

(c) To change from measurements in real life to measurements on the drawing you have to divide by 2 and change the units.

Height of driver in real life = 1.8 m

Height of driver on the drawing = 1.8 ÷ 2 = 0.9 cm

Exercise 11.7

1 Measure each of these lines as accurately as possible.
Using a scale of 1 cm to 4 m, work out the length that each line represents.

   (a) ——————————

   (b) ——————————————

   (c) ——————————————————

   (d) —————————

2 Measure each of these lines as accurately as possible.
Using a scale of 1 cm to 10 km, work out the length that each line represents.

   (a) ——————————————

   (b) ——————————————————————————

   (c) ———————————————————

   (d) —————————

3 Draw accurately the line to represent these actual lengths.
Use the scale given.

   (a) 5 m   Scale: 1 cm to 1 m   (b) 10 km   Scale: 1 cm to 2 km

   (c) 30 km   Scale: 2 cm to 5 km   (d) 750 m   Scale: 1 cm to 100 m

4 Here is a plan of a bungalow.
The scale of the drawing is 1 cm to 2 m.

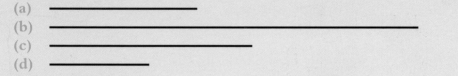

   (a) How long is the hall in real life?

   (b) Work out the length and width of each of the six rooms in real life.

   (c) The bungalow is on a plot of land measuring 26 m by 15 m.
What will the measurements of the plot of land be on this scale drawing?

Exercise continues …

5 The map shows some towns and cities in the south-east of England.
   The scale of the map is 1 cm to 20 km.

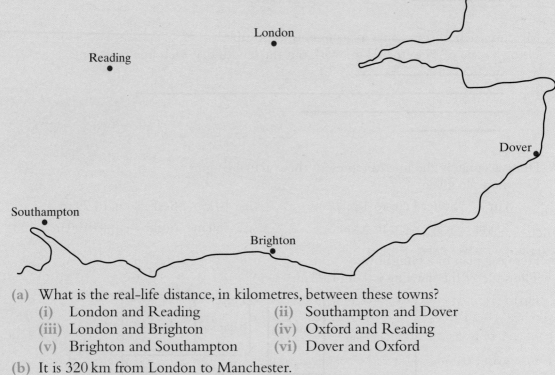

(a) What is the real-life distance, in kilometres, between these towns?
   (i)   London and Reading   (ii)   Southampton and Dover
   (iii) London and Brighton   (iv) Oxford and Reading
   (v)   Brighton and Southampton   (vi) Dover and Oxford

(b) It is 320 km from London to Manchester.
   How many centimetres will this be on the map?

Check up 11.4

Make a scale drawing of your classroom.

Use a scale of 1 cm to 1 m.

You will need to measure the length and the width of the room and
the size of the windows and doors.

You could also include items that are in the room, such as tables and
cupboards.

Discovery 11.6

The diagram is a scale drawing of a lake.

The scale of the drawing is 2 cm to 1 km.

There is a path running all the way round the outside of the lake.

How long is the path in kilometres?

Discuss with a partner how you can find the curved length accurately.

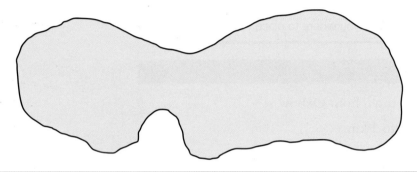

Scale drawings and bearings

When you want to describe accurately the direction in which something is travelling you use a **bearing**.

A bearing is an angle measured in degrees clockwise from a North line.

All bearings are written as three-figure angles. So, when a bearing is less than 100° you have to put one or two zeros in front of the figures. For example 045° or 008°.

It is easiest to measure a bearing with an angle measurer. This is a 360° protractor.

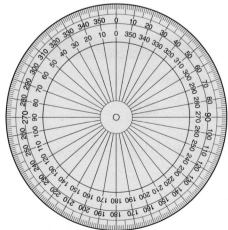

When you are measuring bearings, you always use the outside scale. You can ignore the inside scale because this is anticlockwise.

The centre of your angle measurer must be placed on the point where the bearing is being measured from. The zero line should go straight up, on top of the North line. Then, using the outside scale, work clockwise around to the angle.

TIP

Look for the word *from* as this will tell you where to place your protractor.

Example 11.10

Measure the bearing of each town from Oxbow.

The scale of the map is 1 cm to 1 km.

Work out the distance of eachtown from Oxbow.

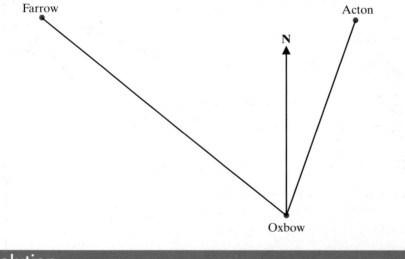

Solution

You place the centre of your angle measurer on Oxbow.

The angle between the North line and Acton is 20° so the bearing is 020°.

The length of the line on the map from Oxbow to Acton is 5.5 cm so the distance between the two towns is 5.5 km.

Acton is 5.5 km from Oxbow on a bearing of 020°.

Similarly, Farrow is 8.5 km from Oxbow on a bearing of 308°.

Example 11.11

A pilot flies for 25 km on a bearing of 070°.

She then changes direction and flies 15 km on a bearing of 220°.

(a) Make an accurate drawing of the flight. Use a scale of 1 cm to 5 km.

(b) Use your drawing to find how far the pilot is from her starting point.

(c) Use your drawing to find the bearing on which she needs to fly to get back to her starting point.

Solution

(a) Mark a point and draw a vertical line.

This is the North line.

Draw an angle of 70° clockwise from the North line.

The first stage of the journey is 25 km.

The scale is 1 cm to 5 km, so you divide the distance by 5 and change the units to find the length on the map.

Length of first stage on the map = 25 ÷ 5 = 5 cm.

Measure 5 cm along the line you drew on a bearing of 070°.

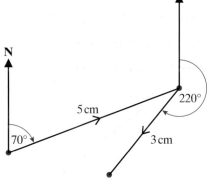

Mark a point and draw a vertical North line from this point.

The second stage of the journey is 15 km long on a bearing of 220°.

Length of second stage on the map = 15 ÷ 5 = 3 cm.

Draw a bearing of 220° and mark a point 3 cm along the line.

(b) To find the distance back to the start point you draw a line from the end of the second stage back to the start.

Then you measure the distance on the drawing and change it to a real-life distance.

Distance on map = 2.8 cm

Distance in real life = 2.8 × 5 = 14 km

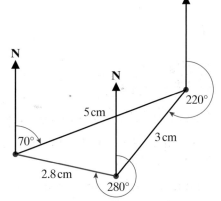

(c) To find the bearing, you draw a North line at the end of the second stage.

Then you measure the bearing back to the start point.

The bearing is 280°.

Exercise 11.8

1 Measure the bearing of each of the points in the diagram from O.

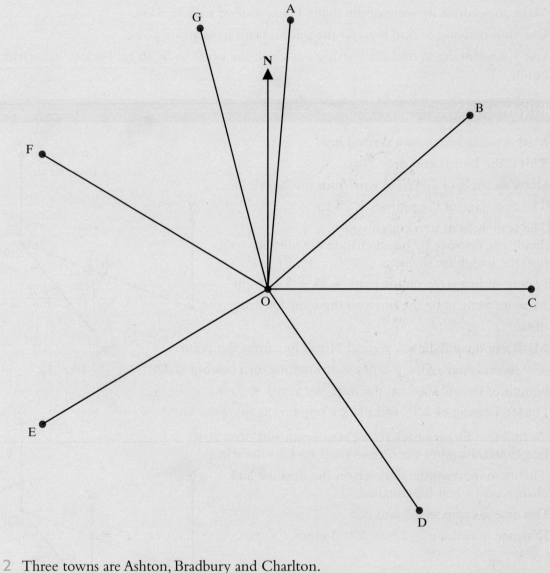

2 Three towns are Ashton, Bradbury and Charlton.
Bradbury is 10 km from Ashton on a bearing of 085°.
Charlton is 8 km from Ashton on a bearing of 150°.
Make a scale drawing showing these three towns.
Use a scale of 1 cm to 2 km

Exercise continues …

3　Measure the bearing of each of the places on the map from home.
How many kilometres is each place from home?
The scale of the map is 1 cm to 1 km.

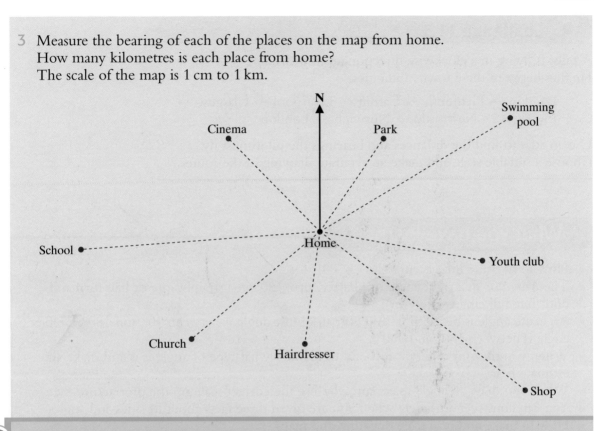

4　John went for a walk in the park.
He started at the car park (C) and walked 1 km on a bearing of 125° to the lake (L).
From the lake he walked 1.5 km on a bearing of 250° to the picnic site (P).

(a)　Draw an accurate scale drawing of John's walk.
Use a scale of 5 cm to 1 km.

(b)　How far is John from the car park?

(c)　On what bearing must he walk to get back to the car park?

5　The coastguard tracks a boat as it passes by.
He keeps a record of the boat's distance away and its bearing from the coastguard station.

Position	A	B	C	D	E
Distance from coastguard	4 km	6 km	5.5 km	6.5 km	5 km
Bearing from coastguard	050°	085°	145°	170°	215°

Use this information to make an accurate scale drawing of the boat's positions from the coastguard station.
Use a scale of 1 cm to 1 km.

Challenge 11.5

A pilot is flying in a clockwise direction around Britain.
He flies between these towns and cities.

London → Plymouth → Cardiff → Liverpool → Glasgow →
Inverness → Newcastle → Norwich → London

Use an atlas to find the distances and bearings the pilot must fly.
Choose a suitable scale and make an accurate drawing of the route.

What you have learned

- How to measure lines accurately
- There are 90° in a right angle or quarter turn, 180° in a straight line or half turn and 360° in a full circle or full turn
- An acute angle is between 0° and 90°; an obtuse angle is between 90° and 180°; a reflex angle is between 180° and 360°
- When you measure angles it is useful to identify what type of angle it is and make an estimate first
- When you draw an angle make sure you use the correct scale on the protractor
- You can construct a triangle when you are given three facts about its sides and angles
- How to make and read scale drawings and maps
- You can describe a direction using a bearing; a bearing is an angle measured in degrees clockwise from a North line
- All bearings are written as three-figure angles

Mixed exercise 11

1 Measure the length of each side of this shape.

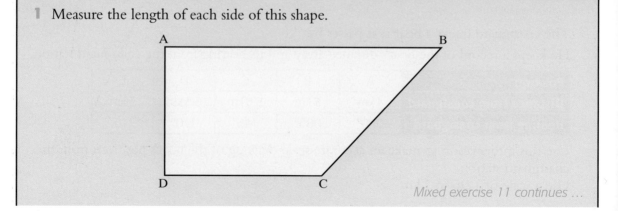

Mixed exercise 11 continues …

2 Look at the clock face.

Is the angle the hour hand turns through acute, right angle, obtuse or reflex when it moves

(a) from 12 noon to 3 p.m.?   (b) from 2 p.m. to 7 p.m.?

(c) from 10 a.m. to 1 p.m.?   (d) from 8 a.m. to 5 p.m.?

You may want to draw a sketch to help you.

3 Measure the size of each of these angles.

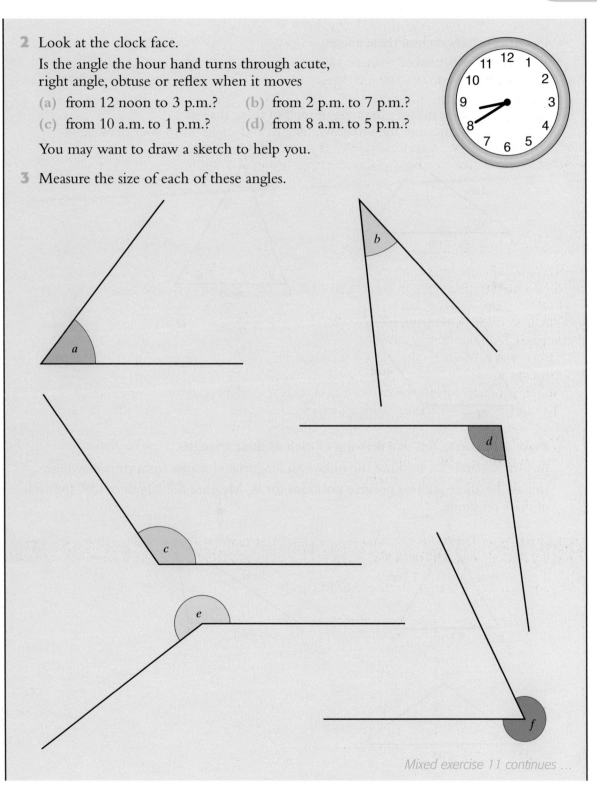

Mixed exercise 11 continues …

4 Draw accurately each of these angles.

(a) 75°   (b) 38°   (c) 104°

(d) 93°   (e) 207°   (f) 316°

5 Make an accurate, full-size drawing of each of these shapes.
For each shape, measure the unknown lengths and angles from your drawing.

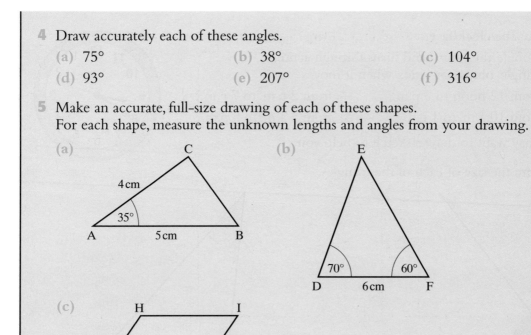

(a)

(b)

(c)

6 Make an accurate, full-size drawing of each of these triangles.

In parts (a) and (b), measure the unknown lengths and angles from your drawings.

In part (c), there are two possible positions for R. Measure the length of QR for each of these positions.

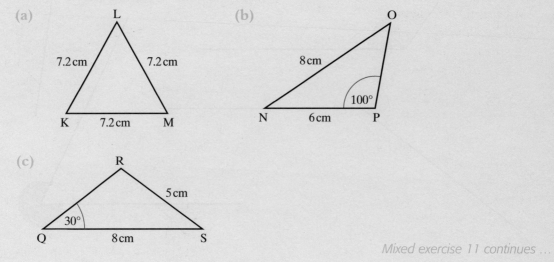

(a)

(b)

(c)

Mixed exercise 11 continues …

7 Here is a plan of the ground floor of a large house. The scale is 1 cm to 2 m.

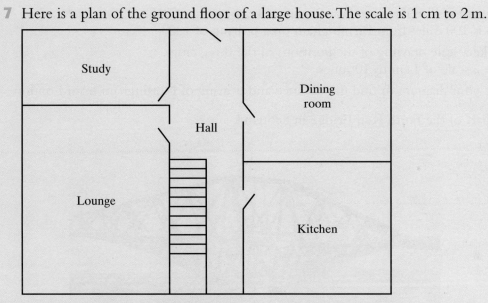

(a) How wide are the stairs?

(b) How long is the hall?

(c) Work out the length and width of each of the four rooms.

(d) The house is on a plot of land measuring 226 m by 105 m.
What will the measurements of the plot of land be on this scale drawing?

8 The diagram shows three boats A, B and C at sea. The scale of the diagram is 1 cm to 50 m.

What is the distance and bearing

(a) of B from A?

(b) of A from C?

(c) of C from B?

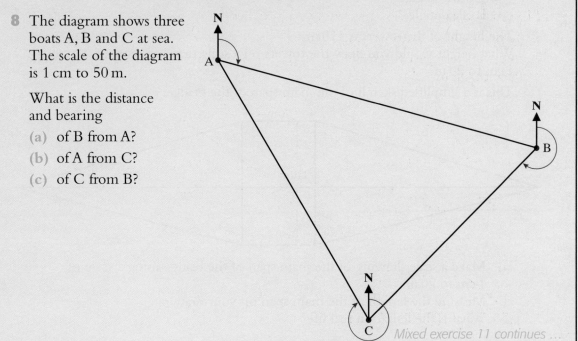

Mixed exercise 11 continues ...

9 Nottingham is 45 miles from Birmingham on a bearing of 045°.
London is 100 miles from Birmingham on a bearing of 130°.

(a) Make a scale drawing of the positions of the three cities.
Use a scale of 1 cm to 10 miles.

(b) Use your diagram to find the distance and bearing of Nottingham from London.

10 This is part of the Forth Rail Bridge in Scotland.

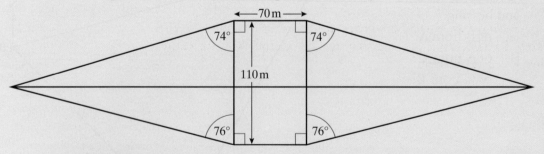

(a) Find an example of each of these angles in the picture.
(i) Right angle
(ii) Acute angle
(iii) Obtuse angle
(iv) Reflex angle

(b) The height of the towers is 110 m.
What height would you draw the towers on a scale drawing with a scale of
1 cm to 20 m?

(c) This is a simplified sketch of the main span of the bridge.

(i) Make a scale drawing of the main span of the bridge using a scale of
1 cm to 20 m.
(ii) Measure the length of the main span on your drawing.
What is the length in real life?

Using a calculator

◉ This chapter is about

- Rounding to the nearest integer, to a given number of decimal places and to 1 significant figure
- Using the square, square root and powers functions on your calculator
- Understanding the order in which your calculator does calculations
- Using your calculator efficiently to do more difficult calculations

◉ You should already know

- How to use the four basic arithmetic functions $+, -, \times, \div$ on your calculator
- The meanings of the terms *square* and *square root*

Rounding numbers

In Chapter 1 you learned a method for rounding numbers to the nearest 10, 100, 1000, … . You can use this method for rounding to other levels of accuracy.

Rounding to the nearest whole number

Look at this number line.

1.7 is nearer to 2 than 1. So 1.7 to the nearest whole number is 2.
2.4 is nearer to 2 than 3. So 2.4 to the nearest whole number is 2.

Look at this number line.

6.5 is halfway between 6 and 7.
You always round up any numbers that are in the middle.
So 6.5 to the nearest whole number is 7.

To round to the nearest whole number, look at the first decimal place.
- If it is less than 5, leave the whole number as it is.
- If it is 5 or more add 1 to the whole number.

You ignore any digits in the second decimal place and further to the right.

Example 12.1

Round these to the nearest whole number.
(a) 4.91   **(b)** 17.32   **(c)** 91.5   **(d)** 4.032   **(e)** 146.9

Solution

(a) 5   The first decimal place is 9 so you add 1 to 4.
(b) 17   The first decimal place is 3 so you leave 17 as it is.
(c) 92   The first decimal place is 5 so you add 1 to 91.
(d) 4   The first decimal place is 0 so you leave 4 as it is.
(e) 147   The first decimal place is 9 so you add 1 to 146.

Rounding to a given number of decimal places

When a number is written in decimal form, the digits on the right-hand side of the decimal point are known as **decimal places**. Numbers can have many different decimal places.

65.3 is written to 1 decimal place.
25.27 is written to 2 decimal places.
98.654 is written to 3 decimal places.
And so on.

You can shorten 'decimal place' to 'd.p.'.

When working with numbers, you may be asked to round a number to a certain number of decimal places. You can adapt the method you learned for rounding in Chapter 1 to do this.

- Count the decimal places from the decimal point and look at the first digit you need to remove.
- If this digit is less than 5, just remove all the unwanted places.
- If this digit is 5 or larger, add 1 to the digit in the last decimal place you want and then remove the unwanted decimal places.

Example 12.2

Round these numbers to the number of decimal places given.

(a) 65.533 to 1 decimal place   **(b)** 21.334 to 2 decimal places

(c) 88.653 to 1 decimal place   **(d)** 327.556 to 2 decimal places

(e) 2.658 97 to 3 decimal places

Solution

(a) 65.5   The second decimal place is less than 5 so you just remove the unwanted decimal places.

(b) 21.33   The third decimal place is less than 5 so you just remove the unwanted decimal places.

(c) 88.7   The second decimal place is 5 so you add 1 to the digit in the first decimal place.

(d) 327.56   The third decimal place is larger than 5 so you add 1 to the digit in the second decimal place.

(e) 2.659   The fourth decimal place is larger than 5 so you add 1 to the digit in the third decimal place.

Rounding to a given number of decimal places is often used in everyday situations.

Example 12.3

£50 is shared equally between seven people.
How much does each receive?

Solution

Using a calculator, $50 \div 7 = 7.142\,857\,143$.

Since a penny is the smallest coin there is, it makes sense to round this answer to 2 decimal places.

This makes the answer £7.14 (since the third decimal place is less than 5).

Rounding to 1 significant figure

To find the first **significant figure** of a number you start at the *left* of the number and look for the first non-zero digit. The second significant figure is the next digit to the right.

In 19 765   the first significant figure is 1 and it represents 10 000 and the second significant figure is 9 and it represents 9000.

In 202 322   the first significant figure is 2 and it represents 200 000 and the second significant figure is 0 and it tells you there are no ten thousands.

Notice that zero can be a significant figure when it is not the first significant figure. You can shorten 'significant figure' to 'sig fig' or 's.f.'.

Numbers are often rounded to 1 significant figure in newspapers.

A headline might read '20 000 attended test match' when the number was actually 19 765.

Another headline might be 'Gang steal £200 000' when the actual amount was £202 322.

You can adapt the method you have used before to round a number to 1 significant figure.
- Starting from the left of the number, find the first and second significant figures.
- If the second significant figure is less than 5, leave the first significant figure as it is and replace all the other digits with zeros.
- If the second significant figure is 5 or larger, add 1 to the first significant figure and replace all the other digits with zeros.

The zeros that you use to replace all but the first significant figure show the size of the number.

Example 12.4

Round these numbers to 1 significant figure.

(a) 5210 **(b)** 69 140 **(c)** 406 **(d)** 45 200

Solution

(a) 5000 The second significant figure is 2 so you leave 5 as it is and replace the other three digits with three zeros.

(b) 70 000 The second significant figure is 9 so you add 1 to 6 and replace the other four digits with four zeros.

(c) 400 The second significant figure is 0 so you leave 4 as it is and replace the other two digits with two zeros.

(d) 50 000   The second significant figure is 5 so you add 1 to 4 and replace the other four digits with four zeros.

> ### TIP
>
> A common error is to put the wrong number of zeros when rounding.
>
> Remember that what is being found is the approximate value of the number so it must be about the same size as the original number.

Exercise 12.1

1   Round these numbers to the nearest whole number.

   (a) 14.2   (b) 16.5   (c) 581.4   (d) 204.6   (e) 8.96
   (f) 28.48   (g) 319.6   (h) 924.23   (i) 1.12   (j) 34.57

2   Round these numbers to 1 decimal place.

   (a) 5.237   (b) 48.124   (c) 0.8945   (d) 7.6666   (e) 9.8876

3   Round the numbers in question 2 to 2 decimal places.

4   Round these numbers to 1 significant figure.

   (a) 1402   (b) 3121   (c) 59 104   (d) 42   (e) 616 312
   (f) 8 546 217   (g) 294   (h) 4092   (i) 631

5   There are 23 214 people at a rally in Hyde Park.
    How many is this correct to 1 significant figure?

● Challenge 12.1

Eight people won £1 842 631 between them in the lottery.
They shared it equally.

How much did each receive?
Give your answer correct to the nearest pound.

● Challenge 12.2

An extension cost £8000 to build, correct to 1 significant figure.

What was the least that the extension could have cost?

Squares

You learned about squares in Chapter 1. Remember 5^2 is read as 5 squared and means 5×5. So $5^2 = 25$.

To square numbers on your calculator you can use the $\boxed{\times}$ key or the $\boxed{x^2}$ key.

◯ Discovery 12.1

Make sure that you can find the $\boxed{x^2}$ key on your calculator.

TIP

Remember, the square key may look different on your calculator.

(a) Work out 3.4^2 by doing $\boxed{3}\boxed{.}\boxed{4}\boxed{\times}\boxed{3}\boxed{.}\boxed{4}\boxed{=}$.

Now do the calculation again by pressing $\boxed{3}\boxed{.}\boxed{4}\boxed{x^2}\boxed{=}$.

Check that the answers are the same.

(b) Use both methods to work out these squares.
   (i) 5.3^2 (ii) 12.7^2 (iii) 29^2
   (iv) 168^2 (v) 0.86^2 (vi) 0.17^2

● Challenge 12.3

Look again at the answers you found in part **(b)** of Discovery 12.1.

Previously, squaring has always given a bigger answer than the original number.

The last two answers are smaller than the original number.

When is the square of a number smaller than the original number?

Are there any numbers whose square is the same size as the original number?

Square roots

To find the square root of a number you look for a number which when multiplied by itself gives the original number.

So the square root of 9 is 3 because $3 \times 3 = 9$.

You write $\sqrt{9} = 3$.

◯ Discovery 12.2

Make sure that you can find the $\boxed{\sqrt{}}$ key on your calculator.

(a) Work out the square root of 60.84 by pressing $\boxed{\sqrt{}}\boxed{6}\boxed{0}\boxed{.}\boxed{8}\boxed{4}\boxed{=}$.

Then check your answer by doing $\boxed{ANS}\boxed{x^2}\boxed{=}$. You should get 60.84.

(b) Find these square roots and check your answers.

   (i) $\sqrt{27.04}$ (ii) $\sqrt{8.41}$ (iii) $\sqrt{21.16}$

   (iv) $\sqrt{2916}$ (v) $\sqrt{0.5184}$ (vi) $\sqrt{0.1849}$

● Challenge 12.4

Look again at the answers you found in part **(b)** of Discovery 12.2.

Which numbers have a square root that is bigger than the original?

When is the square root of a number bigger than the original?

Are there any numbers that have a square root equal to the original?

Accuracy

If you work out 2.63^2 on your calculator the display will show 6.9169.

If you work out $\sqrt{6.8}$ on your calculator the display will show 2.607 680 9... . (The number of decimal places will be different on some calculators.)

You will probably not need to give so many decimal places in your answer. Usually, rounding your answer to 2 or 3 decimal places will be accurate enough. Sometimes the question will tell you how accurately to give your answer.

$2.63^2 = 6.92$ or 6.917 is probably accurate enough.

$\sqrt{6.8} = 2.61$ or 2.608 is probably accurate enough.

Only round your final answer. Do not use rounded numbers in a calculation.

TIPS

- If you round a number, always state the number of decimal places you have rounded to, for example 2.61 correct to 2 d.p. (decimal places).
- If you have rounded a number for an answer to part of a question and need to use that answer for another part of the question, always go back to the more accurate version.
- It is a good idea to keep an answer in your calculator until you know that you are not going to use it in the next part of the question.

Other powers

In Chapter 1 you learned that 5^2 can be read as '5 squared' or as '5 to the power 2'. You also learned that 5^4 can be read as '5 to the power 4' and means $5 \times 5 \times 5 \times 5$.

You can work out higher powers on your calculator using the $\boxed{\times}$ key on your calculator but there is also a key for calculating powers directly. This may be labelled $\boxed{\wedge}$ or $\boxed{x^y}$ or $\boxed{y^x}$ or $\boxed{x^?}$.

Find how to calculate powers on your calculator.

Check up 12.1

Practise using the powers key on your calculator using examples that you can do in your head, for example 4^3 or 2^4.

Example 12.5

Work out these.

(a) 6^4 **(b)** 3.1^7

Solution

(a) $6^4 = 1296$ You press $\boxed{6}\,\boxed{\wedge}\,\boxed{4}\,\boxed{=}$

(b) $3.1^7 = 2751.26$ (2 d.p.) You press $\boxed{3}\,\boxed{.}\,\boxed{1}\,\boxed{\wedge}\,\boxed{7}\,\boxed{=}$

Even if the powers key on your calculator is labelled $\boxed{x^y}$ or $\boxed{y^x}$ or $\boxed{x^?}$ rather than $\boxed{\wedge}$ you may see the symbol $\wedge$ in your display.

Exercise 12.2

1 Work out these squares.
    (a) 2.7^2 (b) 4.7^2 (c) 38^2 (d) 328^2 (e) 0.62^2
    (f) 0.19^2 (g) 0.07^2 (h) 1.8^2 (i) 2.16^2 (j) 31.6^2

2 Work out these square roots.
    (a) $\sqrt{16.81}$ (b) $\sqrt{59.29}$ (c) $\sqrt{2.56}$ (d) $\sqrt{295.84}$ (e) $\sqrt{1296}$
    (f) $\sqrt{0.0961}$ (g) $\sqrt{0.2401}$ (h) $\sqrt{0.7396}$ (i) $\sqrt{14.0625}$ (j) $\sqrt{0.002\,209}$

Exercise continues …

3 Work out these square roots. Give your answers to 2 decimal places.

(a) $\sqrt{17.32}$ (b) $\sqrt{29.8}$ (c) $\sqrt{88}$ (d) $\sqrt{567}$ (e) $\sqrt{2348}$

(f) $\sqrt{0.345}$ (g) $\sqrt{0.9}$ (h) $\sqrt{23\,790}$ (i) $\sqrt{1.87}$ (j) $\sqrt{0.078}$

4 The area of a square is $480\,\text{cm}^2$. Find the length of the side.
   Give your answer to 1 decimal place.

5 Work out these. Where necessary, give your answer to 2 decimal places.

(a) 4.2^4 (b) 0.52^4 (c) 2.01^6 (d) 3.24^5

Order of operations

◉ Discovery 12.3

(a) What are the answers to these calculations?
   (i) $4 + 8 \div 2$ (ii) $2 + 3 \times 4 - 5$

(b) Press these sequences of keys on your calculator.

(i) $\boxed{4}\,\boxed{+}\,\boxed{8}\,\boxed{\div}\,\boxed{2}\,\boxed{=}$ (ii) $\boxed{2}\,\boxed{+}\,\boxed{3}\,\boxed{\times}\,\boxed{4}\,\boxed{-}\,\boxed{5}\,\boxed{=}$

Did you get the answers you expected?

If you have a scientific calculator, it always does multiplication and division before addition and subtraction.

If you have a calculator that only does add, subtract, multiply and divide it will do the calculations from left to right.

Check which your calculator does.

In Discovery 12.3, if your calculator gave the answer 6 to $4 + 8 \div 2$, then it works left to right. If your calculator gave the answer 8, then it does multiplication and division before addition and subtraction. This is the correct way to do calculations.

So the correct answers to Discovery 12.3 are $4 + 8 \div 2 = 8$ and $2 + 3 \times 4 - 5 = 9$.

One way to avoid problems is to use brackets.

For example, in a calculation such as $(3 + 4) \times 2$, the brackets mean do $3 + 4$ first and then multiply by 2. So the answer is $7 \times 2 = 14$.

You always do what is in the brackets first.

If your calculator has brackets then you can use them rather than writing down the middle step.

Example 12.6

Work out $(5.9 + 3.3) \div 2.3$.

Solution

If your calculator has brackets, press this sequence of keys.

$\boxed{(}\boxed{5}\boxed{.}\boxed{9}\boxed{+}\boxed{3}\boxed{.}\boxed{3}\boxed{)}\boxed{\div}\boxed{2}\boxed{.}\boxed{3}\boxed{=}$

If your calculator does not have brackets, work out $5.9 + 3.3$ first.
Write down the answer, 9.2.

Now do $9.2 \div 2.3$.

For both methods the answer is 4.

Example 12.7

Use your calculator to work out these.

(a) $\sqrt{(5.2 + 2.7)}$ **(b)** $5.2 \div (3.7 \times 2.8)$

Solution

(a) The brackets show that you need to work out $5.2 + 2.7$ before
finding the square root.
Press this sequence of keys.

$\boxed{\sqrt{}}\boxed{(}\boxed{5}\boxed{.}\boxed{2}\boxed{+}\boxed{2}\boxed{.}\boxed{7}\boxed{)}\boxed{=}$

The answer is 2.811 correct to 3 decimal places.

(b) You need to do 3.7×2.8 before doing the division.
Press this sequence of keys.

$\boxed{5}\boxed{.}\boxed{2}\boxed{\div}\boxed{(}\boxed{3}\boxed{.}\boxed{7}\boxed{\times}\boxed{2}\boxed{.}\boxed{8}\boxed{)}\boxed{=}$

The answer is 0.502 correct to 3 decimal places.

Exercise 12.3

Work out these on your calculator.
If the answers are not exact, give them correct to 3 decimal places.

1   $(5.2 + 2.3) \div 3.1$ 2 $(127 - 31) \div 25$ 3 $(5.3 + 4.2) \times 3.6$

4   $\sqrt{(15.7 - 3.8)}$ 5 $3.2^2 + \sqrt{5.6}$ 6 $(6.2 + 1.7)^2$

7   $6.2^2 + 1.7^2$ 8 $5.3 \div (4.1 \times 3.1)$ 9 $2.8 \times (5.2 - 3.6)$

10   $6.3^2 - 3.7^2$ 11 $\sqrt{(5.3 \times 9.2)}$ 12 $25.2 \div (6.1 + 3.8)$

Challenge 12.5

(a) You do a calculation on your calculator and the display shows 0.333 333.
(The number of 3s may vary, depending on how many figures your display holds.)
What does this answer mean?
How can you write this answer?

(b) You do another calculation on your calculator and the display shows 0.666 666.
(Again the number of 6s may vary, depending on how many figures your display holds.)
What does this answer mean?
How can you write this answer?

Challenge 12.6

Look at this calculation.

$$4.8 + 3.7 \times 173 \div 2.5$$

(a) Show where to put brackets to give these answers.
(i) 588.2 **(ii)** 260.84 **(iii)** 257.96

(b) Can you find a different answer?

What you have learned

- To round numbers to the nearest whole number you look at the digit in the first decimal place; if it is less than 5 you leave the whole number as it is; if it is 5 or more you add 1 to the whole number
- You use similar methods to round to a given number of decimal places
- The first significant figure of a number is the first non-zero digit when starting from the left of the number and moving right
- To round numbers to 1 significant figure you look at the second significant figure; if it is less than 5 you leave the first significant figure as it is; if it is 5 or more you add 1 to the first significant figure. You use zeros to replace all but the first significant figure in the number so that the size of the number is correct.
- You can work out squares and square roots using the x^2 and $\sqrt{}$ buttons on your calculator
- What order your calculator does operations
- You can change the order of operations by using brackets

Mixed exercise 12

1 Round these numbers to the nearest whole number.
   (a) 47.3   (b) 1.624   (c) 98.37   (d) 104.53   (e) 6.75

2 Round these numbers to 1 decimal place.
   (a) 9.14   (b) 1.64   (c) 68.67   (d) 1.385   (e) 3.879   (f) 16.375

3 Round these numbers to 2 decimal places.
   (a) 1.385 (b) 3.879 (c) 16.375 (d) 43.9543 (e) 1.333 33 (f) 2.666 666

4 Round these numbers to 1 significant figure.
   (a) 41   (b) 29 184   (c) 8162   (d) 756 324   (e) 9871

5 There are 16 478 people at a football match.
How many people are there correct to 1 significant figure?

6 Work out these squares.
   (a) 7.1^2 (b) 6.4^2 (c) 38^2 (d) 521^2 (e) 0.46^2

7 Work out these square roots.
   (a) $\sqrt{23.04}$ (b) $\sqrt{68.89}$ (c) $\sqrt{590.49}$ (d) $\sqrt{0.1089}$ (e) $\sqrt{0.003\,969}$

8 Work out these square roots. Give your answers to 2 decimal places.
   (a) $\sqrt{37.3}$ (b) $\sqrt{537}$ (c) $\sqrt{40\,682}$ (d) $\sqrt{0.389}$ (e) $\sqrt{0.0786}$

9 The area of a square field is $9650\,\text{m}^2$.
Find the length of the side.
Give your answer in metres to 1 decimal place.

10 Work out these. Give your answers to 2 decimal places.
   (a) 2.6^4 (b) 3.8^5 (c) 4.7^6

11 Work out these on your calculator.
If the answers are not exact, give them correct to 3 decimal places.
   (a) $(4.2 + 8.6) \div 1.7$ (b) $\sqrt{(148 - 37)}$ (c) $(6.3 - 1.9)^2$
   (d) $5.7 \times (6.8 + 9.2)$ (e) $4.3^2 + \sqrt{28.3}$ (f) $54.2 \div (5.3 \times 4.1)$

12 Use your calculator to find out how many seconds there are in one week.

Statistical diagrams 1

This chapter is about

- Drawing and interpreting pie charts
- Drawing and interpreting line graphs

You should already know

- How to draw simple graphs
- How to measure and draw angles
- How to calculate the mean, median, mode and range for a set of data

Drawing pie charts

In Chapter 3 you saw that it is possible to take the data from a frequency table and draw a vertical line graph or a bar chart to show the results. Another way of showing these results is to use a **pie chart**.

The table shows some data on the number of children in families. The diagram on the right is the pie chart showing the same information.

Number of children	Frequency
1	12
2	10
3	3
4	2
5	2
More than 5	1
Total	30

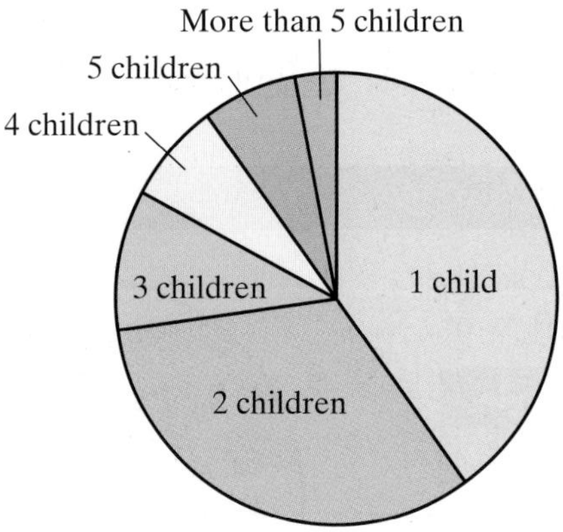

To draw a pie chart you need to work out the angle for each **sector** (slice of the pie) of the circle.

There are 30 families altogether and there are 360° in a circle, so the angle for one family is

$$360° \div 30 = 12°.$$

The working for the angles used in the pie chart is shown in the table.

Number of children	Calculation	Angle
1	12 × 12° =	144°
2	10 × 12° =	120°
3	3 × 12° =	36°
4	2 × 12° =	24°
5	2 × 12° =	24°
More than 5	1 × 12° =	12°

TIP

Check that the angles for your pie chart add up to 360° before you start drawing.

When drawing the sectors you need to measure the angles accurately. Make sure that the next sector starts where the previous one finishes.

It is a good idea to measure the last sector. It provides a check that you have drawn the other angles correctly. If it is not the size you have calculated it should be, check the angles you have drawn for the other sectors.

Example 13.1

Draw a pie chart to represent these data.

Type of pet	Frequency
Cat	18
Dog	14
Horse	3
Rabbit	7
Bird	5
Other	13

Solution

The total number of pets is 60 so the angle for one pet is 360° ÷ 60 = 6°.

Type of pet	Calculation	Angle
Cat	18 × 6° =	108°
Dog	14 × 6° =	84°
Horse	3 × 6° =	18°
Rabbit	7 × 6° =	42°
Bird	5 × 6° =	30°
Other	13 × 6° =	78°

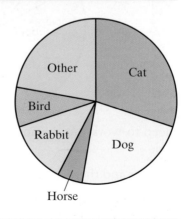

Exercise 13.1

1 Draw a pie chart for each of these sets of data.

(a)

Number of cars	Frequency
None	2
1	12
2	10
3	5
More than 3	1
Total	30

(b)

Number of bedrooms	Frequency
1	2
2	8
3	14
4	11
More than 4	1
Total	36

(c)

Number of pets	Frequency
0	9
1	7
2	2
3	5
More than 3	1
Total	24

2 Draw a pie chart for each of these sets of data.

(a)

Favourite crisps	Frequency
Beef	21
Chicken	8
Cheese and onion	10
Salt and vinegar	12
Other	9

(b)

Eye colour	Frequency
Blue	18
Brown	29
Green	9
Grey	14
Other	2

(c)

Hair colour	Frequency
Black	8
Blonde	12
Brown	22
Red	4
Other	2

3 Draw a pie chart for each of these sets of data.

(a)

Country	Population (millions)
England	50.0
Wales	5.5
Scotland	3.0
N. Ireland	1.5
Total	60.0

(b)

Fruit crumble ingredient	Weight (grams)
Fruit	900
Flour	140
Butter	140
Sugar	140
Oats and nuts	120
Total	1440

Interpreting pie charts

Pie charts are very useful for comparing proportions, for example of votes cast in an election. The slices of the pie are called **sectors**.

Example 13.2

The pie chart shows the makes of mobile phone owned by a group of people.

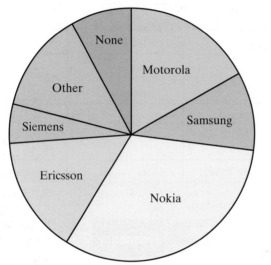

(a) Which is the most popular make?

(b) Which is the least popular make?

Solution

(a) The biggest sector is for Nokia.
Nokia is the most popular make.

(b) The smallest sector is for Siemens.
Siemens is the least popular make.

In Example 13.2 you could not say how many people owned a Nokia mobile phone. If you are told the number of data items represented by the pie, however, you can work out the number of items in each category. To do this you must measure the angles of the sectors. You then write the angle as a fraction of a full turn. Finally, you find this fraction of the total frequency.

Example 13.3

The number of people surveyed in Example 13.2 was 60.

(a) How many people own Nokia mobile phones?

(b) How many people own each of the other makes of phone?

Solution

(a) There are 360° in a full turn.

The angle for Nokia is 114°.

To work out the number of people who own Nokia phones you need to divide the angle for Nokia by 360° and then multiply this fraction by the total number of people in the survey.

$$\frac{114}{360} \times 60 = 19 \text{ people}$$

(b) You can lay out your calculations in a table.

Type of phone	Calculation	Frequency
Motorola	$\frac{60}{360} \times 60$	10
Samsung	$\frac{36}{360} \times 60$	6
Ericsson	$\frac{54}{360} \times 60$	9
Siemens	$\frac{18}{360} \times 60$	3
Other	$\frac{48}{360} \times 60$	8
None	$\frac{30}{360} \times 60$	5

TIP

You can check your answer by adding the frequencies.
The total should equal the number of people in the survey.
$19 + 10 + 6 + 9 + 3 + 8 + 5 = 60$

If you asked a different group of 60 people you might get exactly the same results but it is very unlikely.

If you think about how many people own a mobile phone, 60 is a very small proportion. A different 60 people are likely to have different preferences so the frequencies will be different. It is possible that one of the 'other' phone makes is so much more popular that you would give it a category of its own.

● Challenge 13.1

How many people in your class own mobile phones?
What about other people in your households?

Is Nokia the most popular make of mobile phone?
Do you think the proportions for your class are similar to the proportions for other classes in your school?
What about for a different school or for a group of people you ask in a shopping centre?
What reasons can you think of for your answers?

Draw a pie chart to show the makes of mobile phone owned by people in your class or your households.

Exercise 13.2

1　The pie chart shows the nutritional values of a bag of crisps.

　(a)　What proportion of the crisps is carbohydrate?

　(b)　About what proportion of the crisps is fat?

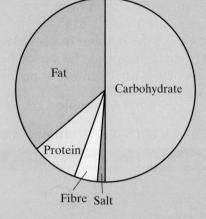

2　The pie chart shows the proportions of the different newspapers and magazines James delivers on his paper round.

　(a)　What does he deliver most of?

　(b)　Which of the categories have the same proportion?

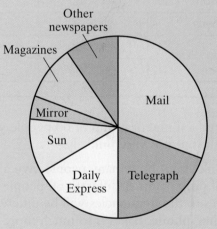

3　The pie chart shows the number of cars owned by the families of 90 people questioned in a survey.

　(a)　How many families owned one car only?

　(b)　How many families did not own a car?

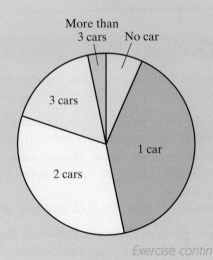

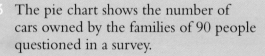

Exercise continues ...

4 The pie chart shows the number of
bedrooms in the homes of 72 people
questioned in a survey.

   (a) How many people had
   three bedrooms?

   (b) How many people had
   two bedrooms?

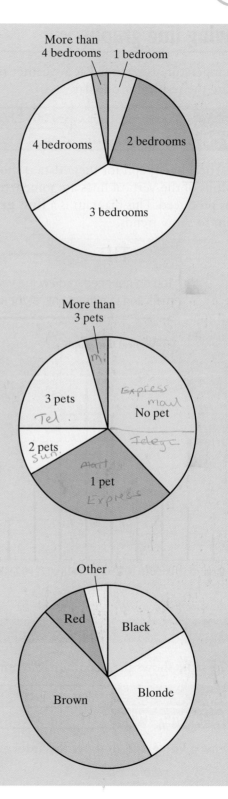

5 The pie chart shows the number of
pets owned by 48 people questioned
in a survey.

   (a) How many people had one pet?

   (b) How many people had three pets?

   (c) How many people did not own
   a pet?

6 The pie chart shows the hair colour of
96 people questioned in a survey.

   (a) How many people had brown hair?

   (b) How many people had black hair?

   (c) How many people had blonde hair?

Drawing line graphs

The temperature is recorded at a weather station every three hours. Here are the results for one day.

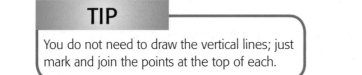

Time	00:00	03:00	06:00	09:00	12:00	15:00	18:00	21:00	24:00
Temp (°C)	10	12	13	15	19	21	16	14	11

The vertical line graph for these data is shown below on the left. If the tops of the vertical lines are joined, the diagram shown on the right is produced. This diagram is a **line graph**. A line graph is a useful diagram to show trends.

TIP

You do not need to draw the vertical lines; just mark and join the points at the top of each.

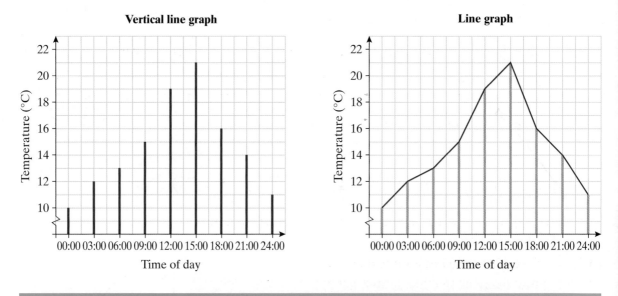

Exercise 13.3

1 The table shows the maximum daytime temperature in Wakefield one week.

Day	Mon	Tues	Wed	Thurs	Fri	Sat	Sun
Temperature (°C)	17	19	22	21	23	20	18

Draw a line graph to show this information.

Exercise continues …

2 The table shows the monthly sales of a company in 2009.

Month	Jan	Feb	Mar	Apr	May	June	July	Aug	Sept	Oct	Nov	Dec
Sales (× £1000)	17	14	12	13	17	16	15	11	14	16	19	22

Draw a line graph to show this information.

3 A container of water is heated.
The table shows the temperature of the water as it was heated.

Minutes after start	0	5	10	15	20	25	30
Temperature (°C)	52	55	60	67	75	86	100

Draw a line graph to show this information.

4 The table shows the numbers of people visiting a museum one week.

Day	Mon	Tues	Wed	Thur	Fri	Sat	Sun
Number of visitors	350	425	475	450	375	700	550

Draw a line graph to show this information.

5 The table shows the sales made by a charity shop over a period of two weeks.

Day	1	2	3	4	5	6	7	8	9	10	11	12	13	14
Sales (£)	75	85	30	80	95	116	0	69	78	27	77	89	108	0

Draw a line graph to show this information.

● Challenge 13.2

Look in newspapers or magazines and find some examples of pie charts and line graphs.

Make a wall display explaining what they show.

Interpreting line graphs

When you draw a line graph you plot points to represent the data given and then join the points with a line.

Sometimes the values between the points you have plotted will not have a meaning and sometimes they will.

Remember in Chapter 3 you learned about discrete and continuous data. If the values on the horizontal scale of your graph are discrete values, such as the days of the week or the months of the year, then values between the plotted points will not have a value. You cannot have Wednesday and a half, so do not read the graph between the points. These graphs usually have the points joined with broken or dotted lines.

However, if the values on the horizontal scale of the graph are continuous values, such as times or time in hours, then the values between the plotted points will have a value. You can have a time of 12:30 or $2\frac{1}{2}$ hours. It is important to know that the values you read from the lines between plotted values are only **estimates**. You do not know the actual value because you did not measure it at that point.

Example 13.4

The temperature was measured every 3 hours in Liverpool one day. The graph shows the results.

(a) What was the temperature at noon?

(b) At what times was the temperature 16°C?

(c) Estimate the temperature at 07:00.

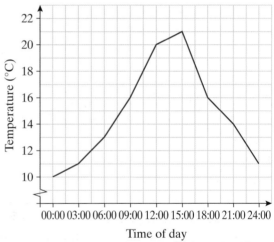

Solution

(a) Find 12:00 on the horizontal axis. Follow the grid line to the graph line. Follow the grid line to the vertical axis and read off the value.
The temperature at noon was 20°C.

(b) Notice the hint in the question: it is asking for more than one time.
This time start at the vertical axis and read off the values from the horizontal axis.
The times when the temperature was 16°C were 09:00 and 18:00.

(c) From the graph you can estimate that the temperature at 07:00 was 14°C.

It is possible, and sometimes very useful, to show more than one set of data on a single diagram.

Example 13.5

The line graph shows how people on a one-week training course had travelled there.

(a) How many people travelled by car on Tuesday?

(b) How many people walked on Thursday?

(c) On which days did the same number of people use the bus?

(d) Which way of travelling was used most?

(e) How many people were on the course?

(f) What was the mean number of people walking each day?

(g) Is there any meaning to the values between the plotted points?

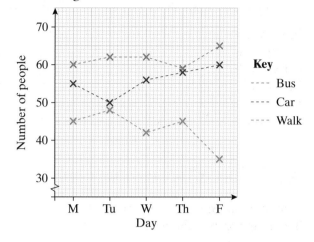

Solution

You read values from the graph in the same way as you did in Example 13.4.
Make sure you understand the scale.
The symbol on the vertical axis shows that the scale does not start at zero.

(a) 50

(b) 45

(c) Tuesday and Wednesday

(d) You do not need to work out the numbers. You can see that the line for the numbers using the bus is higher than the other lines.

(e) Read off the values for each way of travel for any one day.
Then add the numbers together.
For example, using Friday: $65 + 60 + 35 = 160$.

(f) Mean $= \dfrac{\text{Total number walking each day}}{\text{Number of days}}$

$= \dfrac{45 + 48 + 42 + 45 + 35}{5}$

$= \dfrac{215}{5} = 43$

The mean number of people walking each day was 43.

(g) There is no meaning to the values between the plotted points because the days of the week are discrete. The lines on the graph just show the trends.

Exercise 13.4

1 The temperature of a liquid was taken
every 5 minutes as it cooled.
The line graph shows the results.

(a) What was the temperature of the liquid
at the beginning of the experiment?

(b) How many minutes did it take for the
liquid to reach 50°C?

(c) What was the temperature of the liquid
after 25 minutes?

(d) What was the lowest temperature reached?

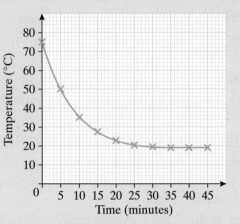

2 The nature of the injuries of the patients attending a casualty department were recorded
each day for one week. The line graph shows the results.

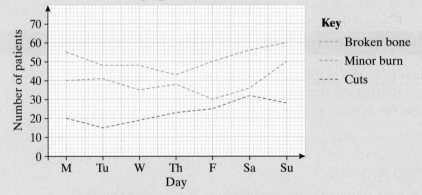

Key
---- Broken bone
---- Minor burn
---- Cuts

(a) How many patients suffered broken bones on Thursday?

(b) On which day were there fewest patients with burns?

(c) How many patients were there in total on Monday?

(d) How many patients with cuts were there in total in the week?

3 The line graph shows the temperature at noon each day in London for one week.

(a) What was the temperature
at noon on Thursday?

(b) What was the lowest
temperature at noon
that week?

(c) What was the range of the
temperatures at noon during
the week?

(d) Calculate the mean temperature at noon for the week.

Exercise continues....

4   The line graph shows the value of
the monthly sales made by an
umbrella company one year.

   (a)   In which month were the
sales greatest?

   (b)   What was the range of the
monthly sales?

   (c)   What was the total value of
sales for the year?

   (d)   What was the mean value of
the monthly sales?

5   The line graph shows the
number of visitors to a theme
park during one week.

   (a)   How many visitors were
there to the theme park
that week?

   (b)   What was the mean
number of visitors per day?

   (c)   What was the range of the
numbers of visitors
attending each day?

6   The line graph shows the value of the
sales of soft drinks in a café during
a 10-day period.

   (a)   What do you think happens on
Wednesday?

   (b)   What was the value of the sales
of soft drinks on Saturday?

   (c)   On which day when the café was
open were sales the lowest?

   (d)   What was the mean value of the sales
for the days when the café was open?

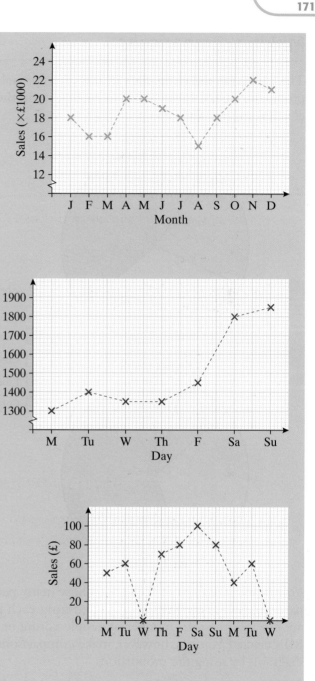

Making comparisons

The pie charts show the share of the vote for each party and the proportions of MPs elected from each party in the 2005 general election in the UK.

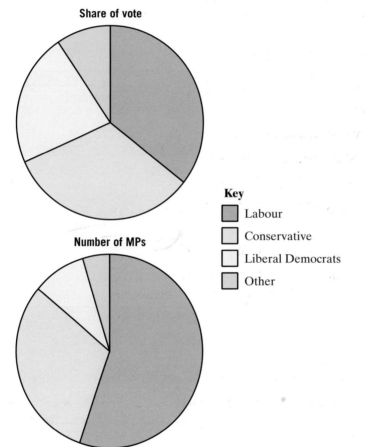

Share of vote

Number of MPs

Key

- Labour
- Conservative
- Liberal Democrats
- Other

You cannot tell from the pie charts how many people voted for each party or how many MPs were elected from each party because you are not told the total number of people voting or the total number of MPs elected. You can, however, make comparisons between the two diagrams by using the proportions.

These are some of the things you can say about these pie charts.
- The share of the vote for the Conservatives and the number of Conservative MPs elected are of about the same proportion.
- The Liberal Democrats and the parties other than the main three gained a smaller number of MPs than their share of the vote.
- Labour had a greater proportion of MPs than their share of the vote.

Challenge 13.3

If you want to be more precise, you need to use numbers.
You can use the angles of the different sectors of the pie.

For example, the share of the vote for the Liberal Democrats is represented by an angle of 80°. The angle for the number of Liberal Democrat MPs elected is 34°.

You could write these as fractions of 360°. They are $\frac{80}{360}$ and $\frac{34}{360}$, respectively. (You would not cancel them because you need the denominators to be the same to make a comparison.) However, this still does not give a very clear picture.

You can get a better picture by converting the fractions to percentages.

The Liberal Democrats' share of the vote as a percentage is

$\frac{80}{360} \times 100 = 22\%$ (to the nearest whole number).

The Liberal Democrats' proportion of the MPs elected as a percentage is

$\frac{34}{360} \times 100 = 9\%$ (to the nearest whole number).

Measure the other angles in the pie charts and compare the share of the vote with the proportion of MPs elected for Labour, Conservative and the other parties.

Example 13.6

The line graphs show the temperatures in Bradford and in Leeds on the same day.
Compare the temperatures of the two cities.

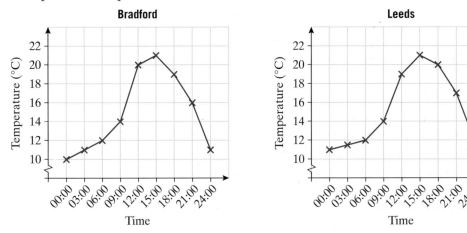

Solution

Similarities

- Both cities have the same maximum temperature (21°C).
- At 09:00 the temperature is the same in both cities (14°C).
- In both cities, the temperature at the end of the day is one degree higher than at the start of the day.

Differences

- Bradford has lower temperatures than Leeds at the start and end of the day.
- Between 09:00 and 12:00, Bradford gets warmer faster than Leeds.
- The range of the temperatures is slightly greater in Bradford.

● Challenge 13.4

How can you illustrate the data in Example 13.6 so that it is easier to compare?

Draw a suitable graph.

Exercise 13.5

1 The pie charts show the population and the land area of the countries in the UK.

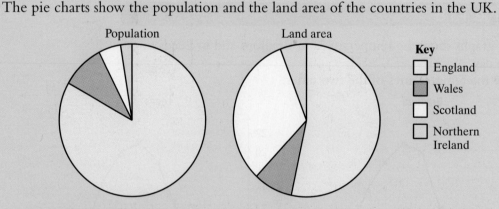

Population   Land area

Key
- England
- Wales
- Scotland
- Northern Ireland

(a) Which country has the largest land area and also has the largest population?

(b) Which country has a land area that is roughly in the same proportion as its population?

(c) Compare the land area and the population of Scotland.

Exercise continues …

2   The bar charts show the marks obtained by boys and by girls in a spelling test.

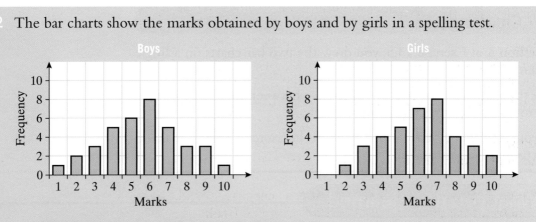

(a) How many boys scored 6 in the test?

(b) How many girls scored 1 in the test?

(c) What was the modal score for the girls?

(d) What was the range of the scores for the boys?

(e) Did the boys or the girls do better in this test? Give a reason for your answer.

3   The wealth of a country can be measured by its Gross Domestic Product (GDP) per person. The diagrams show the GDP per person of five developed and five undeveloped countries in 2003.

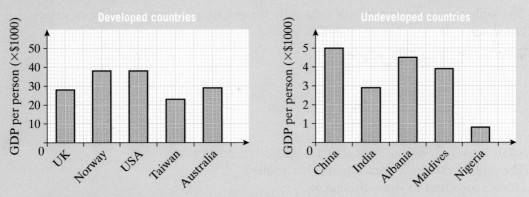

(a) What is the GDP per person of Norway?

(b) What is the GDP per person of China?

(c) What is the mean GDP per person for the developed countries?

(d) What is the mean GDP per person for the undeveloped countries?

(e) What is the range of the GDP per person for all ten countries?

(f) Which group of countries is the wealthier?

(g) Draw both bar charts on a single diagram using an appropriate scale.

Challenge 13.5

In question 3 of Exercise 13.5, you drew the two bar charts on a single diagram.

Using the same scale showed the difference between the GDPs of the two groups of countries much more clearly.

Can you change any of the other diagrams in Exercise 13.5 to make it easier to make comparisons?

What you have learned

- A pie chart is useful for comparing proportions; the angle of each sector represents its size
- To calculate the angle to use for a sector you divide 360° by the total frequency; you then multiply the result by the frequency for the category
- A line graph is useful for showing trends
- The points on the lines between the plotted points of a line graph may or may not have a meaning, depending on what is plotted on the horizontal axis
- You can compare two sets of data by looking at diagrams representing the data

Mixed exercise 13

1 The table shows the nutritional values, by weight, of a 72-gram portion of cereal. Draw a pie chart to show these data.

Food type	Weight (g)
Carbohydrate	67
Protein	3
Other	2

2 Sarah is paid £720 a month after tax and other deductions. The table shows how Sarah spends her money. Draw a pie chart to show these data.

Item	Amount spent (£)
Rent	252
Food	180
Clothes	108
Entertainment	144
Savings	36
Total	720

Mixed exercise 13 continues …

3 The pie chart shows how the money received from council tax payments is spent.

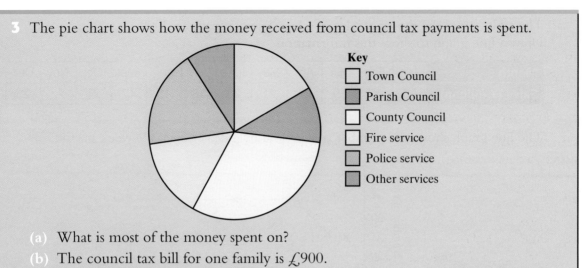

Key
☐ Town Council
▦ Parish Council
☐ County Council
☐ Fire service
▦ Police service
▦ Other services

(a) What is most of the money spent on?

(b) The council tax bill for one family is £900.
Use the pie chart to work out how much money the family contributed to these areas.
(i) The County Council
(ii) The Police service
(iii) Other services

4 120 people were surveyed about the holiday destination they would most like to visit.
The pie chart shows the results of the survey.

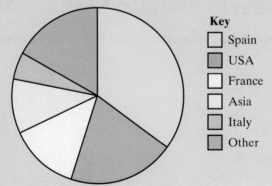

Key
☐ Spain
▦ USA
☐ France
☐ Asia
▦ Italy
▦ Other

(a) Which was the most popular destination?

(b) How many people chose each destination?

5 The table shows the daily temperatures in Norwich one week.
Draw a line graph to show this information.

Day	Mon	Tues	Wed	Thurs	Fri	Sat	Sun
Temperature (°C)	17	19	22	21	23	20	18

Mixed exercise 13 continues …

6 The table shows the monthly sales of a Japanese company in 2009.
Draw a line graph to show this information.

Month	Jan	Feb	Mar	Apr	May	June	July	Aug	Sept	Oct	Nov	Dec
Sales (¥ million)	16	17	19	23	21	20	18	15	17	21	22	25

7 The line graph shows the number of cars sold by a dealer over a 9-week period.

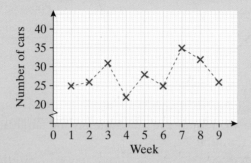

(a) In which week did the dealer sell the most cars?

(b) How many cars were sold in week 5?

(c) How many cars were sold in total over the 9-week period?

(d) What was the mean number of cars sold each week?

8 The line graph shows the mean temperature and the rainfall each month in Muddville during one year.

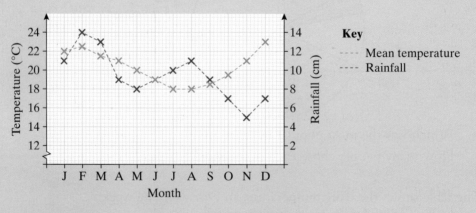

Key
- - - - Mean temperature
- - - - Rainfall

(a) Which months had the lowest mean temperature?

(b) What was the rainfall in February?

(c) What was the range of the monthly mean temperatures?

(d) What was the total rainfall for the year?

Mixed exercise 13 continues …

9 The pie charts show how land is used on two continents.

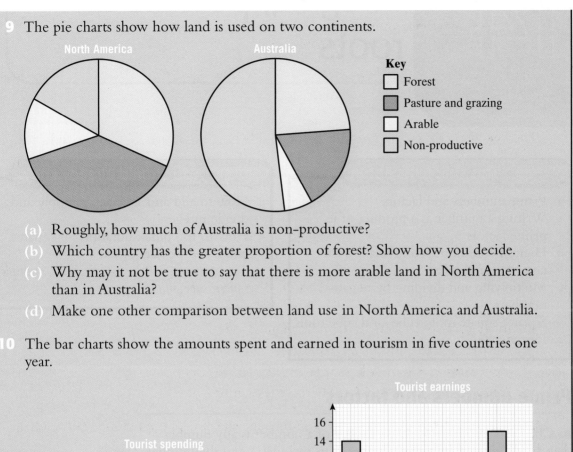

(a) Roughly, how much of Australia is non–productive?

(b) Which country has the greater proportion of forest? Show how you decide.

(c) Why may it not be true to say that there is more arable land in North America than in Australia?

(d) Make one other comparison between land use in North America and Australia.

10 The bar charts show the amounts spent and earned in tourism in five countries one year.

(a) Which country spent most? How much was this?

(b) Which country earned most? How much was this?

(c) Which countries earned more than they spent?

(d) Draw a single diagram, using a suitable scale, to show all this information.

▶ This chapter is about

- Prime numbers and factors
- Writing a number as a product of its prime factors
- Highest common factors and lowest common multiples
- Multiplying and dividing by negative numbers
- Squares, square roots, cubes and cube roots
- Reciprocals

▶ You should already know

- How to add and subtract, multiply and divide integers
- How to use index notation for squares, cubes and powers of ten
- The meaning of the words *factor*, *multiple*, *square*, *cube*, *square root*

Prime numbers and factors

In Chapter 1 you learned that a **factor** of a number is any number that divides exactly into that number. This includes 1 and the number itself.

🔍 Check up 14.1

The factors of 2 are 1 and 2.   The factors of 22 are 1, 2, 11 and 22.
Write down all the factors of these numbers.

(a) 14   **(b)** 16   **(c)** 40

⊙ Discovery 14.1

(a) Write down all the factors of the other numbers from 1 to 20.
(b) Write down all the numbers under 20 that have two, and only two, different factors.

The numbers you found in part **(b)** of Discovery 14.1 are called **prime numbers**. Notice that 1 is not a prime number as it only has one factor.

> **TIP**
>
> It is useful to learn the prime numbers up to 50.

Check up 14.2

Find all the prime numbers up to 50.

If you have time, go further.

Writing a number as a product of its prime factors

When you multiply two or more numbers together the result is a **product**.

When you write a number as a product of its prime factors you work out which prime numbers are multiplied together to give the number.

The number 6 written as a product of its prime factors is 2×3.

2×3 is called the **prime factor decomposition** of 6.

It is easy to write down the prime factors of 6 because it is a small number. To write a larger number as a product of its prime factors, use this method.
- Try dividing the number by 2.
- If it divides by 2 exactly, try dividing by 2 again.
- Continue dividing by 2 until your answer will not divide by 2.
- Next try dividing by 3.
- Continue dividing by 3 until your answer will not divide by 3.
- Then try dividing by 5.
- Continue dividing by 5 until your answer will not divide by 5.
- Continue to work systematically through the prime numbers.
- Stop when your answer is 1.

Example 14.1

(a) Write 12 as a product of its prime factors.

(b) Write 126 as a product of its prime factors.

Solution

(a) $2 \overline{)12}$ $12 = 2 \times 2 \times 3$

    $2 \overline{)\ 6}$ In Chapter 1 you learned that you can write 2×2 as 2^2.

    $3 \overline{)\ 3}$ So you can write $2 \times 2 \times 3$ in a shorter way as $2^2 \times 3$.

     1

(b) $2\overline{)126}$ $\qquad$ $126 = 2 \times 3 \times 3 \times 7 = 2 \times 3^2 \times 7$

 $3\overline{)\,63}$ $\qquad$ Remember 3^2 means 3 squared and this is the special

 $3\overline{)\,21}$ $\qquad$ name for 3 to the power 2.

 $7\overline{)\ \ 7}$ $\qquad$ The power, 2 in this case, is called the **index**.

 1

TIP

Check by multiplying the prime factors together. You should get the original number.

Exercise 14.1

Write each of these numbers as a product of its prime factors.

1 6	2 10	3 15	4 21	5 32
6 36	7 140	8 250	9 315	10 420

● Challenge 14.1

The factors of 24 are 1, 2, 3, 4, 6, 8, 12, 24.

This is eight different factors. You can write this as F(24) = 8.

24 written as a product of its prime factors is $2 \times 2 \times 2 \times 3 = 2^3 \times 3^1$.

(You do not usually include the index if it is 1 but you need it for this activity.)

Now add 1 to each of the indices: (3 + 1) = 4 and (1 + 1) = 2.
Then multiply these numbers: 4 × 2 = 8.

Your answer is the same as F(24), the number of factors of 24.

Here is another example.

The factors of 8 are 1, 2, 4, 8.

This is four different factors so F(8) = 4.

8 written as a product of its prime factors is 2^3.

There is just one power this time.

Add 1 to the index: (3 + 1) = 4.

This is the same as F(8), the number of factors of 8.

(a) Try this for 40.

(b) Investigate if there is a similar connection between the number of factors and powers of the prime factors for some other numbers.

Highest common factors and lowest common multiples

The **highest common factor (HCF)** of a set of numbers is the largest number that will divide exactly into each of the numbers.

The largest number that will divide into both 8 and 12 is 4.

So 4 is the highest common factor of 8 and 12.

You can find the highest common factor of 8 and 12 without using any special methods. You list, perhaps mentally, the factors of 8 and 12 and compare the lists to find the largest number that appears in both lists.

When you can give the answer without using any special methods it is called finding **by inspection**.

Check up 14.3

Find, by inspection, the highest common factor (HCF) of these pairs of numbers.

(a) 12 and 18   **(b)** 27 and 36   **(c)** 48 and 80

TIP

The HCF is never bigger than the smaller of the numbers.

You probably found parts **(a)** and **(b)** of Check up 14.3 fairly easy but part **(c)** more difficult.

This is the method to use when it is not easy to find the highest common factor by inspection.
- Write each number as a product of its prime factors.
- Choose the common factors.
- Multiply them together.

This method is shown in the next example.

Example 14.2

Find the highest common factor of these pairs of numbers.
(a) 28 and 72   **(b)** 96 and 180

Solution

(a) Write each number as the product of its prime factors.

$$28 = ②\times②\times 7 \qquad = 2^2 \times 7$$
$$72 = ②\times②\times 2 \times 3 \times 3 = 2^3 \times 3^2$$

The common factors are 2 and 2.
The highest common factor is $2 \times 2 = 2^2 = 4$.

(b) Write each number as the product of its prime factors.

$$96 = ② \times ② \times 2 \times 2 \times 2 \times ③ = 2^5 \times 3$$
$$180 = ② \times ② \times ③ \times 3 \times 5 \qquad = 2^2 \times 3^2 \times 5$$

The common factors are 2, 2 and 3.
The highest common factor is $2 \times 2 \times 3 = 2^2 \times 3 = 12$.

The **lowest common multiple (LCM)** of a set of numbers is the smallest number into which all the members of the set will divide.

The smallest number into which both 8 and 12 will divide is 24.

So 24 is the lowest common multiple of 8 and 12.

As for the highest common factor, you can find the lowest common multiple of small numbers by inspection. One way is to list the multiples of each of the numbers and compare the lists to find the smallest number that appears in both lists.

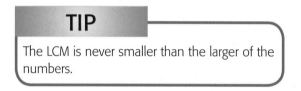

TIP

The LCM is never smaller than the larger of the numbers.

Check up 14.4

Find, by inspection, the lowest common multiple (LCM) of these pairs of numbers.

(a) 3 and 5   **(b)** 12 and 16   **(c)** 48 and 80

You probably found parts **(a)** and **(b)** of Check up 14.4 fairly easy but part **(c)** more difficult.

This is the method to use when it is not easy to find the lowest common multiple by inspection.
- Write each number as a product of its prime factors.
- Choose the highest power of each of the factors that occur in either of the lists.
- Multiply the numbers you choose together.

This method is shown in the next example.

Example 14.3

Find the lowest common multiple of these pairs of numbers.

(a) 28 and 42   **(b)** 96 and 180

Solution

(a) Write each number as the product of its prime factors.

$$28 = 2 \times 2 \times 7 = (2^2) \times 7$$
$$42 = 2 \times (3) \times (7)$$

The highest power of 2 is 2^2.

The highest power of 3 is $3^1 = 3$.

The highest power of 7 is $7^1 = 7$.

The lowest common multiple is $2^2 \times 3 \times 7 = 84$.

Notice that a number can be written as that number to the power 1. For example 3 was written as 3^1. A number to the power 1 is equal to the number. $3^1 = 3$.

(b) Write each number as the product of its prime factors.

$$96 = 2 \times 2 \times 2 \times 2 \times 2 \times 3 = (2^5) \times 3$$
$$180 = 2 \times 2 \times 3 \times 3 \times 5 \quad = 2^2 \times (3^2) \times (5)$$

The highest power of 2 is 2^5.

The highest power of 3 is 3^2.

The highest power of 5 is $5^1 = 5$.

The lowest common multiple is $2^5 \times 3^2 \times 5 = 1440$.

Summary

- To find the highest common factor (HCF), use the prime numbers that appear in *both* lists and use the *lower* power for each prime.
- To find the lowest common multiple (LCM), use all the prime numbers that appear in the lists and use the *higher* power of each prime.

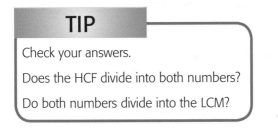

TIP

Check your answers.

Does the HCF divide into both numbers?

Do both numbers divide into the LCM?

Exercise 14.2

For each of these pairs of numbers
- write the numbers as products of their prime factors.
- state the highest common factor.
- state the lowest common multiple.

1 4 and 6   2 12 and 16   3 10 and 15   4 32 and 40   5 35 and 45

6 27 and 63   7 20 and 50   8 48 and 84   9 50 and 64   10 42 and 49

Challenge 14.2

The students in Year 11 at a school are to be split into groups of equal size.
Two possible sizes for the groups are 16 and 22.
What is the smallest number of students that there can be in Year 11?

Multiplying and dividing by negative numbers

Discovery 14.2

(a) Work out this sequence of calculations.

$5 \times 5 = 25$
$5 \times 4 = 20$
$5 \times 3 =$
$5 \times 2 =$
$5 \times 1 =$
$5 \times 0 =$

What is the pattern in the answers?

Use the pattern to continue the sequence.

$5 \times -1 =$
$5 \times -2 =$
$5 \times -3 =$
$5 \times -4 =$

(b) Work out this sequence of calculations.

$5 \times 4 =$
$4 \times 4 =$
$3 \times 4 =$
$2 \times 4 =$
$1 \times 4 =$
$0 \times 4 =$

Spot the pattern and continue the sequence.

You should have found in Discovery 14.2 that a positive number multiplied by a negative number gives a negative answer.

◉ Discovery 14.3

Work out this sequence of calculations.

$$-3 \times 5 =$$
$$-3 \times 4 =$$
$$-3 \times 3 =$$
$$-3 \times 2 =$$
$$-3 \times 1 =$$
$$-3 \times 0 =$$

What is the pattern in the answers?
Use the pattern to continue the sequence.

$$-3 \times -1 =$$
$$-3 \times -2 =$$
$$-3 \times -3 =$$
$$-3 \times -4 =$$
$$-3 \times -5 =$$

Your answers to Discoveries 14.2 and 14.3 suggest these rules.

$$+ \times - = -$$
and
$$- \times + = -$$

$$+ \times + = +$$
and
$$- \times - = +$$

Example 14.4

Work out these.

(a) 6×-4 **(b)** -7×-3 **(c)** -5×8

Solution

(a) $+ \times - = -$
$6 \times 4 = 24$
So $6 \times -4 = -24$

(b) $- \times - = +$
$7 \times 3 = 21$
So $-7 \times -3 = +21 = 21$

(c) $- \times + = -$
$5 \times 8 = 40$
So $-5 \times 8 = -40$

Discovery 14.4

$4 \times 3 = 12$ From this calculation you can say that $12 \div 4 = 3$ and $12 \div 3 = 4$.

$10 \times 6 = 60$ From this calculation you can say that $60 \div 6 = 10$ and $60 \div 10 = 6$.

In Example 14.4 you saw that $6 \times -4 = -24$.

In the same way as for the calculations above you can write down these two division sums.

$$-24 \div 6 = -4 \qquad \text{and} \qquad -24 \div -4 = 6$$

(a) Work out 2×-9.
     Then write two division sums in the same way as above.
(b) Work out -7×-4.
     Then write two division sums in the same way as above.

Your answers to Discovery 14.4 suggest these rules.

$$+ \div - = -$$
and
$$- \div + = -$$

$$+ \div + = +$$
and
$$- \div - = +$$

You now have a complete set of rules for multiplying and dividing positive and negative numbers.

$$+ \times - = -$$
$$+ \div - = -$$
$$- \times + = -$$
$$- \div + = -$$

$$+ \times + = +$$
$$+ \div + = +$$
$$- \times - = +$$
$$- \div - = +$$

Here is another way of thinking of these rules.

Signs different: answer negative   Signs the same: answer positive

Example 14.5

Work out these.

(a) 5×-3 (b) -2×-3 (c) $-10 \div 2$ (d) $-15 \div -3$

Solution

First work out the signs. Then work out the numbers.

(a) -15 $(+ \times - = -)$ (b) $+6 = 6$ $(- \times - = +)$
(c) -5 $(- \div + = -)$ (d) $+5 = 5$ $(- \div - = +)$

You can extend the rules to calculations with more than two numbers.

If there is an even number of negative signs the answer is positive.
If there is an odd number of negative signs the answer is negative.

Example 14.6

Work out $-2 \times 6 \div -4$.

Solution

You can work this out by taking each part of the calculation in turn.

$$-2 \times 6 = -12 \qquad (- \times + = -)$$
$$-12 \div -4 = 3 \qquad (- \div - = +)$$

Or you can count the number of negative signs and then work out the numbers.

There are two negative signs so the answer is positive.

$$-2 \times 6 \div -4 = 3$$

Exercise 14.3

Work out these.

1 4×3

2 -5×4

3 -6×-5

4 -9×6

5 4×-7

6 -2×8

7 -3×-6

8 $24 \div -6$

9 $-25 \div -5$

10 $-32 \div 4$

11 $18 \div 6$

12 $-14 \div -7$

13 $-45 \div 5$

14 $49 \div -7$

15 $36 \div -9$

16 $6 \times 10 \div -5$

17 $-84 \div -12 \times -3$

18 $4 \times 9 \div -6$

19 $-3 \times -6 \div -2$

20 $-6 \times 2 \times -5 \div -3$

● Challenge 14.3

Find the value of these expressions when $x = -3$, $y = 4$ and $z = -1$.

(a) $5xy$ (b) $x^2 + 2x$ (c) $2y^2 - 2yz$

(d) $3xz - 2xy + 3yz$ (e) $4xyz$

Squares, square roots, cubes and cube roots

In Chapter 1 you learned about squares and cubes. The **square** of a number is the number multiplied by itself.

For example 2 squared is written 2^2 and equals $2 \times 2 = 4$.
The **cube** of a number is the number $\times$ number $\times$ number.
For example 2 cubed is written 2^3 and equals $2 \times 2 \times 2 = 8$.

It is useful to know the squares of the numbers 1 to 15 and the cubes of the numbers 1 to 5 and of 10.

Check up 14.5

(a) What are the squares of the numbers 1 to 15?

(b) What are the cubes of the numbers 1 to 5 and of 10?

The squares of integers are called **square numbers**.

The cubes of integers are called **cube numbers**.

Because $4^2 = 4 \times 4 = 16$, the **square root** of 16 is 4.

This is written as $\sqrt{16} = 4$.

But $(-4)^2 = -4 \times -4 = 16$. So the square root of 16 is also -4.

This is often written as $\sqrt{16} = \pm 4$. Similarly $\sqrt{81} = \pm 9$ and so on.

In many practical problems where the answer is a square root, the negative answer has no meaning and should be left out.

Because $5^3 = 5 \times 5 \times 5 = 125$, the **cube root** of 125 is 5.

This is written as $\sqrt[3]{125} = 5$. It can only be positive.

$(-5)^3 = -5 \times -5 \times -5 = -125$. So $\sqrt[3]{-125} = -5$.

Finding the square root is the reverse operation to squaring. Finding the cube root is the reverse operation to cubing. So you can find the square roots and cube roots of the squares and cubes you know.

Make sure you know how to use your calculator to work out squares, cubes, square roots and cube roots.

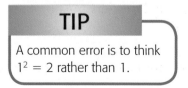

TIP

A common error is to think that $1^2 = 2$ rather than 1.

Example 14.7

Work out these giving your answers to 2 decimal places.

(a) The square of 3.4

(b) The cube of 9.2

(c) The square root of 57

(d) The cube root of 86

Solution

(a) $3.4^2 = 11.56$ You press ③ . ④ x^2 = on your calculator.

(b) $9.2^3 = 778.69$ You press ⑨ . ② x^3 = on your calculator.

(c) $\sqrt{57} = \pm 7.55$ You press $\sqrt{}$ ⑤ ⑦ = on your calculator.

(d) $\sqrt[3]{86} = 4.41$ You press $\sqrt[3]{}$ ⑧ ⑥ = on your calculator.

> **TIP**
>
> Calculators vary. You may need to use the SHIFT key to find some of these functions on your calculator. Look at the labels on and above the keys.

Exercise 14.4

1   Write down the value of each of these.

   (a) 7^2 (b) 11^2 (c) $\sqrt{36}$ (d) $\sqrt{144}$

   (e) 2^3 (f) 10^3 (g) $\sqrt[3]{64}$ (h) $\sqrt[3]{1}$

2   A square has an area of 36 cm². What is the length of one side?

3   Find the square of each of these numbers.

   (a) 25   (b) 40   (c) 35   (d) 32   (e) 1.2

4   Find the cube of each of these numbers.

   (a) 12   (b) 2.5   (c) 6.1   (d) 30   (e) 5.4

5   Find the square root of each of these numbers.
   Where necessary, give your answer correct to 2 decimal places.

   (a) 400   (b) 575   (c) 1284   (d) 3684   (e) 15 376

6   Find the cube root of each of these numbers.
   Where necessary, give your answer correct to 2 decimal places.

   (a) 512   (b) 676   (c) 8000   (d) 9463   (e) 10 000

7   Find two numbers less than 200 which are both a square number and a cube number.

Challenge 14.4

(a) The side of a square is 2.2 m long.
What is the area of the square?

(b) The volume of a cube is 96 cm³.
What is the length of an edge, correct to 1 decimal place?

Reciprocals

The **reciprocal** of a number is $\dfrac{1}{\text{the number}}$.

For example, the reciprocal of 2 is $\frac{1}{2}$.

The reciprocal of n is $\dfrac{1}{n}$.

The reciprocal of $\dfrac{1}{n}$ is n.

The reciprocal of $\dfrac{a}{b}$ is $\dfrac{b}{a}$.

0 does not have a reciprocal.

To find the reciprocal of a number without a calculator you divide 1 by the number.

To find the reciprocal of a number with a calculator you use the $\boxed{x^{-1}}$ key.

Example 14.8

Without using a calculator, find the reciprocal of each of these.

(a) 5 **(b)** $\frac{5}{8}$ **(c)** $1\frac{1}{8}$

Solution

(a) To find the reciprocal of a number, divide 1 by the number.
The reciprocal of 5 is $\frac{1}{5}$ or 0.2.

(b) The reciprocal of $\frac{5}{8}$ is $\frac{8}{5} = 1\frac{3}{5}$.
Note: You should always convert improper fractions to mixed numbers unless you are told not to.

(c) First convert $1\frac{1}{8}$ to an improper fraction.
$1\frac{1}{8} = \frac{9}{8}$
The reciprocal of $\frac{9}{8} = \frac{8}{9}$.

Example 14.9

Use your calculator to find the reciprocal of 1.25.

Give your answer as a decimal.

Solution

This is the sequence of keys to press.

The display should read 0.8.

🔍 Check up 14.6

Write down the reciprocal of each of these numbers.

(a) 2   (b) 5   (c) 10   (d) $\frac{3}{5}$

◯ Discovery 14.5

(a) Multiply each of the numbers in Check up 14.6 by its reciprocal. What do you notice about your answers?

(b) Now try these products on your calculator.

    (i) 55×2 (press $=$) $\times \frac{1}{2}$ (press $=$)

    (ii) 15×4 (press $=$) $\times \frac{1}{4}$ (press $=$)

    (iii) 8×10 (press $=$) $\times 0.1$ (press $=$)

         What do you notice about your answers?

(c) Try some more calculations and explain what is happening.

● Challenge 14.5

An **inverse operation** takes you back to the previous number.

Multiplying by a number and multiplying by its reciprocal are inverse operations.

Write down as many operations and their inverse operation as you can.

Exercise 14.5

1 Write down the reciprocal of each of these numbers.
   (a) 3   (b) 6   (c) 49   (d) 100   (e) 640

2 Write down the numbers of which these are the reciprocals.
   (a) $\frac{1}{16}$ (b) $\frac{1}{9}$ (c) $\frac{1}{52}$ (d) $\frac{1}{67}$ (e) $\frac{1}{1000}$

3 Find the reciprocal of each of these numbers.
   Give your answers as fractions or mixed numbers.
   (a) $\frac{4}{5}$ (b) $\frac{3}{8}$ (c) $1\frac{3}{5}$ (d) $3\frac{1}{3}$ (e) $\frac{2}{25}$

4 Find the reciprocal of each of these numbers.
   Give your answers as decimals.
   (a) 2.5   (b) 0.5   (c) 125   (d) 0.16   (e) 3.2

What you have learned

- A prime number has two factors only, 1 and itself
- How to write a number as the product of its prime factors
- The highest common factor (HCF) of a set of numbers is the largest number that will divide exactly into each of the numbers
- How to find the highest common factor of a pair of numbers using prime factors
- The lowest common multiple (LCM) of a set of numbers is the smallest number into which all the members of the set will divide
- How to find the lowest common multiple using prime factors
- That, when multiplying or dividing positive and negative numbers,

$$+ \times + = + \qquad - \times - = + \qquad + \times - = - \qquad - \times + = -$$
$$+ \div + = + \qquad - \div - = + \qquad + \div - = - \qquad - \div + = -$$

- $5^3 = 5 \times 5 \times 5 = 125$, so the cube root of 125 is 5
- The reciprocal of a number is 1 divided by the number: the reciprocal of n is $\frac{1}{n}$
- The reciprocal of $\frac{a}{b}$ is $\frac{b}{a}$
- 0 does not have a reciprocal

Mixed exercise 14

1   Write each of these numbers as a product of its prime factors.
    (a) 75   (b) 140   (c) 420

2   For each of these pairs of numbers
    • write the numbers as products of their prime factors.
    • state the highest common factor.
    • state the lowest common multiple.
    (a) 24 and 60   (b) 100 and 150   (c) 81 and 135

3   Work out these.
    (a) 4×-3 (b) -2×8 (c) $-48 \div -6$ (d) $2 \times -6 \div -4$

4   Write down the square and the cube of each of these numbers.
    (a) 4   (b) 6   (c) 10

5   Write down the square root of each of these numbers.
    (a) 64   (b) 196

6   Write down the cube root of each of these numbers.
    (a) 125   (b) 27

7   Find the square and the cube of each of these numbers.
    (a) 4.6   (b) 21   (c) 2.9

8   Find the square root and the cube root of each of these numbers.
    Give your answers correct to 2 decimal places.
    (a) 89   (b) 124   (c) 986

9   Find the reciprocal of each of these numbers.
    (a) 5   (b) 8   (c) $\frac{1}{8}$ (d) 0.1   (e) 1.6

Chapter 15 — Algebra 2

▶ This chapter is about

- Expanding brackets in algebra
- Collecting like terms
- Factorising expressions
- Index notation in algebra

▶ You should already know

- Letters can be used to stand for numbers
- How to use index notation for numbers

Expanding brackets

Joe has a job making cheese sandwiches.
He uses two slices of bread and one slice of cheese for each sandwich.
He wants to know how much it will cost to make 25 sandwiches.

Joe uses 50 slices of bread and 25 slices of cheese to make 25 sandwiches.

You could use letters to represent the cost of the sandwich ingredients.
Let b represent the cost of each slice of bread in pence.
Let c represent the cost of each slice of cheese in pence.

You could then write one sandwich costs $2b + c$.

(You write $1c$ as just c.)

25 sandwiches costs 25 times this amount. You could write $25(2b + c)$.

$2b$ and c are called **terms**. $25(2b + c)$ is called an **expression**.

To work out the cost of 25 sandwiches you could write

$$25(2b + c) = 50b + 25c.$$

This is called **expanding the bracket**. You multiply *each* term inside the bracket by the number outside the bracket.

Example 15.1

Work out these.

(a) $10(2b + c)$

(b) $35(2b + c)$

(c) $16(2b + c)$

(d) $63(2b + c)$

Solution

(a) $10(2b + c) = 10 \times 2b + 10 \times c = 20b + 10c$

(b) $35(2b + c) = 35 \times 2b + 35 \times c = 70b + 35c$

(c) $16(2b + c) = 16 \times 2b + 16 \times c = 32b + 16c$

(d) $63(2b + c) = 63 \times 2b + 63 \times c = 126b + 63c$

There may be more than two terms in a bracket. You expand the bracket in the same way.

Example 15.2

Expand these.

(a) $12(2b + 7g)$ **(b)** $6(3m - 4n)$

(c) $8(2x - 5)$ **(d)** $3(4p + 2v - c)$

(e) $x(3 + 2y)$ **(f)** $x(x - 6)$

Solution

(a) $12(2b + 7g) = 12 \times 2b + 12 \times 7g$
$$= 24b + 84g$$

(b) $6(3m - 4n) = 6 \times 3m - 6 \times 4n$
$$= 18m - 24n$$

(c) $8(2x - 5) = 8 \times 2x - 8 \times 5$
$$= 16x - 40$$

(d) $3(4p + 2v - c) = 3 \times 4p + 3 \times 2v - 3 \times c$
$$= 12p + 6v - 3c$$

(e) $x(3 + 2y) = x \times 3 + x \times 2y$ You write the numbers first then
$$= 3x + 2xy$$ the letters in alphabetical order.

(f) $x(x - 6) = x \times x - x \times 6$
$$= x^2 - 6x$$ You write $x \times x$ as x^2.

Exercise 15.1

Expand these.

1 $10(2a + 3b)$ 2 $3(2c + 7d)$ 3 $5(3e - 8f)$ 4 $7(4g - 3h)$

5 $5(2u + 3v)$ 6 $6(5w + 3x)$ 7 $7(3y + z)$ 8 $8(2v + 5)$

9 $6(2 + 7w)$ 10 $4(3 - 8a)$ 11 $2(4g - 3)$ 12 $5(7 - 4b)$

13 $2(3i + 4j - 5k)$ 14 $4(5m - 3n + 2p)$ 15 $6(2r - 3s - 4t)$ 16 $a(b + c)$

17 $p(q - 2)$ 18 $y(3z + 4)$ 19 $d(5 - 2e)$ 20 $b(a - c + 3d)$

21 $x(x + 3)$

Challenge 15.1

Expand these.

(a) $6(5a + 4b - 3c - 2d)$ (b) $4(3w - 5x + 7y - 9z)$

(c) $9(4p - 7q - 8r + 3s)$ (d) $12(7e - 9f + 12g - 16h)$

Collecting like terms

In Chapter 2 you learned that terms with the same letter are called **like terms** and that like terms can be added.

One weekend Joe works on both Saturday and Sunday.

On Saturday he makes 75 cheese sandwiches and on Sunday he makes 45 cheese sandwiches.

You can calculate the total cost of the bread and the total cost of the cheese he uses as follows.

Write expressions for the two days and expand the brackets.

$$75(2b + c) + 45(2b + c) = 150b + 75c + 90b + 45c$$

Simplify the expression by collecting like terms.

$$150b + 75c + 90b + 45c = 150b + 90b + 75c + 45c$$
$$= 240b + 120c$$

Example 15.3

Expand the brackets and simplify these.

(a) $10(2b + c) + 5(2b + c)$ (b) $35(b + c) + 16(2b + c)$

(c) $16(2b + c) + 63(b + 2c)$ (d) $30(b + 2c) + 18(b + c)$

Solution

(a) $10(2b + c) + 5(2b + c) = 20b + 10c + 10b + 5c$
$$= 20b + 10b + 10c + 5c$$
$$= 30b + 15c$$

(b) $35(b + c) + 16(2b + c) = 35b + 35c + 32b + 16c$
$$= 35b + 32b + 35c + 16c$$
$$= 67b + 51c$$

(c) $16(2b + c) + 63(b + 2c) = 32b + 16c + 63b + 126c$
$$= 32b + 63b + 16c + 126c$$
$$= 95b + 142c$$

(d) $30(b + 2c) + 18(b + c) = 30b + 60c + 18b + 18c$
$$= 30b + 18b + 60c + 18c$$
$$= 48b + 78c$$

You have to take particular care when there are minus signs in the expression.

Example 15.4

Expand the brackets and simplify these.

(a) $2(3c + 4d) + 5(3c + 2d)$ **(b)** $5(4e + f) - 3(2e - 4f)$

(c) $5(4g + 3) - 2(3g + 4)$

Solution

(a) Multiply all terms in the first bracket by the number in front of that bracket. Multiply all the terms in the second bracket by the number in front of that bracket. Then collect like terms together.

$$2(3c + 4d) + 5(3c + 2d) = 6c + 8d + 15c + 10d$$
$$= 6c + 15c + 8d + 10d$$
$$= 21c + 18d$$

(b) Take care with the signs.

Think of the second part of the expression, $-3(2e - 4f)$, as $+ (-3) \times (2e + (-4f))$. You need to multiply both the terms in the second bracket by -3. You learned the rules for calculating with negative numbers in Chapter 14.

$$5(4e + f) - 3(2e - 4f) = 5(4e + f) + (-3) \times (2e + (-4f))$$
$$= 5 \times 4e + 5 \times f + (-3) \times 2e + (-3) \times (-4f)$$
$$= 20e + 5f + (-6e) + (+12f)$$
$$= 20e + (-6e) + 5f + 12f$$
$$= 20e - 6e + 5f + 12f$$
$$= 14e + 17f$$

(c) Again you need to take care with the signs.

Think of the second part of the expression, $-2(3g + 4)$, as $+ (-2) \times (3g + 4)$. You need to multiply both the terms in the second bracket by -2.

$$5(4g + 3) - 2(3g + 4) = 5(4g + 3) + (-2) \times (3g + 4)$$
$$= 20g + 15 + (-2) \times 3g + (-2) \times 4$$
$$= 20g + 15 + (-6g) + (-8)$$
$$= 20g + (-6g) + 15 + (-8)$$
$$= 20g - 6g + 15 - 8$$
$$= 14g + 7$$

Exercise 15.2

Expand the brackets and simplify these.

1 (a) $8(2a + 3) + 2(2a + 7)$ (b) $5(3b + 7) + 6(2b + 3)$

  (c) $2(3 + 8c) + 3(2 + 7c)$ (d) $6(2 + 3a) + 4(5 + a)$

2 (a) $5(2s + 3t) + 4(2s + 7t)$ (b) $2(2v + 7w) + 5(2v + 7w)$

  (c) $7(3x + 8y) + 3(2x + 7y)$ (d) $3(2v + 5w) + 4(8v + 3w)$

3 (a) $4(3x + 5) + 3(3x - 4)$ (b) $2(4y + 5) + 3(2y - 3)$

  (c) $5(2 + 7z) + 4(3 - 8z)$ (d) $3(2 + 5x) + 5(6 - x)$

4 (a) $3(2n + 7p) + 2(5n - 6p)$ (b) $5(3q + 8r) + 3(2q - 9r)$

  (c) $7(2d + 3e) + 3(3d - 5e)$ (d) $4(2f + 7g) + 3(2f - 9g)$

  (e) $3(3h - 8j) - 5(2h - 7j)$ (f) $6(2k - 3m) - 3(2k - 7m)$

Factorising

You learned in Chapter 14 that the highest common factor of a set of numbers is the largest number that will divide into all of the numbers in the set.

Remember that in algebra you use letters to stand for numbers. For example, $2x = 2 \times x$ and $3x = 3 \times x$.

You do not know what x is but you do know that $2x$ and $3x$ both divide by x. So x is the highest common factor of $2x$ and $3x$.

When you have unlike terms, for example $2x$ and $4y$, you must assume that x and y do not have any common factors. However, you can look for common factors in the numbers. The highest common factor of 2 and 4 is 2 so the highest common factor of $2x$ and $4y$ is 2.

Factorising is the reverse of expanding a bracket. You divide each of the terms in the bracket by the highest common factor and write this common factor outside the bracket.

Example 15.5

Factorise these.

(a) $12x + 16$ **(b)** $x - x^2$ **(c)** $8x^2 - 12x$

Solution

(a) $12x + 16$ The highest common factor of $12x$ and 16 is 4.
$4($ $)$ You write this factor outside the bracket.
You then divide each term inside the original bracket by the highest common factor, 4.
$12x \div 4 = 3x$ and $16 \div 4 = 4$
$4(3x + 4)$ You write the new terms inside the bracket.
$12x + 16 = 4(3x + 4)$

> **TIP**
>
> Check that your answer is correct by expanding it.
> $4(3x + 4) = 4 \times 3x + 4 \times 4 = 12x + 16$

(b) $x - x^2$ The highest common factor of x and x^2 is x.
  (Remember that x^2 is $x \times x$.)
$x($ $)$ You write this factor outside the bracket.
You then divide each term inside the original bracket by the highest common factor, x.
$x \div x = 1$ and $x^2 \div x = x$
$x - x^2 = x(1 - x)$

(c) $8x^2 - 12x$ Think about the numbers and the letters separately and then combine them. The highest common factor of 8 and 12 is 4 and the highest common factor of x^2 and x is x. Therefore, the highest common factor of $8x^2$ and $12x$ is $4 \times x = 4x$.
$4x($ $)$ You write this factor outside the bracket.
You then divide each term inside the original bracket by the highest common factor, $4x$.
$8x^2 \div 4x = 2x$ and $12x \div 4x = 3$
$8x^2 - 12x = 4x(2x - 3)$

Exercise 15.3

Factorise these.

1 (a) $10x + 15$ (b) $2x + 6$ (c) $8x - 12$ (d) $4x - 20$

2 (a) $14 + 7x$ (b) $8 + 12x$ (c) $15 - 10x$ (d) $9 - 12x$

3 (a) $3x^2 + 5x$ (b) $5x^2 + 20x$ (c) $12x^2 - 8x$ (d) $6x^2 - 8x$

● Challenge 15.2

Factorise these.

(a) $24x + 32y$ (b) $15ab - 20ac$

(c) $30f^2 - 18fg$ (d) $42ab + 35a^2$

Index notation

In Chapter 14 you learned that you can write 2 to the power 4 as 2^4 and that the power, 4 in this case, is called the index. You can say that 2^4 is written in **index notation**.

You can use index notation in algebra too. You have already met x^2. This means x squared or x to the power 2.

y^5 is another example of an expression written using index notation. It means y to the power 5 or $y \times y \times y \times y \times y$. The index is 5.

Example 15.6

Write these using index notation.

(a) $5 \times 5 \times 5 \times 5 \times 5 \times 5$ (b) $x \times x \times x \times x \times x \times x \times x$

(c) $p \times p \times p \times p \times r \times r \times r$ (d) $3w \times 4w \times 5w$

Solution

(a) $5 \times 5 \times 5 \times 5 \times 5 \times 5 = 5$ to the power $6 = 5^6$

(b) $x \times x \times x \times x \times x \times x \times x = x$ to the power $7 = x^7$

(c) $p \times p \times p \times p \times r \times r \times r = p^4 \times r^3 = p^4 r^3$

(d) You multiply the numbers together first, then the letters.

$$3w \times 4w \times 5w = (3 \times 4 \times 5) \times (w \times w \times w)$$
$$= 60 \times w^3$$
$$= 60w^3$$

Exercise 15.4

Simplify each of the following, writing your answer using index notation.

1   (a) $3 \times 3 \times 3 \times 3$
    (b) $7 \times 7 \times 7$
    (c) $10 \times 10 \times 10 \times 10 \times 10$

2   (a) $x \times x \times x \times x \times x$
    (b) $y \times y \times y \times y$
    (c) $z \times z \times z \times z \times z \times z \times z$

3   (a) $m \times m \times n \times n \times n \times n$
    (b) $f \times f \times f \times f \times g \times g \times g \times g \times g$
    (c) $p \times p \times p \times r \times r \times r \times r$

4   (a) $2k \times 4k \times 7k$
    (b) $3y \times 5y \times 8y$
    (c) $4d \times 2d \times d$

● Challenge 15.3

Simplify each of the following, writing your answer using index notation.

(a) $m^2 \times m^4$ (b) $x^3 \times 5x^6$
(c) $5y^4 \times 3y^3$ (d) $2b^3 \times 3b^2 \times 4b$

What you have learned

- When you expand a bracket such as $25(2b + c)$ or $p(q + r)$ you multiply each of the terms inside the bracket by the number or term outside the bracket
- Like terms use exactly the same letters but can use different numbers, for example a and $5a$ are like terms, ab and $4ab$ are like terms but a and ab are not like terms
- Factorising an expression is the inverse of expanding a bracket
- To factorise an expression you take the highest common factor outside the bracket
- You can use index notation in algebra, for example you can write $x \times x \times x \times x$ as x^4

Mixed exercise 15

1 Expand these.

(a) $8(3a + 2b)$ (b) $5(4a + 3b)$ (c) $12(3a - 5b)$

(d) $9(a - 2b)$ (e) $3(4x + 5y)$ (f) $6(3x - 2y)$

(g) $4(5x - 3y)$ (h) $2(4x + y)$ (i) $5(3f - 4g)$

(j) $3(2j + 5k)$ (k) $7(r + 2s)$ (l) $4(3v - w)$

2 Expand the brackets and simplify these.

(a) $2(3x + 4) + 3(2x + 1)$ (b) $4(2x + 3) + 3(4x + 5)$

(c) $2(2x + 3) + 3(x + 2)$ (d) $5(2y + 3) + 2(3y - 5)$

(e) $3(3y + 5) + 2(3y - 4)$ (f) $3(5y + 2) + 2(3y - 1)$

(g) $3(2a + 4) - 3(a + 2)$ (h) $2(6m + 2) - 3(2m + 1)$

(i) $6(3p + 4) - 3(4p + 2)$ (j) $4(5t + 3) - 3(2t - 4)$

(k) $2(4j + 8) - 3(3j - 5)$ (l) $6(2w + 5) - 4(3w - 4)$

3 Factorise these.

(a) $4x + 8$ (b) $6x + 12$ (c) $9x - 6$

(d) $12x - 18$ (e) $6 - 10x$ (f) $10 - 15x$

(g) $24 + 8x$ (h) $16x + 12$ (i) $6x + 8$

(j) $32x - 12$ (k) $20 - 16x$ (l) $15 + 20x$

(m) $2x - x^2$ (n) $3y - 7y^2$ (o) $5z^2 + 2z$

4 Simplify each of the following, writing your answer using index notation.

(a) $4 \times 4 \times 4 \times 4 \times 4 \times 4$ (b) $5 \times 5 \times 5 \times 5$

(c) $2 \times 2 \times 2 \times 2 \times 2$ (d) $a \times a \times a \times a \times a \times a \times a$

(e) $j \times j \times j$ (f) $t \times t \times t \times t \times t \times t$

(g) $v \times v \times v \times w \times w \times w$ (h) $d \times d \times d \times e \times e \times e \times e \times e \times e$

(i) $x \times x \times x \times y \times y \times y \times y \times y$ (j) $5p \times 4p \times 3p$

Statistical diagrams 2

▸ This chapter is about

- Frequency diagrams
- Frequency polygons
- Stem-and-leaf diagrams

▸ You should already know

- That a large amount of data can be displayed in a diagram, such as a bar graph
- The difference between discrete and continuous data
- The meaning of *mode* and *median*

Frequency diagrams

You learned in Chapter 3 that when you have a lot of data it is often more convenient to group the data into bands or intervals. You have already drawn bar charts to display grouped discrete data.

To display **grouped continuous data**, you can use a **frequency diagram**. This is very like a bar chart: the main difference is that there are no gaps between the bars.

TIP

Remember that the intervals should usually be of equal size.

Example 16.1

Saul measured the heights of 34 students.
He grouped the data into intervals of 5 cm.
Here is his table of values.

(a) Draw a grouped frequency diagram to show these data.

(b) Which of the intervals is the modal class?

(c) Which of the intervals contains the median value?

TIP

$145 < h \leqslant 150$ means all heights, h, which are bigger than 145 cm (but not equal to 145 cm) and up to and including 150 cm.

Height (h cm)	Frequency
$140 < h \leqslant 145$	3
$145 < h \leqslant 150$	8
$150 < h \leqslant 155$	8
$155 < h \leqslant 160$	9
$160 < h \leqslant 165$	2
$165 < h \leqslant 170$	4

Solution

(a)

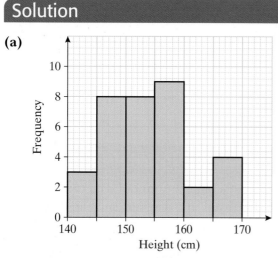

Don't forget to label the axes.

The horizontal axis shows the type of data being collected.

The vertical axis shows the **frequency**, or how many data items there are in each of the intervals.

(b) $155 < h \leqslant 160$ The modal class is the one with the highest frequency.
It has the highest number in the 'frequency' row of the table, and the highest bar in the frequency diagram.

(c) The median value is the value halfway along the ordered list.
As there are 34 values, the median will lie between the 17th and 18th values. Add on the frequency for each interval until you find the interval containing the 17th and 18th values:

	3 is smaller than 17.
	The 17th and 18th values do not lie in interval $140 < h \leqslant 145$.
$3 + 8 = 11$	11 is smaller than 17.
	The 17th and 18th values do not lie in interval $145 < h \leqslant 150$.
$11 + 8 = 19$	19 is larger than 18.
	The 17th and 18th values must lie in interval $150 < h \leqslant 155$.

Interval $150 < h \leqslant 155$ contains the median value.

Exercise 16.1

1   The manager of a leisure centre recorded the ages of the women who used the swimming pool one morning. The table shows his results.

Draw a frequency diagram to show these data.

Age (a years)	Frequency
$15 \leqslant a < 20$	4
$20 \leqslant a < 25$	12
$25 \leqslant a < 30$	17
$30 \leqslant a < 35$	6
$35 \leqslant a < 40$	8
$40 \leqslant a < 45$	3
$45 \leqslant a < 50$	12

Exercise continues …

2   In a survey, the annual rainfall was measured at
    100 different towns.
    The table shows the results of the survey.

    (a) Draw a frequency diagram to show these data.
    (b) Which of the intervals is the modal class?
    (c) Which of the intervals contains the median value?

Rainfall (r cm)	Frequency
$50 \leqslant r < 70$	14
$70 \leqslant r < 90$	33
$90 \leqslant r < 110$	27
$110 \leqslant r < 130$	8
$130 \leqslant r < 150$	16
$150 \leqslant r < 170$	2

3   As part of a fitness campaign, a business measured
    the weight of all of its workers.
    The table shows the results.

    (a) Draw a frequency diagram to show these data.
    (b) Which of the intervals is the modal class?
    (c) Which of the intervals contains the median value?

Weight (w kg)	Frequency
$60 \leqslant w < 70$	3
$70 \leqslant w < 80$	18
$80 \leqslant w < 90$	23
$90 \leqslant w < 100$	7
$100 \leqslant w < 110$	2

4   Here is a frequency diagram.

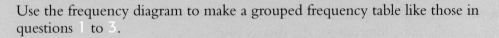

Use the frequency diagram to make a grouped frequency table like those in
questions 1 to 3.

● Challenge 16.1

Lisa checked the price of kettles on the internet.
Here are the prices of the first 30 she saw.

£6.54	£8.90	£9.60	£12.95	£13.90	£13.95
£14.25	£16.75	£16.90	£17.75	£17.90	£19.50
£19.50	£21.75	£22.40	£23.25	£24.50	£24.95
£26.00	£26.75	£27.00	£27.50	£29.50	£29.50
£29.50	£29.50	£32.25	£34.50	£35.45	£36.95

Complete a tally chart and draw a frequency diagram to show these data.
Use appropriate intervals for your groups.

TIP

When you choose the size of the interval, make sure
you don't end up with too many, or too few, groups.
Between five and ten intervals is usually about right.
Remember that the intervals should be equal.

● Challenge 16.2

(a) Measure the height of everyone in your class and record the data
in two lists, one for boys and one for girls.

(b) Choose suitable intervals for the data.

(c) Draw two frequency diagrams, one for the boys' data and one for
the girls'.
Use the same scales for both diagrams so that they can be
compared easily.

(d) Compare the two diagrams.
What do the shapes of the graphs tell you, in general, about the
heights of the boys and girls in your class?

(e) Compare your frequency diagrams with others in your class.
Have they used the same intervals for the data as you?
If they haven't, has this made a difference to their answers to part **(d)**?
Which of the diagrams looks the best? Why?

Frequency polygons

A **frequency polygon** is another way of representing grouped continuous data.

A frequency polygon is formed by joining, with straight lines, the midpoints of the tops of the bars in a frequency diagram. The bars are not drawn. This means that several frequency polygons can be drawn on one grid, which makes them easier to compare.

To find the midpoint of each interval, add the bounds of each interval and divide the sum by 2.

Example 16.2

The grouped frequency table shows the number of days that students in a tutor group were absent one term.

Draw a frequency polygon to show these data.

Days absent (d)	Frequency
$0 \leqslant d < 5$	11
$5 \leqslant d < 10$	8
$10 \leqslant d < 15$	6
$15 \leqslant d < 20$	0
$20 \leqslant d < 25$	5

Solution

First find the midpoint of each class.

$$\frac{0 + 5}{2} = 2.5 \qquad \frac{5 + 10}{2} = 7.5 \qquad \frac{10 + 15}{2} = 12.5$$

$$\frac{15 + 20}{2} = 17.5 \qquad \frac{20 + 25}{2} = 22.5$$

TIP

Notice that the midpoints go up in fives: this is because the interval size is five.

Now you can draw your frequency polygon.

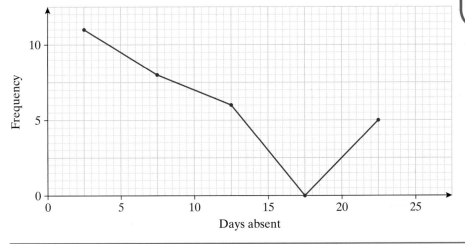

Exercise 16.2

1   The table shows the weight loss of people in a
    slimming club over 6 months.

    Draw a frequency polygon to show these data.

Weight (w kg)	Frequency
$0 \leqslant w < 6$	8
$6 \leqslant w < 12$	14
$12 \leqslant w < 18$	19
$18 \leqslant w < 24$	15
$24 \leqslant w < 30$	10

2   The table shows the length of time that cars stayed
    in a car park one day.

    Draw a frequency polygon to show these data.

Time (t mins)	Frequency
$15 \leqslant t < 30$	56
$30 \leqslant t < 45$	63
$45 \leqslant t < 60$	87
$60 \leqslant t < 75$	123
$75 \leqslant t < 90$	67
$90 \leqslant t < 105$	22

3   The table shows the heights of 60 students.

    Draw a frequency polygon to show these data.

Height (h cm)	Frequency
$168 \leqslant h < 172$	2
$172 \leqslant h < 176$	6
$176 \leqslant h < 180$	17
$180 \leqslant h < 184$	22
$184 \leqslant h < 188$	10
$188 \leqslant h < 192$	3

4   The table shows the number
    of words per sentence in the
    first 50 sentences of two
    books.

No of words (w)	Frequency Book 1	Frequency Book 2
$0 < w \leqslant 10$	2	27
$10 < w \leqslant 20$	9	11
$20 < w \leqslant 30$	14	9
$30 < w \leqslant 40$	7	0
$40 < w \leqslant 50$	4	3
$50 < w \leqslant 60$	8	0
$60 < w \leqslant 70$	6	0

    (a) On the same grid, draw a frequency polygon for each book.

    (b) Use the frequency polygons to compare the number of words per sentence in each
         book.

Stem-and-leaf diagrams

Another way of presenting data is to show them in a **stem-and-leaf diagram**.

Stem–and–leaf diagrams are useful because the data is grouped but the individual values are not lost.

Often the stem shows the tens digit of each value and the leaves show the units digits. If you put them together you get the original value. For example, 3|1 represents 31.

Or the stem might show the units digit of each value and the leaves might show the tenths digits. For example, 1|7 represents 1.7.

A stem–and–leaf diagram should have a key. The key tells you how to read the numbers.

Example 16.3

Here are the marks gained by a group of students in a maths test.

63	58	63	52	59	65	69	75	70	54
57	63	76	81	63	68	59	40	65	74
80	44	47	53	70	81	68	49	57	61

(a) Construct a stem-and-leaf diagram to represent these data.

(b) What is the mode of these data?

(c) What is the modal class?

(d) What is the median?

Solution

(a) First draw the stem of the diagram.

```
4|
5|
6|
7|
8|
```

The smallest value in the list is 40 and the largest is 81. The stem of the diagram will be the tens digits from 4 to 8.

Now work through the data values and put the second digit on the appropriate row.

```
4|0 4 7 9
5|8 2 9 4 7 9 3 7
6|3 3 5 9 3 3 8 5 8 1
7|5 0 6 4 0
8|1 0 1
```

For the first value, 63, the 3 will go on the 6 row.

The numbers on the right of the diagram are the leaves.

Finally rewrite the diagram with all the leaves in order, with the smallest nearest to the stem.
Remember to include a key.

```
4|0 4 7 9
5|2 3 4 7 7 8 9 9
6|1 3 3 3 3 5 5 8 8 9
7|0 0 4 5 6
8|0 1 1
```

Key: 8|0 represents a mark of 80

(b) The mode is 63.   It is easy to see from the stem-and-leaf diagram that this is the value that occurs most often.

(c) The modal class is the 60 to 69 class.   Even though there is no value 60, the 6 row of the table covers all values from 60 to 69.

(d) As there are 30 values, the median will lie between the 15th and 16th values.
Count the values in the stem-and-leaf diagram to find the 15th and 16th values.
Since these are both 63, the median is 63.

Exercise 16.3

1 The data shows the weights, in kilograms, of 30 children.

41	30	40	30	34	37	55	27	45	39
65	47	36	44	32	31	58	41	41	28
49	53	49	37	33	30	34	59	34	68

Construct a stem-and-leaf diagram to show these weights.

2 The data shows the ages of 25 employees in a business.

19	17	26	33	31	41	41	27	25	28	24	28	32
40	31	29	29	39	26	36	26	35	38	42	38	

(a) Construct a stem-and-leaf diagram to show these ages.

(b) What is the modal age?

(c) What is the median age?

Exercise continues …

3 The data shows the times, in seconds, 45 students took to swim one length of a swimming pool.

14.6	15.2	15.7	19.4	18.5	14.7	16.1	18.5	15.9	19.5	16.2	15.6
21.7	20.0	18.4	17.1	17.8	16.2	17.7	15.1	17.0	15.8	16.1	
18.1	16.2	14.9	15.5	18.6	17.4	18.3	16.3	19.2	16.4	15.5	
18.7	14.9	18.3	14.1	17.4	19.8	16.4	16.3	21.2	15.8	16.8	

(a) Construct a stem-and-leaf diagram to show these times.
Use the key: 14|6 represents 14.6 seconds.

(b) Students who swam the length in less than 15 seconds qualified for the school team. How many students qualified?

(c) What was the median time?

4 For their mock examinations, a group of 41 students take two maths papers. Here are their results.

Paper 1

25	46	33	76	55	40	86	83	83	92	36
38	69	45	87	65	53	52	95	94	80	
49	78	57	31	74	65	63	61	70	92	
24	23	66	42	86	74	73	71	82	91	

Paper 2

38	26	34	32	20	69	64	75	61	87	48
49	35	46	45	32	31	52	50	72	83	
26	48	58	57	45	42	52	63	89	95	
37	22	21	69	56	52	64	73	70	91	

(a) Construct a stem-and-leaf diagram for each set of scores.

(b) Which paper do you think was the harder? Give a reason for your answer.

(c) Find the median mark for each paper.

(d) The pass mark was 40 on each paper.
How many students passed each paper?

What you have learned

- How to construct and interpret frequency diagrams, frequency polygons and stem-and-leaf diagrams
- The modal class is the class or group with the highest frequency
- The median is the value halfway through the data; the frequency of the classes can be added to find the class in which the median lies when data is grouped

Mixed exercise 16

1 Emma kept a record of the time, in minutes, that she had to wait for the school bus each morning for 4 weeks.

11	5	7	4	2	18	3	10	8	1
13	4	9	10	14	4	5	17	6	7

(a) Make a frequency table for these values using the groups
$0 \leqslant t < 5, 5 \leqslant t < 10, 10 \leqslant t < 15$ and $15 \leqslant t < 20$.

(b) Draw a frequency diagram for these data.

(c) Which of the intervals is the modal class?

(d) Which of the intervals contains the median value?

2 The table shows the marks gained by students in an examination.

Mark	$30 \leqslant m < 40$	$40 \leqslant m < 50$	$50 \leqslant m < 60$	$60 \leqslant m < 70$	$70 \leqslant m < 80$	$80 \leqslant m < 90$
Frequency	8	11	18	13	8	12

(a) Draw a frequency polygon to show these data.

(b) Describe the distribution of the marks.

(c) Which is the modal class?

(d) How many students took the examination?

(e) What fraction of students scored 70 or more in the examination?
Give your answer in its simplest form.

3 The data show the number of seats empty at a local theatre during each performance over the pantomime season.

19	1	8	11	15	19	21	43	17	1	23	19	11
12	15	40	21	11	8	4	27	21	20	14	18	7
11	23	31	16	8	2	8	6	10	11	15	9	8
6	2	3	8	21	41	27	32	37	4			

(a) Construct a stem-and-leaf diagram to show these data.

(b) There is one performance each day.
For how many weeks did the pantomime season run?

(c) What was the median number of empty seats?

(d) The theatre makes a loss if the number of empty seats is over 40.
For how many performances does the theatre make a loss?

Equations 2

▶ This chapter is about

- Solving equations

▶ You should already know

- How to collect like terms
- How to add, subtract, multiply and divide with negative numbers
- The squares of whole numbers up to 10

Solving equations

Sometimes the x term in the equation is squared (x^2).

If there is an x squared term and no other x term in the equation, you can solve it using the method you learned in Chapter 6. However, you must remember that if you square a negative number, the result is positive. For example, $(-6)^2 = 36$.

When you solve an equation involving x^2, there will usually be two values that satisfy the equation.

Example 17.1

Solve these equations.

(a) $5x + 1 = 16$ **(b)** $x^2 + 3 = 39$

Solution

(a) $5x + 1 = 16$

$5x + 1 - 1 = 16 - 1$ First subtract 1 from each side.

$5x = 15$

$5x \div 5 = 15 \div 5$ Now divide each side by 5.

$x = 3$

(b) $x^2 + 3 = 39$

$x^2 = 36$ First subtract 3 from each side.

$x = 6 \text{ or } x = -6$ Now find the square root of each side.

TIP

Remember that you must always do each operation to the whole of both sides of the equation.

Exercise 17.1

Solve these equations.

1 $2x - 1 = 13$	2 $2x - 1 = 0$	3 $2x - 13 = 1$	4 $3x - 2 = 19$
5 $6x + 12 = 18$	6 $3x - 7 = 14$	7 $4x - 8 = 12$	8 $4x + 12 = 28$
9 $3x - 6 = 24$	10 $5x - 10 = 20$	11 $x^2 + 3 = 28$	12 $x^2 - 4 = 45$
13 $y^2 - 2 = 62$	14 $m^2 + 3 = 84$	15 $m^2 - 5 = 20$	16 $x^2 + 10 = 110$
17 $x^2 - 4 = 60$	18 $20 + x^2 = 36$	19 $16 - x^2 = 12$	20 $200 - x^2 = 100$

Solving equations with a bracket

You learned how to **expand a bracket** in Chapter 15.

If you are solving an equation with a bracket in it, expand the bracket first.

> **TIP**
>
> Remember to multiply *each* term inside the bracket by the number outside the bracket.

Example 17.2

Solve these equations.

(a) $3(x + 4) = 24$ **(b)** $4(p - 3) = 20$

Solution

(a) $3(x + 4) = 24$
$3x + 12 = 24$ Multiply each term inside the bracket by 3.
$3x = 12$ Subtract 12 from each side.
$x = 4$ Divide each side of the equation by 3.

(b) $4(p - 3) = 20$
$4p - 12 = 20$ Multiply each term inside the bracket by 4.
$4p = 32$ Add 12 to each side.
$p = 8$ Divide each side by 4.

Exercise 17.2

Solve these equations.

1 $3(p - 4) = 36$	2 $3(4 + x) = 21$	3 $6(x - 6) = 6$
4 $4(x + 3) = 16$	5 $2(x - 8) = 14$	6 $2(x + 4) = 10$

Exercise continues …

7 $2(x - 4) = 20$

8 $5(x + 1) = 30$

9 $3(x + 7) = 9$

10 $2(x - 7) = 6$

11 $5(x + 2) = 5$

12 $5(x - 6) = 20$

13 $7(a + 3) = 28$

14 $8(2x + 3) = 40$

15 $5(3x - 1) = 40$

16 $2(5x - 3) = 14$

17 $4(3x - 2) = 28$

18 $2(a + 4) = 1$

19 $7(x - 4) = 28$

20 $3(2x + 7) = 6$

21 $3(5x - 12) = 24$

22 $2(4x + 2) = 20$

23 $2(2x - 5) = 12$

24 $10(3y + 7) = 30$

Equations with x on both sides

Some equations, such as $2x + 5 = 4 + 3x$, have x on both sides.

You could start to solve the equation in two different ways.

Subtract $3x$ from both sides   or   Subtract $2x$ from both sides

$2x + 5 - 3x = 4 + 3x - 3x$ $2x + 5 - 2x = 4 + 3x - 2x$

     $-x + 5 = 4$ $5 = 4 + x$

Subtracting $3x$ gives a negative x term. It is much easier to work with a positive x term, so for this equation it is better to start by subtracting $2x$.

Example 17.3

Solve these equations.

(a) $8x - 3 = 3x + 7$

(b) $18 - 6x = 3x + 9$

Solution

(a) $8x - 3 = 3x + 7$

     $8x - 3 - 3x = 3x + 7 - 3x$ Subtract $3x$ from each side.

          $5x - 3 = 7$

              $5x = 10$ Add 3 to each side.

               $x = 2$ Divide each side by 5.

(b) $18 - 6x = 3x + 9$

   $18 - 6x + 6x = 3x + 9 + 6x$ Add $6x$ to each side.

           $18 = 9x + 9$

             $9 = 9x$ Subtract 9 from each side.

           $1 = x$ or $x = 1$ Divide each side by 9.

Exercise 17.3

Solve these equations.

1 $7x - 4 = 3x + 8$

2 $5x + 4 = 2x + 13$

3 $6x - 2 = x + 8$

4 $5x + 1 = 3x + 21$

5 $5x + 7 = 3x - 3$

6 $9x - 10 = 3x + 8$

7 $5x - 12 = 2x - 6$

8 $4x - 23 = x + 7$

9 $8x + 8 = 3x - 2$

10 $11x - 7 = 6x + 8$

11 $3x - 2 = x - 7$

12 $5 + 3x = x + 9$

13 $2x - 3 = 7 - 3x$

14 $4x - 1 = 2 + x$

15 $2x - 7 = x - 4$

16 $3x - 2 = x + 7$

17 $x - 5 = 2x - 9$

18 $x + 9 = 3x - 3$

19 $3x - 4 = 2 - 3x$

20 $5x - 6 = 16 - 6x$

21 $3(x + 1) = 2x$

22 $49 - 3x = x + 21$

23 $2(x + 5) = 6 - x$

24 $5x + 6 = 3(1 - x)$

● Challenge 17.1

The length of a rectangular field is 10 metres more than its width.

The perimeter of the field is 220 metres.

What are the width and length of the field?

Hint: let x represent the width and draw a sketch of the rectangle.

● Challenge 17.2

A rectangle measures $(2x + 1)$ cm by $(x + 9)$ cm.

Find the value of x for which the rectangle is a square.

Fractions in equations

You know that $k \div 6$ can be written as $\frac{k}{6}$.

In Chapter 6 you learned to solve an equation like $\frac{k}{6} = 2$ by multiplying both sides of the equation by the denominator of the fraction.

Check up 17.1

Solve these equations.

(a) $\dfrac{x}{3} = 10$ (b) $\dfrac{m}{4} = 2$ (c) $\dfrac{m}{2} = 6$ (d) $\dfrac{p}{3} = 9$ (e) $\dfrac{y}{7} = 4$

Some equations involving fractions take more than one step to solve. These are solved using the same method as equations without fractions. You can get rid of the fraction, by multiplying both sides of the equation by the denominator of the fraction, at the end.

Example 17.4

Solve the equation $\dfrac{x}{8} + 3 = 5$.

Solution

$\dfrac{x}{8} + 3 = 5$

$\qquad \dfrac{x}{8} = 2 \qquad$ Subtract 3 from each side.

$\qquad x = 16 \qquad$ Multiply each side by 8.

Exercise 17.4

Solve these equations.

1   $\dfrac{x}{4} + 3 = 7$ 2 $\dfrac{a}{5} - 2 = 6$ 3 $\dfrac{x}{4} - 2 = 3$ 4 $\dfrac{y}{5} - 5 = 5$

5   $\dfrac{y}{6} + 3 = 8$ 6 $\dfrac{p}{7} - 4 = 1$ 7 $\dfrac{m}{3} + 4 = 12$ 8 $\dfrac{x}{8} + 8 = 16$

9   $\dfrac{x}{9} + 7 = 10$ 10 $\dfrac{y}{3} - 9 = 2$

What you have learned

- When solving equations involving a squared term, remember to include the negative solution if appropriate
- To solve equations involving a bracket, expand the bracket first
- To solve equations with x on both sides, avoid ending up with a negative x term
- Solve equations involving fractions in the same way as equations without fractions, and deal with the fraction at the end

Mixed exercise 17

Solve these equations.

1 $3x^2 = 48$ **2** $2x^2 = 72$ **3** $5p^2 + 1 = 81$

4 $4x^2 - 3 = 61$ **5** $2a^2 - 3 = 47$ **6** $6x + 14 = 2$

7 $2(m - 4) = 10$ **8** $5(p + 6) = 40$ **9** $7(x - 2) = 42$

10 $3(4 + x) = 9$ **11** $4(p - 3) = 20$ **12** $3x + 23 = 7x - 5$

13 $4x + 7 = 3x + 12$ **14** $5x - 13 = 9x + 11$ **15** $2(3x + 4) = 7x - 1$

16 $\dfrac{x}{5} - 1 = 4$ **17** $\dfrac{x}{6} + 5 = 10$ **18** $\dfrac{y}{3} + 7 = 13$

19 $\dfrac{y}{7} - 6 = 1$ **20** $\dfrac{a}{4} - 8 = 1$

Ratio and proportion

What is a ratio?

A ratio is used to compare two or more quantities.

If you have three sweets and decide to keep one and give two to your best friend, you and your friend have sweets in the ratio $1:2$. You say this as '1 to 2'.

Larger numbers can also be compared in a ratio. If you have six sweets and decide to keep two and give four to your best friend, you and your friend have sweets in the ratio $2:4$.

You write a ratio in its **lowest terms** or **simplest form** by dividing all the parts of the ratio by any common factors.

$2:4 = 1:2$ 2 and 4 are both multiples of 2. So you can divide each part of the ratio by 2.

Example 18.1

The salaries of three people are £16 000, £20 000 and £32 000.
Write this as a ratio in its lowest terms.

Solution

	16 000 : 20 000 : 32 000			First write the salaries as a ratio.
=	16 : 20 : 32			Divide each part of the ratio by 1000.
=	8 : 10 : 16			Divide each part by 2.
=	4 : 5 : 8			Divide each part by 2.

Notice that your answer should not include units. £4 : £5 : £8 would
be wrong.

To write a ratio in its lowest terms in one step, find the highest
common factor (HCF) of the numbers in the ratio. Then divide each
part of the ratio by the highest common factor.

Example 18.2

Write these ratios in their lowest terms.

(a) 20 : 50   **(b)** 16 : 24   **(c)** 9 : 27 : 54

Solution

(a) 20 : 50 = 2 : 5   Divide each part by 10.
(b) 16 : 24 = 2 : 3   Divide each part by 8.
(c) 9 : 27 : 54 = 1 : 3 : 6   Divide each part by 9.

Check up 18.1

(a) Jane is 4 years old and Petra is 8 years old.
Write the ratio of their ages in its lowest terms.

(b) A recipe uses 500 g of flour, 300 g of sugar and 400 g of raisins.
Write the ratio of these amounts in its lowest terms.

Sometimes you have to change the units of one part of the ratio first.
This is shown in the next example.

Example 18.3

Write each of these ratios in its lowest terms.

(a) 1 millilitre : 1 litre   **(b)** 1 kilogram : 200 grams

Solution

(a) 1 millilitre : 1 litre = 1 millilitre : 1000 millilitres   Write each part in the same units.
$\qquad\qquad\qquad$ = 1 : 1000   When the units are the same, you do
$\qquad\qquad\qquad\qquad\qquad$ not include them in the ratio.

(b) 1 kilogram : 200 grams = 1000 grams : 200 grams   Write each part in the same units.
$\qquad\qquad\qquad\qquad$ = 5 : 1   Divide each part by 200.

Example 18.4

Write each of these ratios in its lowest terms.

(a) 50p : £2   **(b)** 2 cm : 6 mm   **(c)** 600 g : 2 kg : 750 g

Solution

(a) 50p : £2 = 50p : 200p   Write each part in the same units.
$\qquad\quad$ = 1 : 4   Divide each part by 50.

(b) 2 cm : 6 mm = 20 mm : 6 mm   Write each part in the same units.
$\qquad\qquad$ = 10 : 3   Divide each part by 2.

(c) 600 g : 2 kg : 750 g = 600 g : 2000 g : 750 g   Write each part in the same units.
$\qquad\qquad\qquad$ = 12 : 40 : 15   Divide each part by 50.

Exercise 18.1

1 Write each of these ratios in its lowest terms.

   (a) 6 : 3   (b) 25 : 75   (c) 30 : 6   (d) 5 : 15 : 25   (e) 6 : 12 : 8

2 Write each of these ratios in its lowest terms.

   (a) 50 g : 1000 g   (b) 30p : £2   (c) 2 minutes : 30 seconds
   (d) 4 m : 75 cm   (e) 300 ml : 2 litres

3 At a concert there are 350 men and 420 women.
   Write the ratio of men to women in its lowest terms.

4 Al, Peta and Dave invest £500, £800 and £1000 respectively in a business.
   Write the ratio of their investments in its lowest terms.

5 A recipe for vegetable soup uses 1 kg of potatoes, 500 g of leeks and 750 g of celery.
   Write the ratio of the ingredients in its lowest terms.

● Challenge 18.1

(a) Explain why the ratio 20 minutes : 1 hour is not 20 : 1.

(b) What should it be?

Writing a ratio in the form $1 : n$

It is sometimes useful to have a ratio with 1 on the left.
A common scale for a scale model is 1 : 24.
The scale of a map or enlargement is often given as $1 : n$.

To change a ratio to this form, divide both numbers by the one on the left. This can be written in a general form as $1 : n$.

Example 18.5

Write these ratios in the form $1 : n$.

(a) 2 : 5

(b) 8 mm : 3 cm

(c) 25 mm : 1.25 km

Solution

(a) 2 : 5 = 1 : 2.5   Divide each side by 2.

(b) 8 mm : 3 cm = 8 mm : 30 mm   Write each side in the same units.
              = 1 : 3.75   Divide each side by 8.

(c) 25 mm : 1.25 km = 25 : 1 250 000   Write each side in the same units.
                  = 1 : 50 000   Divide each side by 25.

1 : 50 000 is a common map scale. It means that 1 cm on the map represents 50 000 cm, or 500 m, on the ground.

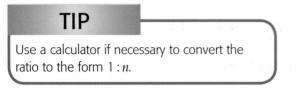

TIP

Use a calculator if necessary to convert the ratio to the form $1 : n$.

Exercise 18.2

1 Write each of these ratios in the form $1:n$.

    (a) $2:6$ (b) $3:15$ (c) $6:15$ (d) $4:7$

    (e) 20p : £1.50   (f) 4 cm : 5 m   (g) $10:2$ (h) 2 mm : 1 km

2 On a map a distance of 8 mm represents a distance of 2 km.
   What is the scale of the map in the form $1:n$?

3 Emily is harvesting sweetcorn in her virtual world.
   A virtual corn cob is 35 mm long.
   In real-life the corn cob is 21 cm long.
   What is the ratio of the virtual corn cob to the real-life corn cob in the form $1:n$?

Using ratios

Sometimes you know one of the quantities in the ratio, but not the other.

If the ratio is in the form $1:n$, you can work out the second quantity
by multiplying the first by n.

You can work out the first quantity by dividing the second quantity by n.

Example 18.6

Jake is building a model Jurassic Park.

The ratio of the models to the size in real life is $1:20$.

(a) A model dragonfly has a wingspan of 36 mm and a length of 24 mm.
    What size was the dragonfly in real life?

(b) A raptor has a height of 1000 mm.
    What height should he make the model?

Solution

(a) $36 \times 20 = 720$ The real-life size of the dragonfly is 20 times
    $24 \times 20 = 480$ bigger than the model so multiply the
                         dimensions by 20.

    The real-life dragonfly had a wingspan of 720 mm and a length of
    480 mm.

(b) $1000 \div 20 = 50$ The model is 20 times smaller than the raptor in
                         real life so divide the height by 20.

    He should make the model raptor with a height of 50 mm.

Example 18.7

A map is drawn to a scale of 1 cm : 2 km.

(a) On the map, the distance between Amhope and Didburn is 5.4 cm. What is the actual distance in kilometres?

(b) The length of a straight railway track between two stations is 7.8 km. How long is this track on the map in centimetres?

Solution

(a) 2 × 5.4 = 10.8   The actual distance, in kilometres,
     Real distance = 10.8 km   is twice as large as the map distance, in centimetres. So multiply by 2.

(b) 7.8 ÷ 2 = 3.9   The map distance, in centimetres,
     Map distance = 3.9 cm   is half as large as the actual distance, in kilometres. So divide by 2.

● Challenge 18.2

What would the answer to part **(a)** of Example 18.7 be in centimetres?

What ratio could you use to work this out?

Sometimes you have to work out quantities using a ratio that is not in the form 1 : n.

To work out an unknown quantity, you multiply each part of the ratio by the same number to get an equivalent ratio which contains the quantity you know. This number is called the **multiplier**.

Example 18.8

To make jam, fruit and sugar are mixed in the ratio 2 : 3.

This means that if you have 2 kg of fruit, you need 3 kg of sugar; if you have 4 kg of fruit, you need 6 kg of sugar.

How much sugar do you need if your fruit weighs

(a) 6 kg?   **(b)** 10 kg?   **(c)** 500 g?

Solution

(a) $6 \div 2 = 3$ Divide the quantity of fruit by the fruit part of the ratio to find the multiplier.

      $2:3 = 6:9$ Multiply each part of the ratio by the multiplier, 3.

      9 kg of sugar

(b) $10 \div 2 = 5$ Divide the quantity of fruit by the fruit part of the ratio to find the multiplier.

      $2:3 = 10:15$ Multiply each part of the ratio by the multiplier, 5.

      15 kg of sugar

(c) $500 \div 2 = 250$ Divide the quantity of fruit by the fruit part of the ratio to find the multiplier.

      $2:3 = 500:750$ Multiply each part of the ratio by the multiplier, 250.

      750 g of sugar

Example 18.9

Two photos are in the ratio $2:5$.

(a) What is the height of the larger photo?

(b) What is the width of the smaller photo?

5 cm

9 cm

Solution

(a) $5 \div 2 = 2.5$ Divide the height of the smaller photo by the smaller part of the ratio to find the multiplier.

      $2:5 = 5:12.5$ Multiply each part of the ratio by the multiplier, 2.5.

      Height of the larger photo $= 12.5$ cm

(b) $9 \div 5 = 1.8$ Divide the width of the larger photo by the larger part of the ratio to find the multiplier.

      $2:5 = 3.6:9$ Multiply each part of the ratio by the multiplier, 1.8.

      Width of the smaller photo $= 3.6$ cm

Example 18.10

To make grey paint, white paint and black paint are mixed in the ratio 5:2.

(a) How much black paint is mixed with 800 ml of white paint?

(b) How much white paint is mixed with 300 ml of black paint?

Solution

A table is often useful for this sort of question.

Paint	White	Black
Ratio	5	2
(a) Amount	800 ml	2 × 160 = 320 ml
Multiplier	800 ÷ 5 = 160	
(b) Amount	5 × 150 = 750 ml	300 ml
Multiplier		300 ÷ 2 = 150

(a) Black paint = 320 ml

(b) White paint = 750 ml

TIP

Make sure you haven't made a silly mistake by checking that the bigger side of the ratio has the bigger quantity.

Example 18.11

To make stew for four people, a recipe uses 1.6 kg of beef.

How much beef is needed using the recipe for six people?

Solution

The ratio of people is 4:6.

4:6 = 2:3   Write the ratio in its lowest terms.

1.6 ÷ 2 = 0.8   Divide the quantity of beef needed for four people by the first part of the ratio to find the multiplier.

0.8 × 3 = 2.4   Multiply the second part of the ratio by the multiplier, 0.8.

Beef needed for six people = 2.4 kg

Exercise 18.3

1　The ratio of the lengths of two squares is 1 : 6.

   (a)　The length of the side of the small square is 2 cm.
   What is the length of the side of the large square?

   (b)　The length of the side of the large square is 21 cm.
   What is the length of the side of the small square?

2　The ratio of helpers to babies in a crèche must be 1 : 4.

   (a)　There are six helpers on a Tuesday.
   How many babies can there be?

   (b)　There are 36 babies on a Thursday.
   How many helpers must there be?

3　Sanjay is mixing pink paint.
   To get the shade he wants, he mixes red and white paint in the ratio 1 : 3.

   (a)　How much white paint should he mix with 2 litres of red paint?

   (b)　How much red paint should he mix with 12 litres of white paint?

4　A photo is 35 mm long. An enlargement of 1 : 4 is made.
   What is the length of the enlargement?

5　A road atlas of Great Britain is to a scale of 1 inch to 4 miles.

   (a)　On the map the distance between Forfar and Montrose is 7 inches.
   What is the actual distance between the two towns in miles?

   (b)　It is 40 miles from Newcastle to Middlesbrough.
   How far is this on the map?

6　For a recipe, Chelsy mixes water and lemon curd in the ratio 2 : 3.

   (a)　How much lemon curd should she mix with 20 ml of water?

   (b)　How much water should she mix with 15 teaspoons of lemon curd?

7　To make a solution of a chemical a scientist mixes 3 parts chemical with 20 parts water.

   (a)　How much water should he mix with 15 ml of chemical?

   (b)　How much chemical should he mix with 240 ml of water?

8　An alloy is made by mixing 2 parts silver with 5 parts nickel.

   (a)　How much nickel must be mixed with 60 g of silver?

   (b)　How much silver must be mixed with 120 g of nickel?

Exercise continues …

9   Sachin and Rehan share a flat. They agree to share the rent in the same ratio as their wages.
Sachin earns £600 a month and Rehan earns £800 a month.
If Sachin pays £90, how much does Rehan pay?

10 A recipe for hotpot uses onions, carrots and stewing steak in the ratio, by mass, of 1 : 2 : 5.
    (a)   What quantity of steak is needed if 100 g of onion is used?
    (b)   What quantity of carrots is needed if 450 g of steak is used?

Dividing a quantity in a given ratio

○ Discovery 18.1

Maya has an evening job making up party bags for a children's party organiser.

She shares out lemon sweets and raspberry sweets in the ratio 2 : 3.

Each bag contains 5 sweets.

(a) On Monday Maya makes up 10 party bags.
    (i)   How many sweets does she use in total?
    (ii)   How many lemon sweets does she use?
    (iii) How many raspberry sweets does she use?
(b) On Tuesday Maya makes up 15 party bags.
    (i)   How many sweets does she use in total?
    (ii)   How many lemon sweets does she use?
    (iii) How many raspberry sweets does she use?
What do you notice?

A ratio represents the number of shares in which a quantity is divided. The total quantity divided in a ratio is found by adding the parts of the ratio together.

To find the quantities shared in a ratio:
● Find the total number of shares.
● Divide the total quantity by the total number of shares to find the multiplier.
● Multiply each part of the ratio by the multiplier.

TIP

The multiplier may not be a whole number. Work with the decimal or fraction and round the final answer if necessary.

Example 18.12

To make fruit punch, orange juice and grapefruit juice are mixed in the ratio 5:3.

Jo wants to make 1 litre of punch.

(a) How much orange juice does she need in millilitres?

(b) How much grapefruit juice does she need in millilitres?

Solution

$5 + 3 = 8$ First work out the total number of shares.

$1000 \div 8 = 125$ Convert 1 litre to millilitres and divide by 8 to find the multiplier.

A table is often helpful for this sort of question.

Punch	Orange	Grapefruit
Ratio	5	3
Amount	5 × 125 = 625 ml	3 × 125 = 375 ml

(a) Orange juice = 625 ml   **(b)** Grapefruit juice = 375 ml

---

TIP

To check your answers, add the parts together: they should equal the total quantity. For example, 625 ml + 375 ml = 1000 ml ✓

Exercise 18.4

1   Share £20 between Dave and Sam in the ratio 2:3.

2   Paint is mixed in the ratio 3 parts red to 5 parts white to make 40 litres of pink paint.
    (a) How much red paint is used?   (b) How much white paint is used?

3   Asif is making mortar by mixing sand and cement in the ratio 5:1.
    How much sand is needed to make 36 kg of mortar?

4   To make a solution of a chemical a scientist mixes 1 part chemical with 5 parts water.
    She makes 300 ml of the solution.
    (a) How much chemical does she use?   (b) How much water does she use?

5   Amit, Bree and Chris share £1600 between them in the ratio 2:5:3.
    How much does each receive?

Exercise continues …

6 In a local election, 5720 people vote.
They vote for Labour, Conservative and other parties in the ratio 6 : 3 : 2.
How many people vote Conservative?

7 St Anthony's College Summer Fayre raised £1750.
The school councillors decided to share the money between the college and a local charity in the ratio 5 to 1.
How much did the local charity receive?
Give your answer correct to the nearest pound.

8 Sally makes breakfast cereal by mixing bran, currants and wheatgerm in the ratio 8 : 3 : 1 by mass.
   (a) How much bran does she use to make 600 g of the cereal?
   (b) One day, she only has 20 g of currants.
       How much cereal can she make? She has plenty of bran and wheatgerm.

● Challenge 18.3

Okera has a photograph which measures 13 cm by 17 cm.
He wants to have it enlarged.
Supa Print offer two sizes: 24 inches by 32 inches and 20 inches by 26.5 inches.
Okera wants the enlargement to be in the same proportions as the original, as nearly as possible.

(a) (i) For the photograph and for each of the enlargements, work out the ratio of the width to the length in the form 1 : n.
    (ii) Which of the two enlargements is closer to the proportions of the photograph? Explain how you make your decision.

(b) Why might Okera choose the other enlargement?

Best value

○ Discovery 18.2

Two packets of cornflakes are available at a supermarket.

Which is the better value for money?

To compare value, you need to compare either
- how much you get for a certain amount of money or
- how much a certain quantity (for example, volume or mass) costs.

In each case you are comparing **proportions**, either of size or of cost.

The better value item is the one with the **lower unit cost** or the **greater number of units per penny** (or pound).

Example 18.13

Sunflower oil is sold in 700 ml bottles for 95p and in 2 litre bottles for £2.45. Show which bottle is the better value.

Solution

Method 1

Work out the price per millilitre for each bottle.

Size	Small	Large
Capacity	700 ml	2 litre = 2000 ml
Price	95p	£2.45 = 245p
Price per ml	95 ÷ 700 = 0.14p	245 ÷ 2000 = 0.1225p

Use the same units for each bottle.

Round your answers to 2 decimal places if necessary.

The price per ml of the 2 litre bottle is lower. It has the lower unit cost. In this case the unit is a millilitre.

The 2 litre bottle is the better value.

Method 2

Work out the amount per penny for each bottle.

Size	Small	Large
Capacity	700 ml	2 litre = 2000 ml
Price	95p	£2.45 = 245p
Amount per penny	700 ÷ 95 = 7.37 ml	2000 ÷ 245 = 8.16 ml

Again, use the same units for each bottle.

Round your answers to 2 decimal places if necessary.

The amount per penny is greater for the 2 litre bottle. It has the greater number of units per penny.

The 2 litre bottle is the better value.

TIP

Make it clear whether you are working out the cost per unit or the amount per penny, and include the units in your answers. Always show your working.

Exercise 18.5

1   A 420 g bag of Choco bars costs £1.59 and a 325 g bag of Choco bars costs £1.09.
    Which is the better value for money?

2   PowerJuice is sold in 2 litre bottles for 85p and in 5 litre bottles for £1.79.
    Show which is the better value.

3   Wallace bought two packs of cheese, a 680 g pack for £3.20 and a 1.4 kg pack for £5.40.
    Which was the better value?

4   One-inch nails are sold in packets of 50 for £1.25 and in packets of 144 for £3.80.
    Which packet is the better value?

5   Toilet rolls are sold in packs of 12 for £1.79 and in packs of 50 for £7.20.
    Show which is the better value.

6   Brillo white toothpaste is sold in 80 ml tubes for £2.79 and in 150 ml tubes for £5.00.
    Which tube is the better value?

7   A supermarket sells cola in three different sized bottles: a 3 litre bottle costs £1.99,
    a 2 litre bottle costs £1.35 and a 1 litre bottle costs 57p.
    Which bottle gives the best value?

8   Crispy cornflakes are sold in three sizes: 750 g for £1.79, 1.4 kg for £3.20 and
    2 kg for £4.89.
    Which packet gives the best value?

What you have learned

- To write a ratio in its lowest terms or simplest form, divide all parts of the ratio by their highest common factor (HCF)
- To write the ratio in the form $1:n$, divide both numbers by the one on the left
- If the ratio is in the form $1:n$, you can work out the second quantity by multiplying the first by n, and you can work out the first quantity by dividing the second quantity by n
- To find an unknown quantity, each part of the ratio must be multiplied by the same number, called the multiplier
- To find the quantities shared in a given ratio, first find the total number of shares, then divide the total quantity by the total number of shares to find the multiplier, then multiply each part of the ratio by the multiplier
- To compare value, work out the cost per unit or the number of units per penny (or pound)
- The better value item is the one with the lower cost per unit or the greater number of units per penny (or pound)

Mixed exercise 18

1 Write each ratio in its simplest form.
(a) 50:35   (b) 30:72   (c) 1 minute:20 seconds
(d) 45 cm:1 m   (e) 600 ml:1 litre

2 Write these ratios in the form 1:n.
(a) 2:8   (b) 5:12   (c) 2 mm:10 cm
(d) 2 cm:5 km   (e) 100:40

3 A notice is enlarged in the ratio 1:20.
(a) The original is 3 cm wide.
How wide is the enlargement?
(b) The enlargement is 100 cm long.
How long is the original?

4 To make 12 scones Maureen uses 150 g of flour.
How much flour does she use to make 20 scones?

5 To make a fruit and nut mixture, raisins and nuts are mixed in the ratio 5:3, by mass.
(a) What mass of nuts is mixed with 100 g of raisins?
(b) What mass of raisins is mixed with 150 g of nuts?

6 Panache made a fruit punch by mixing orange, lemon and grapefruit juice in the ratio 5:1:2.
(a) He made a 2 litre bowl of fruit punch.
How many millilitres of grapefruit juice did he use?
(b) How much fruit punch could he make with 150 ml of orange juice?

7 Show which is the better buy: 5 litres of oil for £18.50 or 2 litres of oil for £7.00.

8 Supershop sells milk in pints at 43p and in litres at 75p.
A pint is equal to 568 ml.
Which is the better buy?

Statistical calculations 2

The mean from a frequency table

In Chapter 8 you learned that the **mean** of a set of data is found by adding the values together and dividing the total by the number of values used.

For example, the following data shows the number of pets owned by nine Year 10 students.

8	4	4	6	3	7	3	2	8

The mean is $45 \div 9 = 5$.

What you are working out is $\dfrac{\text{total number of pets}}{\text{total number of students surveyed}}$.

If you surveyed 150 people you would have a list of 150 numbers. You could find the mean by adding them all up and dividing by 150, but this would take a long time.

Instead, you can put the data in a frequency table and work out the mean using a different method.

Example 19.1

Skye asked all the students in Year 10 at her local girls' school how many brothers they had. The table shows her results.

Work out the mean number of brothers for these students.

Number of brothers	Frequency (number of girls)
0	24
1	60
2	47
3	11
4	5
5	2
6	0
7	0
8	1
Total	150

Solution

The mean of this data is (the total number of brothers) ÷ (the total number of girls surveyed).

First you need to work out the total number of brothers.

You can see from the table that
- 24 girls do not have any brothers. They have 24 × 0 = 0 brothers between them.
- 60 girls have one brother each. They have 60 × 1 = 60 brothers between them.
- 47 girls have two brothers each. They have 47 × 2 = 94 brothers between them.
and so on.

If you add the results for each row of the table together, you will get the total number of brothers.

You can add some more columns to the table to show this.

Number of brothers (x)	Number of girls (f)	Number of brothers × frequency	Total number of brothers (fx)
0	24	0 × 24	0
1	60	1 × 60	60
2	47	2 × 47	94
3	11	3 × 11	33
4	5	4 × 5	20
5	2	5 × 2	10
6	0	6 × 0	0
7	0	7 × 0	0
8	1	8 × 1	8
Total	150		225

> ### TIP
>
> The 'Number of brothers' column is the variable and is usually labelled x.
> The 'Number of girls' column is the frequency and is usually labelled f.
> The 'Total number of brothers' column is usually labelled fx because it
> represents (Number of brothers) $\times$ (Number of girls) $= x \times f$.

The total number of brothers $= 225$
The total number of girls surveyed $= 150$
So the mean $= 225 \div 150 = 1.5$ brothers.

> ### TIP
>
> You can enter the calculations into your calculator as a chain of numbers and
> then press the $=$ key to find the total before dividing by 150.
>
> Input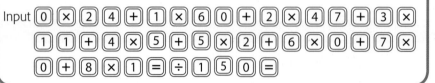

You can also work out the **mode, median** and **range** from the table.

The mode of the number of brothers is 1.
This is the number of brothers with the highest frequency (60).

The median number of brothers is 1.
As there are 150 values, the median will lie between the 75th and 76th
values.
Add on the frequency for each number of brothers (row) until you
find the interval containing the 75th and 76th values:

                    24 is smaller than 75. The 75th and 76th values do not
                    lie in row 0.
$24 + 60 = 84$ 84 is larger than 76. The 75th and 76th values must lie
                    in row 1.

The range of the number of brothers is 8.
This is (the largest number of brothers) $-$ (the smallest number of
brothers) $= 8 - 0 = 8$.

Using a spreadsheet to find the mean

You can also calculate the mean using a computer spreadsheet. Follow
these steps to work out the mean for the data in Example 19.1.

1 Open a new spreadsheet.

2 In cell A1 type the title 'Number of brothers (x)'.
In cell B1 type the title 'Number of girls (f)'.
In cell C1 type the title 'Total number of brothers (fx)'.

> **TIP**
>
> Type the bold text carefully: do not put in any spaces.

3 In cell A2 type the number 0. Then type the numbers 1 to 8 in cells A3 to A10.

4 In cell B2 type the number 24. Then type the other frequencies in cells B3 to B10.

5 In cell C2 type **=A2*B2** and press the enter key.
Click on cell C2, click on Edit in the toolbar and select Copy.
Click on cell C3, and hold down the mouse key and drag down to cell C10. Then click on Edit in the toolbar and select Paste.

6 In cell A11 type the word 'Total'.

7 In cell B11 type **=SUM(B2:B10)** and press the enter key.
In cell C11 type **=SUM(C2:C10)** and press the enter key.

8 In cell A12 type the word 'Mean'.

9 In cell B12 type **=C11/B11** and press the enter key.

Your spreadsheet should look like this.

	A	**B**	**C**
1	Number of brothers (x)	Number of girls (f)	Total number of brothers (fx)
2	0	24	0
3	1	60	60
4	2	47	94
5	3	11	33
6	4	5	20
7	5	2	10
8	6	0	0
9	7	0	0
10	8	1	8
11	Total	150	225
12	Mean	1.5	

Answer one of the questions in the next exercise using a computer spreadsheet.

Exercise 19.1

1 For each of these sets of data
   (i)   find the mode.   (ii) find the median.
   (iii) find the range.   (iv) calculate the mean.

(a)

Score on dice	Number of times thrown
1	89
2	77
3	91
4	85
5	76
6	82
Total	500

(b)

Number of matches	Number of boxes
47	78
48	82
49	62
50	97
51	86
52	95
Total	500

(c)

Number of accidents	Number of drivers
0	65
1	103
2	86
3	29
4	14
5	3
Total	300

(d)

Number of cars per house	Number of students
0	15
1	87
2	105
3	37
4	6
Total	250

2 Calculate the mean for each of these sets of data.

(a)

Number of passengers in taxi	Frequency
1	84
2	63
3	34
4	15
5	4
Total	200

(b)

Number of pets owned	Frequency
0	53
1	83
2	23
3	11
4	5
Total	175

Exercise continues …

(c)

Number of books read in a month	Frequency
0	4
1	19
2	33
3	42
4	29
5	17
6	6
Total	150

(d)

Number of drinks in a day	Frequency
3	81
4	66
5	47
6	29
7	18
8	9
Total	250

3 Calculate the mean for each of these sets of data.

(a)

x	Frequency
1	47
2	36
3	28
4	57
5	64
6	37
7	43
8	38

(b)

x	Frequency
23	5
24	9
25	12
26	15
27	13
28	17
29	14
30	15

(c)

x	Frequency
10	5
11	8
12	6
13	7
14	3
15	9
16	2

(d)

x	Frequency
0	12
1	59
2	93
3	81
4	43
5	67
6	45

Exercise continues …

4   In Barnsfield, bus tickets cost 50p, £1.00, £1.50 or £2.00 depending on the length of the journey. The frequency table shows the numbers of tickets sold on one Friday. Calculate the mean fare paid on that Friday.

Price of ticket (£)	0.50	1.00	1.50	2.00
Number of tickets	140	207	96	57

5   800 people were asked how many newspapers they had bought one week. The table shows the data.

Number of newspapers	0	1	2	3	4	5	6	7	8	9	10	11	12	13	14
Frequency	20	24	35	26	28	49	97	126	106	54	83	38	67	21	26

Calculate the mean number of newspapers bought.

● Challenge 19.1

(a) Design a data collection sheet for the number of pairs of trainers owned by the students in your class.

(b) Collect the data for your class.

(c) (i)   Find the mode of your data.

   (ii)   Find the range of your data.

   (iii) Calculate the mean number of pairs of trainers owned by the students in your class.

Grouped data

The table shows the number of CDs bought in January by a group of 75 people.

Grouping data makes it easier to work with, but it also causes some problems when calculating the mode, median, mean or range.

For example, the modal class of these data is 0–4, because that is the class with the highest frequency.

However, it is impossible to say which number of CDs was the mode because you do not know exactly how many people in this class bought what number of CDs.

Number of CDs purchased	Number of people
0–4	35
5–9	21
10–14	12
15–19	5
20–24	2

It is possible (though not very likely) that seven people bought no
CDs, seven people bought one CD, seven people bought two CDs,
seven people bought three CDs and seven people bought four CDs.
If eight or more people bought nine CDs, then the mode would
actually be 9, even though the modal class is 0–4!

The median presents a similar problem: you can see which class
contains the median value, but you cannot work out what the median
value actually is.

It is also impossible to work out the exact mean from a grouped
frequency table. You can, however, calculate an estimate using a single
value to represent each class: it is usual to use the middle value.

These middle values can also be used to calculate an estimate for the
range. You cannot find the exact range because it is impossible to say
for certain what the highest and lowest numbers of CDs purchased are.
The maximum possible purchase is 24 but you cannot tell whether
anyone did actually buy 24. The minimum possible purchase is 0 but,
again, you cannot tell whether anyone did actually buy no CDs.

Example 19.2

Use the data in the table on the previous page to calculate

(a) an estimate of the mean number of CDs purchased.

(b) an estimate of the range of the number of CDs purchased.

(c) which of the classes contains the median value.

Solution

(a)

Number of CDs purchased (x)	Number of people (f)	Middle (x) value	$f \times$ middle x	fx
0–4	35	2	35 × 2	70
5–9	21	7	21 × 7	147
10–14	12	12	12 × 12	144
15–19	5	17	5 × 17	85
20–24	2	22	2 × 22	44
Total	75			490

TIP

There are five groups
in the table but the
total number of
people is 75.

Do not be tempted
to divide by 5.

The estimate of the mean number of CDs purchased is
$490 \div 75 = 6.5$ (to 1 d.p.).

(b) The estimate of the range of the number of CDs purchased is
$22 - 2 = 20$ but it could be as high as 24 or as low as 16.

(c) As there are 75 values, the median will be the 38th value.
Add on the frequency for each class until you find the class containing the 38th value.
35 is smaller than 38.   The 38th value does not lie in class 0–4.
35 + 21 = 56   56 is larger than 38. The 38th value must lie in class 5–9.
So the 5–9 class contains the median value.

Using a spreadsheet to find the mean of grouped data

An estimate of the mean of grouped data can also be calculated using a computer spreadsheet. The method is the same as before, except for the addition of a 'Middle (x) value' column. This column can then also be used to calculate an estimate of the range.

Follow these steps to calculate estimates of the mean and range of the data in Example 19.2.

1 Open a new spreadsheet.

2 In cell A1 type the title 'Number of CDs purchased (x)'.
In cell B1 type the title 'Number of people (f)'.
In cell C1 type the title 'Middle (x) value'.
In cell D1 type the title 'Total number of CDs purchased (fx)'.

3 In cell A2 type 0–4. Then type the other classes in cells A3 to A6.

4 In cell B2 type the number 35. Then type the other frequencies in cells B3 to B6.

5 In cell C2 type **=(0+4)/2** and press the enter key.
In cell C3 type **=(5+9)/2** and press the enter key.
In cell C4 type **=(10+14)/2** and press the enter key.
In cell C5 type **=(15+19)/2** and press the enter key.
In cell C6 type **=(20+24)/2** and press the enter key.

> **TIP**
>
> Type the bold text carefully: do not put in any spaces.

6 In cell D2 type **=B2*C2** then press the enter key.
Click on cell D2, click on Edit in the toolbar and select Copy.
Click on cell D3, and hold down the mouse key and drag down to cell D6. Then click on Edit in the toolbar and select Paste.

7 In cell A7 type the word 'Total'.

8 In cell B7 type **=SUM(B2:B6)** and press the enter key.
In cell D7 type **=SUM(D2:D6)** and press the enter key.

9 In cell A8 type the word 'Mean'.

10 In cell B8 type **=D7/B7** and press the enter key.

11 In cell A9 type the word 'Range'.

12 In cell B9 type **=C6-C2** and press the enter key.

Your spreadsheet should look like this.

		A	B	C	D
	1	Number of CDs purchased (*x*)	Number of people (*f*)	Middle (*x*) value	Total number of CDs purchased (*fx*)
	2	0-4	35	2	70
	3	5-9	21	7	147
	4	10-14	12	12	144
	5	15-19	5	17	85
	6	20-24	2	22	44
	7	Total	75		490
	8	Mean	6.533333333		
	9	Range	20		

Answer one of the questions in the next exercise using a computer spreadsheet.

Exercise 19.2

1 For each of these sets of data calculate an estimate of
   (i) the range.   (ii) the mean.

(a)

Number of texts received in a day	Number of people	Middle value
0–9	99	4.5
10–19	51	14.5
20–29	28	24.5
30–39	14	34.5
40–49	7	44.5
50–59	1	54.5
Total	200	

(b)

Number of telephone calls made in a day	Number of people	Middle value
0–4	118	2
5–9	54	7
10–14	39	12
15–19	27	17
20–24	12	22
Total	250	

(c)

Number of texts sent	Number of people	Middle value
0–9	79	4.5
10–19	52	14.5
20–29	31	24.5
30–39	13	34.5
40–49	5	44.5
Total	180	

(d)

Number of calls received	Frequency	Middle value
0–4	45	2
5–9	29	7
10–14	17	12
15–19	8	17
20–24	1	22
Total	100	

Exercise continues …

2 For each of these sets of data
   (i)   find the modal class.
   (ii) find the class which cntains the median value.
   (iii) calculate an estimate of the range.
   (iv) calculate an estimate of the mean.

(a)

Number of DVDs owned	Number of people
0–4	143
5–9	95
10–14	54
15–19	26
20–24	12
Total	330

(b)

Number of books owned	Number of people
0–9	54
10–19	27
20–29	19
30–39	13
40–49	7
Total	120

(c)

Number of train journeys in a year	Number of people
0–49	118
50–99	27
100–149	53
150–199	75
200–249	91
250–299	136

(d)

Number of flowers on a plant	Frequency
0–14	25
15–29	52
30–44	67
45–59	36

3 For each of these sets of data
   (i) find the modal class.
   (ii) calculate an estimate of the mean.

(a)

Number of eggs in a nest	Frequency
0–2	97
3–5	121
6–8	43
9–11	7
12–14	2

(b)

Number of peas in a pod	Frequency
0–3	15
4–7	71
8–11	63
12–15	9
16–19	2

Exercise continues …

(c)

Number of leaves on a branch	Frequency
0–9	6
10–19	17
20–29	27
30–39	34
40–49	23
50–59	10
60–69	3

(d)

Number of bananas in a bunch	Frequency
0–24	1
25–49	29
50–74	41
75–99	52
100–124	24
125–149	3

4   A company record the number of complaints they receive each week about their products.
The table shows the data for one year.

Number of complaints	Frequency
1–10	12
11–20	5
21–30	10
31–40	8
41–50	9
51–60	5
61–70	2
71–80	1

Calculate an estimate of the mean number of complaints each week.

5   An office manager records the number of photocopies made by his staff each day in September.
The data is shown in the following table.

Number of photocopies	Frequency
0–99	13
100–199	8
200–299	3
300–399	0
400–499	5
500–599	1

Calculate an estimate of the mean number of copies each day.

Continuous data

So far all the data in this chapter has been **discrete data** (the result of objects beings counted).

When dealing with **continuous data** (the result of measurement) you estimate the mean in the same way as for grouped discrete data.

Example 19.3

A manager records the lengths of telephone calls made by her employees. The table shows the results for one week.

Duration of telephone call in minutes (x)	Frequency (f)
$0 \leqslant x < 5$	86
$5 \leqslant x < 10$	109
$10 \leqslant x < 15$	54
$15 \leqslant x < 20$	27
$20 \leqslant x < 25$	16
$25 \leqslant x < 30$	8
Total	300

> **TIP**
>
> Remember that $15 \leqslant x < 20$ means all times, x, which are greater than or equal to 15 minutes but less than 20 minutes.

(a) State the modal class

(b) Estimate the mean length of a telephone call made during this week.

Solution

(a) The class with the highest frequency (109) is the $5 \leqslant x < 10$ class.

The modal class is $5 \leqslant x < 10$.

Duration of telephone call in minutes (x)	Frequency (f)	Middle (x) value	$f \times$ middle x
$0 \leqslant x < 5$	86	2.5	215
$5 \leqslant x < 10$	109	7.5	817.5
$10 \leqslant x < 15$	54	12.5	675
$15 \leqslant x < 20$	27	17.5	472.5
$20 \leqslant x < 25$	16	22.5	360
$25 \leqslant x < 30$	8	27.5	220
Total	300		2760

(b) The estimate of the mean is $2760 \div 300 = 9.2$ minutes or 9 minutes and 12 seconds.

TIP

Remember that there are 60 seconds in 1 minute. $60 \times 0.2 = 12$ seconds.

Exercise 19.3

Use a spreadsheet to answer one of the questions in this exercise.

1 For each of these sets of data calculate an estimate of
   (i) the range.   **(ii)** the mean.

(a)

Height of plant in centimetres (x)	Number of plants (f)
$0 \leqslant x < 10$	5
$10 \leqslant x < 20$	11
$20 \leqslant x < 30$	29
$30 \leqslant x < 40$	26
$40 \leqslant x < 50$	18
$50 \leqslant x < 60$	7
Total	96

(b)

Weight of egg in grams (x)	Number of eggs (f)
$0 \leqslant x < 8$	3
$8 \leqslant x < 16$	18
$16 \leqslant x < 24$	43
$24 \leqslant x < 32$	49
$32 \leqslant x < 40$	26
$40 \leqslant x < 48$	5
Total	144

(c)

Length of string in centimetres (x)	Frequency (f)
$60 \leqslant x < 64$	16
$64 \leqslant x < 68$	28
$68 \leqslant x < 72$	37
$72 \leqslant x < 76$	14
$76 \leqslant x < 80$	5
Total	100

(d)

Rainfall per day in millimetres (x)	Frequency (f)
$0 \leqslant x < 10$	151
$10 \leqslant x < 20$	114
$20 \leqslant x < 30$	46
$30 \leqslant x < 40$	28
$40 \leqslant x < 50$	17
$50 \leqslant x < 60$	9
Total	365

Exercise continues ...

2 For each of these sets of data
   (i)   write down the modal class.
   (ii) calculate an estimate of the mean.

(a)

Age of chick in days (x)	Number of chicks (f)
0 ≤ x < 3	61
3 ≤ x < 6	57
6 ≤ x < 9	51
9 ≤ x < 12	46
12 ≤ x < 15	44
15 ≤ x < 18	45
18 ≤ x < 21	46

(b)

Weight of apple in grams (x)	Number of apples (f)
90 ≤ x < 100	5
100 ≤ x < 110	24
110 ≤ x < 120	72
120 ≤ x < 130	81
130 ≤ x < 140	33
140 ≤ x < 150	10

(c)

Length of runner bean in centimetres (x)	Frequency (f)
10 ≤ x < 14	16
14 ≤ x < 18	24
18 ≤ x < 22	25
22 ≤ x < 26	28
26 ≤ x < 30	17
30 ≤ x < 34	10

(d)

Time to complete race in minutes (x)	Frequency (f)
40 ≤ x < 45	1
45 ≤ x < 50	8
50 ≤ x < 55	32
55 ≤ x < 60	26
60 ≤ x < 65	5
65 ≤ x < 70	3

3 The table shows the weekly wages of the manual workers in a factory.

   (a) What is the modal class?

   (b) In which class is the median wage?

   (c) Calculate an estimate of the mean wage.

Wage in £ (x)	Frequency (f)
150 ≤ x < 200	4
200 ≤ x < 250	14
250 ≤ x < 300	37
300 ≤ x < 350	15

4 The table shows the masses, in grams, of the first 100 letters posted one day.

   Calculate an estimate of the mean mass of a letter.

Mass in grams (x)	Frequency (f)
0 ≤ x < 15	48
15 ≤ x < 30	36
30 ≤ x < 45	12
45 ≤ x < 60	4

Exercise continues …

5   The table shows the prices paid for birthday cards sold in one day by a greetings card shop.

Price of birthday card in pence (x)	Frequency (f)
$100 \leqslant x < 125$	18
$125 \leqslant x < 150$	36
$150 \leqslant x < 175$	45
$175 \leqslant x < 200$	31
$200 \leqslant x < 225$	17
$225 \leqslant x < 250$	9

Calculate an estimate of the mean price paid for a birthday card that day.

Challenge 19.2

(a) Design a data collection sheet, using appropriate groups, and carry out one of the following tasks. Use the students in your class as the source of your data.
   ● Ask each person the amount of money they spent on lunch on a particular day.
   ● Obtain a piece of string, arrange it in a non–straight line and ask each person to estimate its length.

(b) (i)   Estimate the range of your data.
     (ii)   Calculate an estimate of the mean of your data.

What you have learned

■ To find the mean for discrete data from a frequency table you multiply each data value (x) by its frequency (f) and add these to find the total; you then divide by the number in the survey
■ To find the estimate of the mean for grouped data you use the middle value of each class as the value of x for each class and then proceed as for discrete data
■ To estimate the range for grouped data you subtract the middle value of the first class from the middle value of the last class
■ The modal class is the class with the highest frequency
■ To find the class in which the median value lies you add on the frequency of each class until you find the class containing the value that is halfway through the data
■ You can use a calculator or spreadsheet to calculate the mean and range of large data sets

Mixed exercise 19

1 For each of these sets of data
   (i)   find the mode.
   (ii) find the range.
   (iii) calculate the mean.

(a)

Score on octagonal dice	Number of times thrown
1	120
2	119
3	132
4	126
5	129
6	142
7	123
8	109
Total	1000

(b)

Number of marbles in a bag	Number of bags
47	11
48	25
49	47
50	63
51	54
52	38
53	17
54	5
Total	260

(c)

Number of pets per house	Frequency
0	64
1	87
2	41
3	26
4	17
5	4
6	1

(d)

Number of broad beans in a pod	Frequency
4	17
5	36
6	58
7	49
8	27
9	13

(e)

x	f
1	242
2	266
3	251
4	252
5	259
6	230

(f)

x	f
15	9
16	13
17	18
18	27
19	16
20	7

Mixed exercise 19 continues …

2 Crazyphone top-ups cost £5, £10, £20 or £50 depending upon the amount of credit bought.
The frequency table shows the numbers of each value of top-up sold in one shop on a Saturday.

Price of top-up (£)	5	10	20	50
Number of top-ups	34	63	26	2

Calculate the mean value of top-up bought in the shop that Saturday.

3 A sample of 350 people were asked how many magazines they had bought in September.
The table below shows the data.

Number of magazines	0	1	2	3	4	5	6	7	8	9	10
Frequency	16	68	94	77	49	27	11	5	1	0	2

Calculate the mean number of magazines bought in September.

4 For each of these sets of data, calculate an estimate of
   (i)   the range.   (ii)   the mean.

(a)

Height of cactus in centimetres (x)	Number of plants (f)
$10 \leqslant x < 15$	17
$15 \leqslant x < 20$	49
$20 \leqslant x < 25$	66
$25 \leqslant x < 30$	38
$30 \leqslant x < 35$	15
Total	185

(b)

Wind speed at noon in km/h (x)	Number of days (f)
$0 \leqslant x < 20$	164
$20 \leqslant x < 40$	98
$40 \leqslant x < 60$	57
$60 \leqslant x < 80$	32
$80 \leqslant x < 100$	11
$100 \leqslant x < 120$	3
Total	365

(c)

Time holding breath in seconds (x)	Frequency (f)
$30 \leqslant x < 40$	6
$40 \leqslant x < 50$	29
$50 \leqslant x < 60$	48
$60 \leqslant x < 70$	36
$70 \leqslant x < 80$	23
$80 \leqslant x < 90$	8

(d)

Mass of student in kilograms (x)	Frequency (f)
$40 \leqslant x < 45$	5
$45 \leqslant x < 50$	13
$50 \leqslant x < 55$	26
$55 \leqslant x < 60$	31
$60 \leqslant x < 65$	17
$65 \leqslant x < 70$	8

Mixed exercise 19 continues …

5 The table below shows the lengths, to the nearest minute, of 304 telephone calls.

(a) What is the modal class?

(b) In which class is the median length of call?

(c) Calculate an estimate of the mean length of call.

Length in minutes (x)	Frequency (f)
$0 \leqslant x < 10$	53
$10 \leqslant x < 20$	124
$20 \leqslant x < 30$	81
$30 \leqslant x < 40$	35
$40 \leqslant x < 50$	11

6 The table shows the annual wages of the workers in a company.

Annual wage in thousands of £ (x)	Frequency (f)
$10 \leqslant x < 15$	7
$15 \leqslant x < 20$	18
$20 \leqslant x < 25$	34
$25 \leqslant x < 30$	12
$30 \leqslant x < 35$	9
$35 \leqslant x < 40$	4
$40 \leqslant x < 45$	2
$45 \leqslant x < 50$	1
$50 \leqslant x < 55$	2
$55 \leqslant x < 60$	0
$60 \leqslant x < 65$	1

Calculate an estimate of the mean annual wage of these workers.

Pythagoras' theorem

▶ This chapter is about

- Calculating the length of a side of a right-angled triangle when you know the other two
- Deciding whether or not a triangle is right-angled
- Finding the coordinates of the midpoint and calculating the length of a line segment

▶ You should already know

- How to find squares and square roots on your calculator
- How to round numbers to a given number of decimal places
- The formula for the area of a triangle
- How to use coordinates in two dimensions
- How to work with negative numbers

Pythagoras' theorem

◉ Discovery 20.1

Measure all three sides of the right-angled triangle in the diagram.

Use the lengths to work out the area of each of the three coloured squares.

What do you notice?

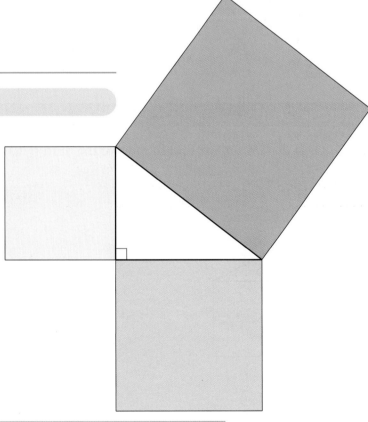

The area of the yellow square added to the area of the blue square is equal to the area of the red square.

The longest side of a right-angled triangle is called the **hypotenuse**.

What you discovered in Discovery 20.1 is true for all right-angled triangles. It was first 'discovered' by Pythagoras, a Greek mathematician, who lived around 500 BC.

Pythagoras' theorem states that:

> The area of the square on the hypotenuse of a right-angled triangle is equal to the sum of the areas of the squares on the other two sides.

That is:

$$P + Q = R$$

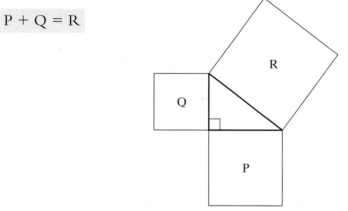

Exercise 20.1

For each of these diagrams, find the area of the third square.

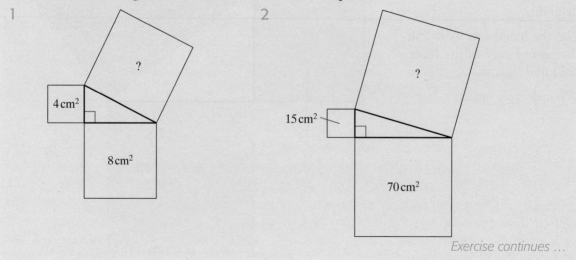

1

4 cm²

?

8 cm²

2

15 cm²

?

70 cm²

Exercise continues …

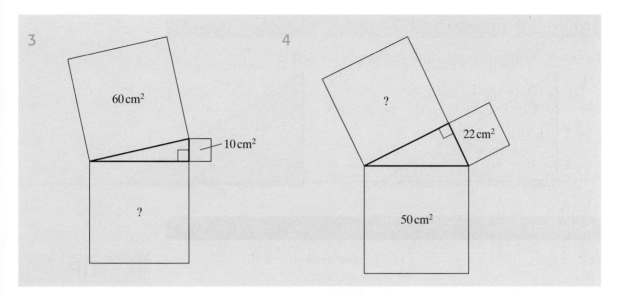

3

60 cm²

10 cm²

?

4

?

22 cm²

50 cm²

Using Pythagoras' theorem

Although the theorem is based on area it is usually used to find the length of a side.

If you drew squares on the three sides of this triangle their areas would be a^2, b^2 and c^2.

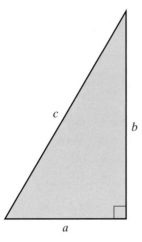

c

b

a

So Pythagoras' theorem can also be written as

$$a^2 + b^2 = c^2.$$

Example 20.1

For each of these triangles, find the length marked x.

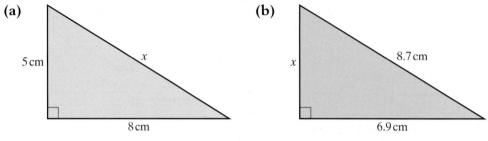

(a)

5 cm

x

8 cm

(b)

8.7 cm

x

6.9 cm

Solution

(a) $c^2 = a^2 + b^2$

    $x^2 = 8^2 + 5^2$

    $x^2 = 64 + 25$
    $x^2 = 89$
    $x = \sqrt{89}$
    $x = 9.43$ cm (to 2 d.p.)

The length marked x is the hypotenuse, or c.
Substitute the numbers into the formula.

Take the square root of both sides.

(b) $a^2 + b^2 = c^2$

    $x^2 + 6.9^2 = 8.7^2$
    $x^2 = 8.7^2 - 6.9^2$
    $x^2 = 75.69 - 47.61$
    $x^2 = 28.08$
    $x = \sqrt{28.08}$
    $x = 5.30$ (to 2 d.p.)

This time the length marked x is the shortest side, or a.

Subtract 6.9^2 from each side.

Take the square root of both sides.

TIP

Always check whether you are finding the longest side (the hypotenuse) or one of the shorter sides.

If you are finding the longest side: add the squares and square root the result.

If you are finding a shorter side: subtract the squares and square root the result.

Exercise 20.2

1 For each of these triangles, find the length of the hypotenuse, marked x.
Where the answer is not exact, give your answer correct to 2 decimal places.

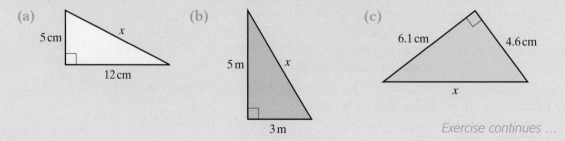

(a)

5 cm

x

12 cm

(b)

5 m

x

3 m

(c)

6.1 cm

4.6 cm

x

Exercise continues …

2 For each of these triangles, find the length of the shorter side marked x.
Where the answer is not exact, give your answer correct to 2 decimal places.

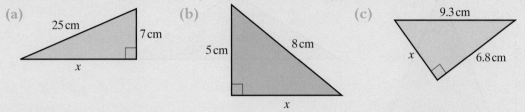

(a) 25 cm, 7 cm, x

(b) 5 cm, 8 cm, x

(c) 9.3 cm, x, 6.8 cm

3 For each of these triangles, find the length marked x.
Where the answer is not exact, give your answer correct to 2 decimal places.

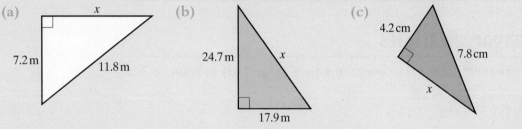

(a) x, 7.2 m, 11.8 m

(b) 24.7 m, x, 17.9 m

(c) 4.2 cm, 7.8 cm, x

4 The diagram shows a ladder standing on horizontal ground
and leaning against a vertical wall.
The ladder is 4.8 m long and the foot of the ladder is 1.6 m
away from the wall.
How far up the wall does the ladder reach?
Give your answer correct to 2 decimal places.

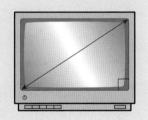

4.8 m

1.6 m

5 The size of a television screen is the length of the diagonal.
The screen size of this television is 27 inches.
If the height of the screen is 13 inches, what is the width?
Give your answer correct to 2 decimal places.

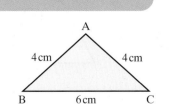

Challenge 20.1

(a) Calculate the area of the
isosceles triangle ABC.

Hint: Draw the height AD
of the triangle.
Calculate the length of AD.

A
4 cm   4 cm
B   6 cm   C

(b) Calculate the area of each of these isosceles triangles.
Give your answers correct to 1 decimal place.

(i)   **(ii)**

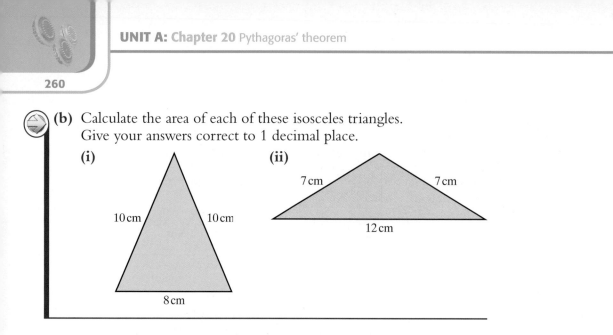

Pythagorean triples

Look again at the answers to question **1** parts **(a)** and **(d)** in Exercise 20.2.

The answers were exact.
In part **(a)** $5^2 + 12^2 = 13^2$
In part **(d)** $7^2 + 24^2 = 25^2$

These are examples of **Pythagorean triples**, or three numbers that exactly fit the Pythagoras relationship.

Another Pythagorean triple is $3, 4, 5$.
You saw this in the diagram at the start of the chapter.

$3, 4, 5$ $5, 12, 13$ and $7, 24, 25$ are the most well-known Pythagorean triples.

You can also use Pythagoras' theorem in reverse.

> If the lengths of the three sides of a triangle form a
> Pythagorean triple then the triangle is right-angled.

Exercise 20.3

Work out whether or not each of these triangles is right-angled. Show your working.

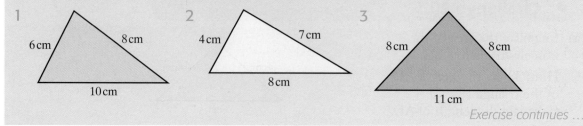

Exercise continues …

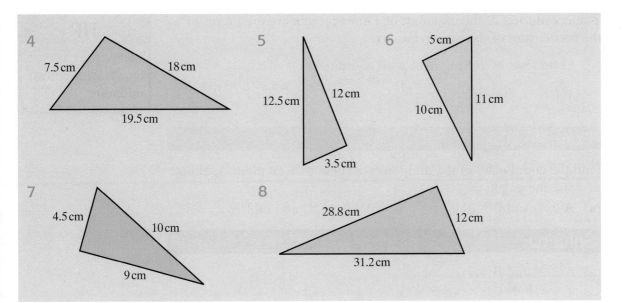

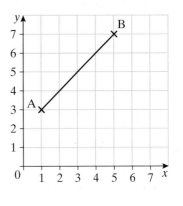

Line segments

A line can be extended at both ends but a **line segment** is the part of a line between two points.

A line has an infinite length but a line segment has a finite length.

○ Discovery 20.2

For each of these pairs of points:

(a) A(1, 3) and B(5, 7)

(b) C(1, 5) and D(7, 1)

(c) E(2, 5) and F(6, 6)

(d) G(3, 7) and H(6, 0)

● Draw a diagram on squared paper. The first one is done for you.

● Find the middle point of the line segment joining the two points and label it M.

● Write down the coordinates of M.

What do you notice?

The coordinates of the midpoint of a line segment are the means of the coordinates of the two endpoints.

> Midpoint of line segment with coordinates
>
> $(a, b), (c, d) = \left(\dfrac{a + c}{2}, \dfrac{b + d}{2}\right)$.

Example 20.2

Find the coordinates of the midpoints of these pairs of points without drawing the graph.

(a) A(2, 1) and B(6, 7)   **(b)** C(−2, 1) and D(2, 5)

Solution

(a) A(2, 1) and B(6, 7)

$a = 2, b = 1, c = 6, d = 7$

$$\text{Midpoint} = \left(\frac{a + c}{2}, \frac{b + d}{2}\right)$$

$$= \left(\frac{2 + 6}{2}, \frac{1 + 7}{2}\right)$$

$$= (4, 4)$$

(b) C(−2, 1) and D(2, 5)

$a = -2, b = 1, c = 2, d = 5$

$$\text{Midpoint} = \left(\frac{a + c}{2}, \frac{b + d}{2}\right)$$

$$\text{Midpoint} = \left(\frac{-2 + 2}{2}, \frac{1 + 5}{2}\right)$$

$$= (0, 3)$$

You can check your answers by drawing the graph of the line segment.

You can use Pythagoras' theorem to find the length of a line segment.

Example 20.3

Find the length AB.

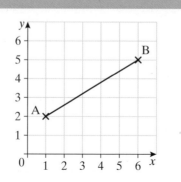

Solution

First make a right-angled triangle by drawing across from A and down from B.

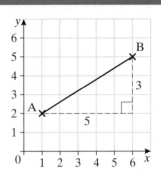

By counting squares, you can see that the lengths of the short sides are 5 and 3.

You can then use Pythagoras' theorem to work out the length of AB.

$AB^2 = 5^2 + 3^2$
$AB^2 = 25 + 9 = 34$
$AB = \sqrt{34} = 5.83$ units (correct to 2 d.p.)

Example 20.4

A is the point $(-5, 4)$ and B is the point $(3, 2)$.
Find the length AB.

Solution

Method 1

Plot the points and complete the right-angled triangle.

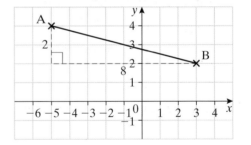

Then use Pythagoras' theorem to work out the length of AB.

$AB^2 = 8^2 + 2^2$
$AB^2 = 64 + 4 = 68$
$AB = \sqrt{68} = 8.25$ units (correct to 2 d.p.)

Method 2

You can also calculate the length of the line segment without drawing a diagram.

Point A has coordinates $(-5, 4)$.
Point B has coordinates $(3, 2)$.

Difference in x values $= 3 - (-5) = 8$
Difference in y values $= 4 - 2 = 2$

$AB^2 = 8^2 + 2^2$
$AB^2 = 64 + 4$
$AB = \sqrt{68} = 8.25$ units (correct to 2 d.p.)

Exercise 20.4

1 For each of the line segments in the diagram
   (i)   find the coordinates of the midpoint.
   (ii) find the length. Where the answer is not exact, give your answer to 2 decimal places.

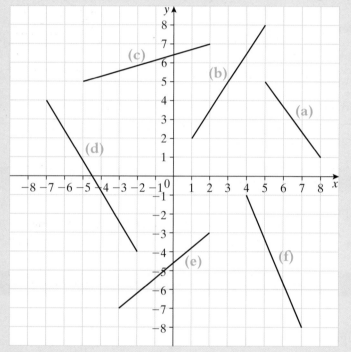

2 For the line segments joining each of the pairs of points below
   (i)   find the coordinates of the midpoint.
   (ii) find the length. Where the answer is not exact, give your answer to 2 decimal places.
   (a) A(1, 4) and B(1, 8)   (b) C(1, 5) and D(7, 3)
   (c) E(2, 3) and F(8, 6)   (d) G(3, 7) and H(8, 2)
   (e) I(−2, 3) and J(4, 1)   (f) K(−4, −3) and L(−6, −11)

● Challenge 20.2

(a) The midpoint of AB is (5, 3).
    A is the point (2, 1).
    What are the coordinates of B?
(b) The midpoint of CD is (−1, 2).
    C is the point (3, 6).
    What are the coordinates of D?

What you have learned

- The longest side of a right-angled triangle is called the hypotenuse
- Pythagoras' theorem states that the area of the square on the hypotenuse of a right-angled triangle is equal to the sum of the areas of the squares on the other two sides; or, using the notation in the diagram, $a^2 + b^2 = c^2$
- To find the length of the longest side using Pythagoras' theorem, add the squares and square root the result
- To find the length of one of the shorter sides using Pythagoras' theorem, subtract the squares and square root the result
- If the lengths of the sides are a Pythagorean triple, the triangle is right-angled
- The three most well-known Pythagorean triples are 3, 4, 5; 5, 12, 13 and 7, 24, 25
- A line segment is the part of a line between two points
- A line segment has a finite length
- The coordinates of the midpoint of the line segment joining (a, b) to (c, d) are $\left(\dfrac{a + c}{2}, \dfrac{b + d}{2}\right)$
- You can find the length of a line segment from the coordinates of the endpoints by plotting the points and counting squares or by finding the difference in the x coordinates and the difference in the y coordinates

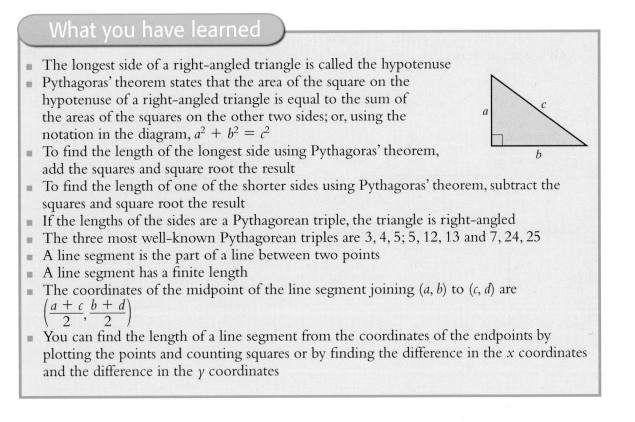

Mixed exercise 20

1 For each of these diagrams, find the area of the third square.

(a)

(b)

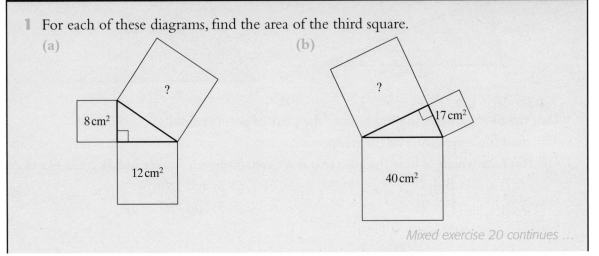

8 cm²

12 cm²

?

17 cm²

40 cm²

?

Mixed exercise 20 continues ...

2 For each of these triangles, find the length marked x.
Give your answers correct to 2 decimal places.

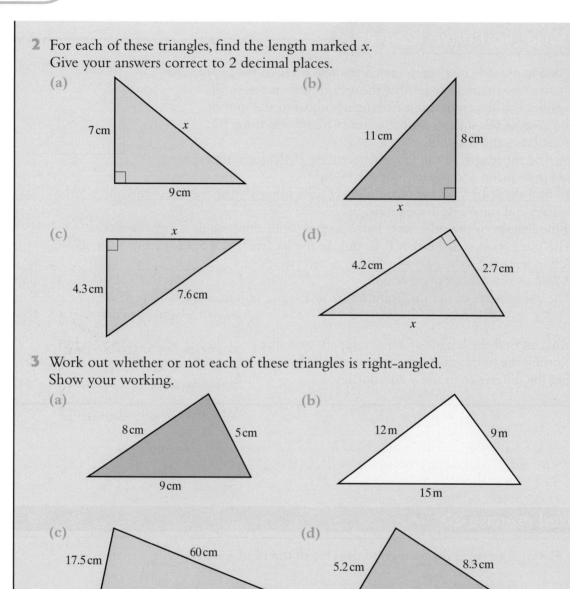

(a)

7 cm
x
9 cm

(b)

11 cm
8 cm
x

(c)

x
4.3 cm
7.6 cm

(d)

4.2 cm
2.7 cm
x

3 Work out whether or not each of these triangles is right-angled.
Show your working.

(a)

8 cm
5 cm
9 cm

(b)

12 m
9 m
15 m

(c)

17.5 cm
60 cm
62.5 cm

(d)

5.2 cm
8.3 cm
9.7 cm

4 For the line segments joining each of the pairs of points below
  (i) find the coordinates of the midpoint
  (ii) find the length. Where the answer is not exact, give your answer to 2 decimal places.
  (a) A(2, 1) and B(4, 7)   (b) C(2, 3) and D(6, 8)
  (c) E(2, 0) and F(7, 9)   (d) G(5, 3) and H(7, −1)

Mixed exercise 20 continues …

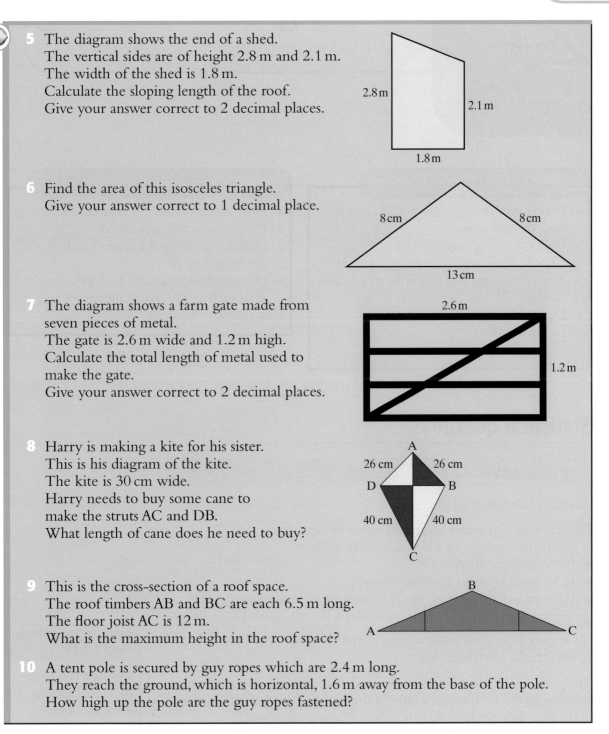

5 The diagram shows the end of a shed.
The vertical sides are of height 2.8 m and 2.1 m.
The width of the shed is 1.8 m.
Calculate the sloping length of the roof.
Give your answer correct to 2 decimal places.

2.8 m

2.1 m

1.8 m

6 Find the area of this isosceles triangle.
Give your answer correct to 1 decimal place.

8 cm

8 cm

13 cm

7 The diagram shows a farm gate made from
seven pieces of metal.
The gate is 2.6 m wide and 1.2 m high.
Calculate the total length of metal used to
make the gate.
Give your answer correct to 2 decimal places.

2.6 m

1.2 m

8 Harry is making a kite for his sister.
This is his diagram of the kite.
The kite is 30 cm wide.
Harry needs to buy some cane to
make the struts AC and DB.
What length of cane does he need to buy?

A

26 cm

26 cm

D

B

40 cm

40 cm

C

9 This is the cross-section of a roof space.
The roof timbers AB and BC are each 6.5 m long.
The floor joist AC is 12 m.
What is the maximum height in the roof space?

B

A

C

10 A tent pole is secured by guy ropes which are 2.4 m long.
They reach the ground, which is horizontal, 1.6 m away from the base of the pole.
How high up the pole are the guy ropes fastened?

Planning and collecting

▶ This chapter is about

- Posing statistical questions and planning how to answer them
- Primary and secondary data
- Choosing a sample and eliminating bias
- The advantages and problems of random samples
- Designing a questionnaire
- Collecting data
- Writing a statistical report

▶ You should already know

- How to make and use tally charts
- How to use two-way tables
- How to calculate the mean, median, mode and range
- How to draw diagrams to represent data, such as bar charts, pie charts and frequency diagrams

Statistical questions

◉ Discovery 21.1

Are boys taller than girls?

Discuss how you could begin to answer this question.
- What information would you need to collect?
- How would you collect it?
- How would you analyse the results?
- How would you present the information in your report?

To answer a question using statistical methods, the first thing you need to do is make a written plan.

You need to decide which statistical calculations and diagrams are relevant to the problem. You should think about this before you start collecting data, so that you can collect it in a useful form.

It is a good idea to rewrite the question as a **hypothesis**. This is a statement such as 'boys are taller than girls'. Your report should present evidence either for or against your hypothesis.

TIP

If you are choosing your own problem, make sure there is more than one aspect to it that you can explore.

Different types of data

When you investigate a statistical problem such as 'boys are taller than girls', there are two types of data which you can use.

- **Primary data** are data which you collect yourself. For example, you could measure the heights of a group of girls and boys.
- **Secondary data** are data which someone else has already collected. For example, you could use the internet database CensusAtSchool, which has a large number of students' heights already collected. Other sources of secondary data include books and newspapers.

Example 21.1

Are these primary or secondary data?

(a) Attendance figures on your class register

(b) The time people have to wait in a queue

(c) The number of hits on a website each day

Solution

(a) Secondary (unless you collected them)

(b) Primary (for the person measuring the times)

(c) Secondary (it is recorded by the website)

Data samples

You are trying to find out whether girls or boys are taller most often. You cannot measure the heights of all boys and all girls, but you can measure the heights of a group of boys and girls and answer the question for that group. In statistics, a group like this is called a **sample**.

The size of your sample is important. If your sample is too small, the results may not be very reliable. In general, the sample size needs to be at least 30. You need to decide what is a reasonable sample size for the hypothesis you are investigating.

You also need to eliminate **bias**. A biased sample is unreliable because it means that certain results are more likely.

Example 21.2

What conclusion might you make about the height of students if your sample consisted of the following? Why?

(a) The boys' basketball team and the girls' swimming team

(b) Only boys under 12 and girls over 15.

Solution

(a) Boys might be taller as basketball players tend to be taller than average.

(b) Boys might be shorter as younger children tend to be shorter than older ones.

It is often a good idea to use a **random** sample, where every person or piece of data has an equal chance of being selected.

Random numbers can be generated by your calculator or a spreadsheet. To select a random sample of five girls from Year 7, for example, you could allocate a random number to each Year 7 girl and then select the five girls with the smallest random numbers.

When you write a report, include reasons for your choice of sample.

Example 21.3

Candace is doing a survey about school meals. She asks every tenth person going into lunch.

Why may this not be a good method of sampling?

Solution

She will not get the opinions of those who dislike school meals and have stopped having them.

◉ Discovery 21.2

A borough council wants to survey public opinion about its library facilities.

How should it choose a sample of people to ask?

Discuss the advantages and disadvantages of each method you suggest.

When you collect large amounts of data, you may need to group it in order to analyse it or to present it clearly. It is usually best to use equal class widths for this. Tally charts are a good way of obtaining a frequency table, or you can use a spreadsheet or other statistics program to help you. Before you collect your data, make sure you design a suitable data collection sheet or spreadsheet.

Example 21.4

Design a data collection sheet to record how far spectators have travelled to a football match.

Solution

Distance (d miles)	Tally	Frequency
$d < 1$		
$1 \leqslant d < 2$		
$2 \leqslant d < 5$		
$5 \leqslant d < 10$		
$d \geqslant 10$		

Designing a questionnaire

A **questionnaire** is often a good way of collecting data.

You need to think carefully about what information you need and how you will analyse the answers to each question. This will help you get the data in the form you need.

For instance, if you are investigating the hypothesis 'boys are taller than girls', you need to know a person's sex as well as their height. If you know their age as well you can find out whether your hypothesis is true for all ages of boys and girls.

Here are some points to bear in mind when you design a questionnaire.
- Make the questions short, clear and relevant to your task.
- Only ask one thing at a time.
- Make sure your questions are not 'leading' questions. Leading questions show bias. They 'lead' the person answering them towards a particular answer: for example, 'do you agree that the cruel practice of testing drugs on animals should be illegal?'

Example 21.5

Suggest a sensible way of asking an adult their age.

Solution

Please tick your age-group:

☐ 18–25 years   ☐ 26–30 years   ☐ 31–40 years
☐ 41–50 years   ☐ 51–60 years   ☐ Over 60 years

This means that the person does not have to tell you their exact age, which many adults don't like doing.

When you have written your questionnaire, test it out on a few people. This is called doing a **pilot survey**. Try also to analyse the data from this pilot, so that you can check whether it is possible. You may then wish to reword one or two questions, regroup your data, or change your method of sampling, before you do the proper survey.

If you encounter practical problems in collecting your data, describe them in your report.

⊙ Discovery 21.3

- Sam decides to test the hypothesis 'fish and chips is the favourite main course at school'. Think of your own survey topic about school lunches. Make sure that it is relevant to your own school or college.
- Write some suitable questions for a survey to test your hypothesis.
- Try them out in a pilot survey. Discuss the results and how you could improve your questions.

Writing up your report

Your report should begin with a clear statement of your aims.

You learned about data collection in Chapter 3; you learned about some statistical calculations in Chapters 8 and 19 and you learned about some statistical diagrams in Chapters 13 and 16. You will learn about some more statistical diagrams in Unit B Chapter 16. You will need to use skills from these chapters to display and analyse your data.

Your report should end with a conclusion. Your conclusion will depend on the results of the statistical calculations you have done with your data and on any differences or similarities illustrated by your statistical diagrams. Throughout the report, you should give your reasons for what you have done and describe any difficulties you encountered, and how you dealt with them.

Use this checklist to make sure that the whole project is clear.
- Use statistical terms whenever possible.
- Make sure you include a written plan.
- Explain how you selected your sample and why you chose to select it that way.
- Show how you found your data.
- Say why you have chosen to draw a particular diagram, and what it shows.
- Relate your findings back to the original problem. Have you proved or disproved your hypothesis?
- Aim to extend the original problem, using ideas of your own.

Exercise 21.1

1   State whether the following are primary data or secondary data.

(a) Measuring people's foot length

(b) Using school records of students' ages

(c) A librarian using a library catalogue to enter new books on the system

(d) A borrower using a library catalogue

2   A borough council wants to survey public opinion about the local swimming pool. Give one disadvantage of each of the following sampling situations.

(a) Selecting people to ring at random from the local phone directory

(b) Asking people who are shopping on Saturday morning

3   Paul plans to ask 50 people at random how long they spent doing homework yesterday evening.

Here is the first draft of his data collection sheet.

Time spent	Tally	Frequency
Up to 1 hour		
1–2 hours		
2–3 hours		

Give two ways in which Paul could improve his collection sheet.

4   For each of these survey questions
   ● state what is wrong with it.
   ● write a better version.

(a) What is your favourite sport: cricket, tennis or athletics?

(b) Do you do lots of exercise each week?

(c) Don't you think this government should encourage more people to recycle waste?

5   Janine is doing a survey about how often people have a meal out in a restaurant. Here are two questions she has written.

Q1. *How often do you eat out?*

☐ *A lot*   ☐ *Sometimes*   ☐ *Never*

Q2. *What food did you eat the last time you ate out?*

(a) Give a reason why each of these questions is unsuitable.

(b) Write a better version of Q1.

6   Design a questionnaire to investigate use of the school library or resource centre. You need to know
   ● which year group the student is in.
   ● how often they use the library.
   ● how many books they usually borrow on each visit.

What you have learned

- Primary data are data which you collect yourself
- Secondary data are data which someone else has already collected, and are found in books or on the internet, for example
- You need to plan how to find evidence for or against your hypothesis, giving evidence of your planning
- Avoid bias when sampling
- In a random sample, every member of the population being considered has an equal chance of being selected
- Make sure the size of the sample is sensible
- In a questionnaire, questions should be short, clear and relevant to your task
- You can do a pilot survey to test out a questionnaire or data collection sheet
- In your report, you should give reasons for what you have done and relate your conclusions back to the original problem, saying whether your hypothesis has been shown to be correct or not

Mixed exercise 21

1 Jan uses train times from the internet.
   Are these data primary or secondary? Give a reason for your answer.

2 Ali wants to test the hypothesis 'older students at secondary school are better at estimating angles than younger ones'.
   (a) What could he ask people to do in order to test this hypothesis?
   (b) How should he choose a suitable sample of people?
   (c) Design a collection sheet for Ali to record his data.

3 Write three suitable questions for a questionnaire asking a sample of people about their favourite music or musicians.
   If you use questions without given categories for responses, show also how you would group the responses to the questions when analysing the data.

Sequences 2

Using rules to find terms of a sequence

You learned how to find the next **term** in a **sequence** in Chapter 9.

For example, for this sequence: 3, 8, 13, 18, 23, 28, … , you find the next term by adding 5.

This is known as the **term-to-term** rule.

You also learned how to find a term given its **position** in the sequence using a **position-to-term** rule.

For example, for the sequence: 3, 8, 13, 18, 23, 28, … , taking the position number (*n*), multiplying it by 5 and then subtracting 2 gives the term.

The term-to-term rule and position-to-term rule for any sequence can be expressed as formulae using the following notation.

T_1 represents the first term of a sequence,
T_2 represents the second term of a sequence,
T_3 represents the third term of a sequence,
and so on.

T_n represents the *n*th term of a sequence.

Example 22.1

Marie makes some matchstick patterns.
Here are her first three patterns.

To get the next pattern from the previous one, Marie adds three more matchsticks to complete another square. She makes this table.

Pattern	1	2	3
Number of matchsticks	4	7	10

(a) Find the term-to-term rule.   **(b)** Find the position-to-term rule.

Solution

(a) First find the rule in words.

The first term is 4.	You must always state the value of the first term
To find the next term, add 3.	when giving a term-to-term rule.

Then write the rule using the notation.

$T_1 = 4$
$T_{n+1} = T_n + 3$ Add 3 to each term to find the next one.
                            For example, $T_4 = T_3 + 3 = 10 + 3 = 13$

(b) First find the rule in words.

Taking the position number (n), multiplying it by 3 and then adding 1 gives the term.

Then write the rule using the notation.

nth term $= 3n + 1$
or $T_n = 3n + 1$ For example, $T_4 = 3 \times 4 + 1 = 12 + 1 = 13$

Example 22.2

For a sequence, $T_1 = 10$ and $T_{n+1} = T_n - 4$.
Find the first four terms of the sequence.

Solution

$T_1 = 10$ $T_2 = T_1 - 4$ $T_3 = T_2 - 4$ $T_4 = T_3 - 4$
                   $= 10 - 4$ $= 6 - 4$ $= 2 - 4$
                   $= 6$ $= 2$ $= -2$

The first four terms are $10, 6, 2$ and -2.

A position-to-term rule is very useful if you need to find a term a long way into the sequence, such as the 100th term. It means that you can find it straight away without having to find the previous 99 terms as you would using a term-to-term rule.

Example 22.3

The nth term of a sequence is $5n + 1$.

(a) Find the first four terms of the sequence.

(b) Find the 100th term of the sequence.

Solution

(a) $T_1 = 5 \times 1 + 1$ $T_2 = 5 \times 2 + 1$ $T_3 = 5 \times 3 + 1$ $T_4 = 5 \times 4 + 1$

     $= 6$ $= 11$ $= 16$ $= 21$

     The first four terms are 6, 11, 16 and 21.

(b) $T_{100} = 5 \times 100 + 1 = 501$

Exercise 22.1

1   Look at this sequence of circles. The first four patterns in the sequence have been drawn.

   (a)   Describe the position-to-term rule for this sequence.

   (b)   How many circles are there in the 100th pattern?

2   Look at this sequence of matchstick patterns.

   (a)   Copy and complete the table.

Pattern number	1	2	3	4	5
Number of matchsticks					

   (b)   What patterns can you see in the numbers?

   (c)   Find the number of matchsticks in the 50th pattern.

3   Here is a sequence of star patterns.

   (a)   Draw the next pattern in the sequence.

   (b)   Without drawing the pattern, find the number of stars in the 8th pattern. Explain how you found your answer.

Exercise continues ...

4 The numbers in a sequence are given by this rule:
Multiply the position number by 3, then subtract 5.
(a) Show that the first term of the sequence is -2.
(b) Find the next four terms in the sequence.

5 Find the first four terms of the sequences with these nth terms.
(a) $6n - 2$ (b) $4n + 1$ (c) $6 - 2n$

6 Find the first five terms of the sequences with these nth terms.
(a) n^2 (b) $n^2 + 2$ (c) n^3

7 The first term of a sequence is 2.
The general rule for the sequence is multiply a term by 2 to get to the next term.
Write down the first five terms of the sequence.

8 For a sequence, $T_1 = 5$ and $T_{n+1} = T_n - 3$.
Write down the first four terms of the sequence.

9 Draw suitable patterns to represent the sequence $1, 5, 9, 13, \ldots$.

10 Draw suitable patterns to represent the sequence $1, 4, 9, 16, \ldots$.

● Challenge 22.1

Draw the first three patterns of a matchstick or dot sequence.
Write on the back of the paper how you would continue the pattern.
Write also how many matchsticks or dots there are in each pattern.

Give the first three patterns to someone else and see if they can continue your sequence.

Give the first three numbers of your sequence to a different person and see if they can continue your sequence.

Finding the nth term of a linear sequence

Look at this linear sequence.

$$4 \quad 9 \quad 14 \quad 19 \quad 24 \quad \ldots$$
$$+5 \quad +5 \quad +5 \quad +5$$

To get from one term to the next, you add 5 each time. Another way of saying this is that there is a **common difference** between the terms, 5.

A sequence like this, which has a common difference, is called a **linear sequence**.

If you plot the terms of a linear sequence on a graph, you get a straight line.

Term (c)	Value (y)
1	4
2	9
3	14
4	19
5	24

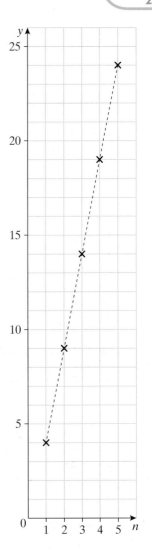

As 5 is added each time,

$$T_2 = T_1 + 5$$
$$= 4 + 5$$
$$T_3 = 4 + 5 \times 2$$
$$T_4 = 4 + 5 \times 3, \text{etc.}$$

So $\quad T_n = 4 + 5(n - 1)$
$$= 5n - 1$$

The nth term of this sequence is $5n - 1$.

Now look at some other linear sequences.

Sequence	Common difference	First term − Common difference	nth term
4, 7, 10, 13, ...	3	$4 - 3 = 1$	$3n + 1$
6, 11, 16, 21, ...	5	$6 - 5 = 1$	$5n + 1$
10, 6, 4, −2, ...	−4	$10 - (-4) = 14$	$-4n + 14$
2, 4, 6, 8, ...	2	$2 - 2 = 0$	$2n$
4, 10, 16, 22, ...	6	$4 - 6 = -2$	$6n - 2$

Looking at the patterns in the table, you can see some evidence for this formula.

nth term of a linear sequence = Common difference × n + (First term − Common difference)

This can be written as

$$n\text{th term} = An + b$$

where A represents the common difference and b is the first term minus A.

You can also find b by comparing An with any term in the sequence.

Example 22.4

Find the nth term of this sequence: $4, 7, 10, 13, \ldots$.

Solution

$$
\begin{array}{cccc}
4 & 7 & 10 & 13 \\
\end{array} \quad \ldots
$$
$$
+3 \quad +3 \quad +3
$$

The common difference (A) is 3, so the formula contains $3n$.

When $n = 1$, $3n = 3$. The first term is actually 4, which is 1 more. So the nth term is $3n + 1$.

You can check your answer using a different term.

When $n = 2$, $3n = 6$. The second term is actually 7, which is 1 more. This confirms that the nth term is $3n + 1$.

You can use sequences and position-to-term rules to solve problems.

Example 22.5

Lucy has ten CDs. She decides to buy three more CDs each month.

(a) Copy and complete the table to show the number of CDs Lucy has after each of the first four months.

Number of months	1	2	3	4
Number of CDs				

(b) Find the formula for the number of CDs she will have after n months.

(c) After how many months will Lucy have 58 CDs?

Solution

(a)

Number of months	1	2	3	4
Number of CDs	13	16	19	22

(b) nth term $= An + b$
   $A = 3$ A is the common difference.
   $b = 10$ b is the first term minus the common difference.
   nth term $= 3n + 10$

(c) $3n + 10 = 58$ Solve the equation to find n when the nth term is 58.
   $3n = 48$
   $n = 16$

   Lucy will have 58 CDs after 16 months.

Some special sequences

You have already met some special sequences.

⊙ Discovery 22.1

Look at each of these sequences.

Even numbers	2, 4, 6, 8, …
Odd numbers	1, 3, 5, 7, …
Multiples of 4	4, 8, 12, 16, …
Powers of 2	2, 4, 8, 16, …
Powers of 10	10, 100, 1000, 10 000, …
Square numbers	1, 4, 9, 16, …
Triangular numbers	1, 3, 6, 10, …

Look for different patterns in each of the sequences.

(a) Describe the term-to-term rule.

(b) Describe the position-to-term rule.

For triangular numbers, you may find it helpful to look at these diagrams.

```
                        *   *   *
            *   *       *   *   *
*           *   *       *   *   *
*           *   *       *   *   *
```

Exercise 22.2

1 Find the nth term for each of these sequences.
   (a) 5, 7, 9, 11, 13, …   (b) 2, 5, 8, 11, 14, …   (c) 7, 8, 9, 10, 11, …

2 Find the nth term for each of these sequences.
   (a) 17, 14, 11, 8, 5, …   (b) 5, 0, −5, −10, −15, …   (c) 0, −1, −2, −3, −4, …

3 Which of these sequences are linear?
   Find the next two terms of each of the sequences that are linear.
   (a) 5, 8, 11, 14, …   (b) 2, 4, 7, 11, …   (c) 6, 12, 18, 24, …   (d) 2, 6, 18, 54, …

4 (a) Write the first five terms of the sequence with nth term $12 - 6n$.
   (b) Write the nth term of this sequence: 8, 2, −4, −10, −16, …

Exercise continues …

5 A theatre agency charges £15 per ticket, plus an overall booking charge of £2.

(a) Copy and complete the table.

Number of tickets	1	2	3	4
Cost in £				

(b) Write an expression for the cost, in pounds, of n tickets.

(c) Jenna pays £107 for her tickets. How many does she buy?

6 Write down the first ten triangular numbers.

7 The nth triangular number is $\dfrac{n(n + 1)}{2}$. Find the 20th triangular number.

8 The nth term of a sequence is 10^n.

(a) Write down the first five terms of this sequence.

(b) Describe this sequence.

9 (a) Write down the first five square numbers.

(b) (i) Compare the following sequence with the sequence of square numbers.

4, 7, 12, 19, 28, …

(ii) Write down the nth term of this sequence.

(iii) Find the 100th term of this sequence.

10 (a) Compare the following sequence with the sequence of square numbers.

3, 12, 27, 48, 75, …

(b) Write down the nth term of this sequence.

(c) Find the 20th term of this sequence.

● Challenge 22.2

Work in pairs.

You are going to use a spreadsheet to explore sequences. Don't let your partner see you input your formula. Make sure that they can't see the formula on the computer screen: click on View in the toolbar and make sure the Formula Bar is not checked.

1 Open a new spreadsheet.

2 Enter the number 1 in cell A1.

Click on cell A1, and hold down the mouse key and drag down the column. Then click on Edit in the toolbar and select Fill, then Series to call up the Series dialogue box. Make sure that the Columns and Linear boxes are checked and that the Step value is 1. Click OK.

3 Enter a formula in cell B1. For example, **=A1*3+5** or **=A1^2**. Press the enter key.
Click on cell B1, click on Edit in the toolbar and select Copy.
Click on cell B2, and hold down the mouse key and drag down the column. Then click on Edit in the toolbar and select Paste.

4 Ask your partner to try to work out the formula and generate the same sequence in column C.

If you have time, you could explore some non-linear sequences as well. For example, enter the formula **=A1^2+A1**.

What you have learned

- Sequences may be described by a list of numbers, diagrams in a pattern, a term-to-term rule (for example, $T_{n+1} = T_n + 3$ when $T_1 = 4$) or a position-to-term rule (for example, nth term $= 3n + 1$ or $T_n = 3n + 1$)
- The nth term of a linear sequence $= An + b$, where A is the common difference and b is the first term minus A
- Here are some important sequences.

Name	Sequence	nth term	Term-to-term rule
Even numbers	2, 4, 6, 8, …	$2n$	Add 2
Odd numbers	1, 3, 5, 7, …	$2n - 1$	Add 2
Multiples e.g. multiples of 6	6, 12, 18, 24, …	$6n$	Add 6
Powers of 2	2, 4, 8, 16, …	2^n	Multiply by 2
Powers of 10	10, 100, 1000, 10 000, …	10^n	Multiply by 10
Square numbers	1, 4, 9, 16, …	n^2	Add 3 then 5 then 7, etc. (the odd numbers)
Triangular numbers	1, 3, 6, 10, …	$\dfrac{n(n + 1)}{2}$	Add 2 then 3 then 4, etc. (consecutive integers)

Mixed exercise 22

1 Look at this sequence of circles. The first four patterns in the sequence have been drawn.

(a) How many circles are there in the 100th pattern?

(b) Describe a rule for this sequence.

Mixed exercise 22 continues …

2 Here is a sequence of star patterns.

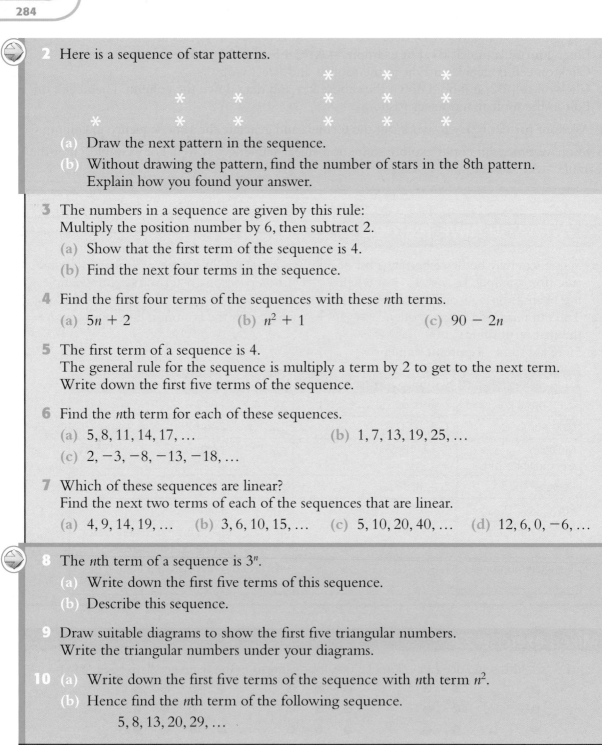

(a) Draw the next pattern in the sequence.

(b) Without drawing the pattern, find the number of stars in the 8th pattern. Explain how you found your answer.

3 The numbers in a sequence are given by this rule:
Multiply the position number by 6, then subtract 2.

(a) Show that the first term of the sequence is 4.

(b) Find the next four terms in the sequence.

4 Find the first four terms of the sequences with these nth terms.

(a) $5n + 2$ (b) $n^2 + 1$ (c) $90 - 2n$

5 The first term of a sequence is 4.
The general rule for the sequence is multiply a term by 2 to get to the next term.
Write down the first five terms of the sequence.

6 Find the nth term for each of these sequences.

(a) $5, 8, 11, 14, 17, \ldots$ (b) $1, 7, 13, 19, 25, \ldots$

(c) $2, -3, -8, -13, -18, \ldots$

7 Which of these sequences are linear?
Find the next two terms of each of the sequences that are linear.

(a) $4, 9, 14, 19, \ldots$ (b) $3, 6, 10, 15, \ldots$ (c) $5, 10, 20, 40, \ldots$ (d) $12, 6, 0, -6, \ldots$

8 The nth term of a sequence is 3^n.

(a) Write down the first five terms of this sequence.

(b) Describe this sequence.

9 Draw suitable diagrams to show the first five triangular numbers.
Write the triangular numbers under your diagrams.

10 (a) Write down the first five terms of the sequence with nth term n^2.

(b) Hence find the nth term of the following sequence.

$5, 8, 13, 20, 29, \ldots$

Constructions 2

▶ This chapter is about

- Constructing the perpendicular bisector of a line segment
- Constructing the perpendicular from a point on a line
- Constructing the perpendicular from a point to a line
- Constructing the bisector of an angle
- Knowing that a locus is a line, a curve or a region of points
- Knowing the four basic loci results
- Locating a locus of points that follow a given rule or rules

▶ You should already know

- How to use a protractor and compasses
- How to make scale drawings
- How to construct a triangle given three facts about its sides and angles

Constructions

You learned in Chapter 11 how to **construct** angles and triangles. You can use these skills in other constructions.

Four important constructions

You need to know four important constructions.

Construction 1: The perpendicular bisector of a line segment

Use the following method to construct the **perpendicular bisector** of line segment AB.

You learned in Chapter 20 that a line segment is the part of a line between two points and has a finite length; however, a line segment is often called a line for short.

TIP

Perpendicular means 'at right angles to'. A *bisector* is something that divides into 'two equal parts'.

1 Open your compasses to a radius more than half the length of the line AB. Put the point of your compasses on A. Draw one arc above the line, and one arc below.

2 Keep your compasses open to the same radius. Put the compass point on B. Draw two more arcs, cutting the first arcs at P and Q.

3 Join the points P and Q. This line divides AB into two equal parts and is at right angles to AB.

⊙ Discovery 23.1

(a) (i) Draw a triangle. Make it big enough to fill about half of your page.
   (ii) Construct the perpendicular bisector of each of the three sides.
   (iii) If you have drawn them accurately enough, the bisectors should meet at one point. Put your compass point on this point, and the pencil on one of the corners of the triangle. Draw a circle.

(b) You have drawn the **circumcircle** of the triangle. What do you notice about this circle?

Construction 2: The perpendicular from a point on a line

Use the following method to construct the perpendicular from point P on the given line.

1 Open your compasses to any radius. Put the compass point on P. Draw an arc on each side of P, cutting the line at Q and R.

2 Open your compasses to a larger radius. Put the compass point on Q. Draw an arc above the line. Now put the compass point on R and draw another arc, with the same radius, cutting the first arc at X.

3 Join the points P and X. This line is at right angles to the original line.

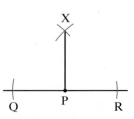

○ Discovery 23.2

(a) Draw a line 10 cm long. Label it AB.

(b) At A, draw a circle of radius 5 cm.
Label the point where the circle crosses the line P.

(c) Construct the perpendicular from P.
This perpendicular is the **tangent** to the circle at P.

Construction 3: The perpendicular from a point to a line

Use the following method to construct the perpendicular from point P to the given line.

1 Open your compasses to any radius.
Put the compass point on P.
Draw two arcs, cutting the line at Q and R.

2 Keep your compasses open to the same radius.
Put the compass point on Q.
Draw an arc below the line.
Now put the compass point on R and draw another arc, cutting the first arc at X.

3 Line up your ruler with points P and X.
Draw the line PM.
This line is at right angles to the original line.

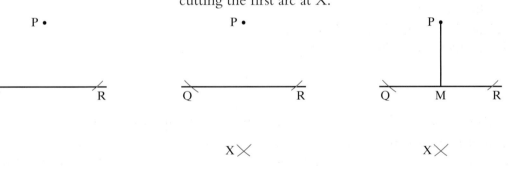

○ Discovery 23.3

(a) (i) Draw a line across your page. Put a cross on one side of the line and label it P.
(ii) Construct the perpendicular from P to the line. Make sure you keep your compasses open to the same radius all the time.
This time, join P to X, don't stop at the original line.

(b) (i) Measure PM and XM.
(ii) What do you notice?
What can you say about P and X?

Construction 4: The bisector of an angle

Use the following method to construct the bisector of the given angle.

1 Open your compasses to any radius.
Put the compass point on A. Draw two arcs, cutting the 'arms' of the angle at P and Q.

2 Open your compasses to any radius.
Put the compass point on P. Draw an arc inside the angle.
Now put the compass point on Q and draw another arc, with the same radius, cutting the first arc at X.

3 Join the points A and X. This line divides the given angle into two equal parts.

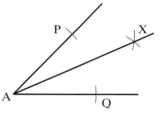

Discovery 23.4

(a) (i)   Draw a triangle. Make it big enough to fill about half of your page.
   (ii)   Construct the bisector of each of the three angles.
   (iii) If you have drawn them accurately enough, the bisectors should meet at one point. Label this point A.
          Construct the perpendicular from A to one of the sides of the triangle. Label the point where the perpendicular meets the side of the triangle B.
   (iv)   Put your compass point on A, and the pencil on point B. Draw a circle.

(b) You have drawn the **incircle** of the triangle.
   (i)   What do you notice about this circle?
   (ii)   What can you say about the side of the triangle to which you have drawn the perpendicular from A?
          Hint: Look at your diagram from Discovery 23.2.
   (iii) What can you say about each of the other two sides of the triangle?

Constructing a locus

A **locus** is a line, curve or region of points that satisfy a certain rule.

The plural of locus is **loci**.

Four important loci

You need to know four important loci.

Locus 1: The locus of points that are the same distance from a given point

The locus of points that are 2 cm from point A is a circle, with centre A and radius 2 cm.

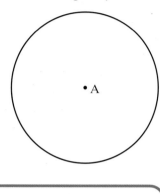

> **TIP**
>
> The locus of points that are less than 2 cm from A is the region inside the circle.
> The locus of points that are more than 2 cm from A is the region outside the circle.

> **TIP**
>
> When trying to identify a particular locus, find several points that satisfy the required rule and see what sort of line, curve or region they form.

Locus 2: The locus of points that are the same distance from two given points

The locus of points that are the same distance from the points A and B is the perpendicular bisector of the line joining A and B.

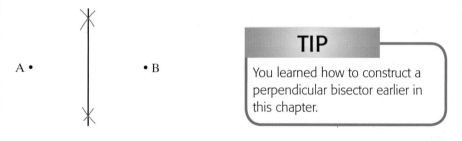

> **TIP**
>
> You learned how to construct a perpendicular bisector earlier in this chapter.

Locus 3: The locus of points that are the same distance from two, given, intersecting lines

The locus of points that are the same distance from AB and AC is the bisector of the angle BAC.

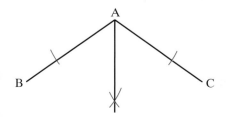

> **TIP**
>
> You learned how to construct the bisector of an angle earlier in this chapter.

Locus 4: The locus of points that are the same distance from a given line

The locus of points that are 3 cm from the line AB is a pair of lines, parallel to AB and 3 cm away from it on either side; at each end of the line there is a semicircle with centre A or B and radius 3 cm.

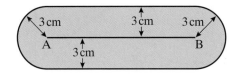

You can use constructions and loci to solve problems.

Example 23.1

Two towns, P and Q, are 5 km apart.
Toby lives exactly the same distance from P as from Q.

(a) Construct the locus of where Toby could live.
   Use a scale of 1 cm to 1 km.

Toby's school is closer to P than to Q.

(b) Shade the region where Toby's school could be.

Solution

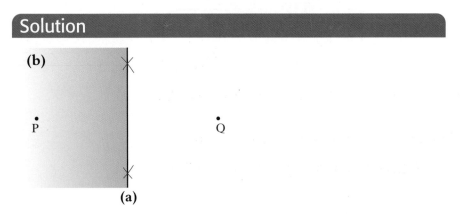

The locus of where Toby lives is the perpendicular bisector of the line joining P and Q.

Any point to the left of the line drawn in part **(a)** is closer to point P than to point Q. Any point to the right of the line is closer to point Q than to point P.

Example 23.2

A security light is attached to a wall.
The light illuminates an area up to 20 m.

Construct the region illuminated by the light.
Use a scale of 1 cm to 5 m.

Solution

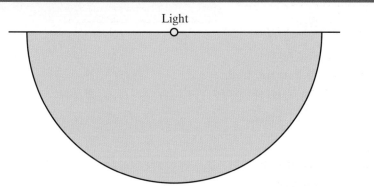

Remember that the light cannot illuminate the area behind the wall.

Example 23.3

The diagram shows a port, P, and some rocks.

To leave the port safely, a boat must keep the same distance from each set of rocks.

Copy the diagram and construct the path of a boat from the port.

Solution

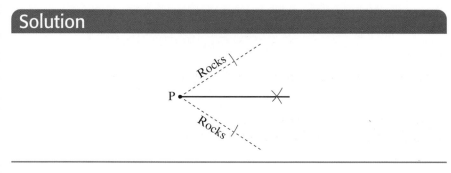

Exercise 23.1

1   Draw the locus of points that are less than 5 cm from a fixed point A.

2   Two rocks are 100 m apart. A boat passes between the rocks so that it is always the same distance from each of them.
Construct the locus of the path of the boat. Use a scale of 1 cm to 20 m.

3   In a farmer's field, a tree is 60 m from a long hedge.
The farmer decides to build a fence between the tree and the hedge.
The fence must be as short as possible.
Make a scale drawing of the tree and the hedge.
Construct the locus of where the fence must be built.
Use a scale of 1 cm to 10 m.

4   Draw an angle of 60°.
Construct the bisector of the angle.

5   Draw a square, ABCD, with side 5 cm.
Draw the locus of points inside the square that are less than 3 cm from corner C.

6   A rectangular shed measures 4 m by 2 m.
A path, 1 m wide and perpendicular to the shed, is to be built from the door of the shed.
Construct the locus showing the edges of the path.
Use a scale of 1 cm to 1 m.

7   Construct a compass like the one opposite.
   • Draw a circle with radius 4 cm.
   • Draw a horizontal diameter of the circle.
   • Construct the perpendicular bisector of the diameter.
   • Bisect each of the four angles.

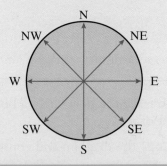

8   Draw a line 7 cm long.
Construct the region of points that are less than 3 cm from the line.

9   Construct a triangle ABC with AB = 8 cm, AC = 7 cm and BC = 6 cm.
Shade the locus of points inside the triangle that are closer to AB than to AC.

10   The owner of a theme park decides to build a moat 20 m wide around a castle.
The castle is a rectangle measuring 80 m long by 60 m wide.
Draw accurately an outline of the castle and the moat.

Intersecting loci

Often a locus is defined by more than one rule.

Example 23.4

Two points, P and Q, are 5 cm apart.

Find the locus of points that are less than 3 cm from P and equidistant from P and Q.

Solution

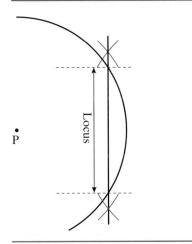

The locus of points that are less than 3 cm from P is within a circle with centre P and radius 3 cm.

The locus of points that are equidistant from P and Q is the perpendicular bisector of the line joining P and Q.

The points that satisfy both rules lie within the circle and on the line.

Example 23.5

A rectangular garden is 25 m long and 15 m wide.

A tree is to be planted in the garden so that it is more than 2.5 m from the boundary and less than 10 m from the south-west corner.

Using a scale of 1 cm to 5 m, make a scale drawing of the garden and find the region where the tree can be planted.

Solution

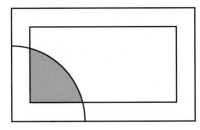

Draw a rectangle measuring 5 cm by 3 cm to represent the garden.

The locus of points more than 2.5 m from the boundary is a smaller rectangle inside the first. Each side of the smaller rectangle is 0.5 cm inside the sides of the larger rectangle.

The locus of points less than 10 m from the south–west corner is a sector with the corner as its centre and radius 2 cm.

The points that satisfy both rules lie within the smaller rectangle and the sector.

Exercise 23.2

1 Draw a point and label it A.
Draw the region of points that are more than 3 cm from A but less than 6 cm from A.

2 Two towns, P and Q, are 7 km apart.
Amina wants to buy a house that is within 5 km of P and also within 4 km of Q.
Make a scale drawing to indicate the region where Amina could buy a house.
Use a scale of 1 cm to 1 km.

3 Draw a square ABCD with side 5 cm.
Find the region of points that are more than 3 cm from AB and AD.

4 Steve is using an old map to locate treasure.
He is searching a rectangular plot of land, EFGH, which measures 8 m by 5 m.

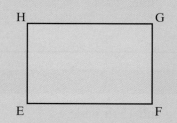

The map says that the treasure is hidden 6 m from E on a line equidistant from F and H.
Using a scale of 1 cm to 1 m, make a scale drawing to locate the treasure.
Mark the position where the treasure is hidden with the letter T.

5 Construct triangle ABC with AB = 11 cm, AC = 7 cm and BC = 9 cm.
Construct the locus of points, inside the triangle, that are closer to AB than to AC and equidistant from A and B.

6 The diagram shows the corner of a farm building, which stands in a field.
A donkey is tethered at a point D with a rope 5 m long.
Shade the region of the field that the donkey can graze.

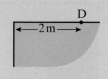

Exercise continues...

7   A sailor is shipwrecked at night. She is 140 m from the straight coastline.
    She swims straight for the shore.

    (a) Make a scale drawing of the path she swims.

    The coastguard is standing on the beach exactly where the sailor will come ashore.
    He has a searchlight that can illuminate up to a distance of 50 m.

    (b) Mark on your diagram the part of the sailor's swim that will be lit up.

8   The positions of three radio stations, A, B and C, form a triangle such that AB = 7 km,
    BC = 9.5 km and angle ABC = 90°.
    The signal from each radio station can be received up to 5 km away.
    Make a scale drawing to locate the region where none of the three radio stations can
    be received.

9   A rectangular garden measures 20 m by 14 m. The house wall is along one of the
    shorter sides of the garden.
    Pete is going to plant a tree. It must be more than 10 m from the house and more than
    8 m from any corner of the garden.
    Find the region of the garden where the tree can be planted.

10 Two towns, H and K, are 20 miles apart.
    A new leisure centre is to be built within 15 miles of H, but nearer to K than to H.
    Using a scale of 1 cm to 5 miles, draw a diagram to show where the leisure centre
    could be built.

What you have learned

- How to construct the perpendicular bisector of a line segment
- How to construct the perpendicular from a point on a line
- How to construct the perpendicular from a point to a line
- How to construct the bisector of an angle
- That the locus of points that are the same distance from a given point is a circle
- That the locus of points that are the same distance from two points is the perpendicular
  bisector of the line segment joining the two points
- That the locus of points which are the same distance from two intersecting lines is the
  bisector of the angle formed by the two lines
- That some loci must satisfy more than one rule

Mixed exercise 23

1   Draw an angle of 100°.
    Construct the bisector of the angle.

2   Draw a line 8 cm long.
    Construct the locus of the points that are the same distance from each end of the line.

3   Draw a line 7 cm long.
    Draw the locus of points which are 3 cm away from this line.

4   Draw the triangle ABC with AB = 9 cm, BC = 8 cm and CA = 6 cm.
    Construct the perpendicular line from C to AB.
    Measure the length of this line and hence work out the area of the triangle.

5   Draw the rectangle PQRS with PQ = 8 cm and QR = 5 cm.
    Shade the region of points which are nearer to QP than to QR.

6   Two radio stations are 40 km apart. Each station can transmit signals up to 30 km.
    Construct a scale drawing to show the region which can receive signals from both
    radio stations.

7   A garden is a rectangle ABCD with AB = 5 m and BC = 3 m.
    A tree is planted so that it is within 5 m of A and within 3 m of C.
    Indicate the region where the tree could be planted.

8   Draw triangle EFG with EF = 8 cm, EG = 6 cm and angle E = 70°.
    Construct the point which is equidistant from F and G and is also 5 cm from G.

9   A lawn is a square with side 5 m.
    A water sprinkler covers a circle of radius 3 m.
    If the gardener puts the sprinkler at each corner, will the whole lawn get watered?

10 The gardener in question 9 loans his water sprinkler to a neighbour.
    The neighbour has a large garden with a rectangular lawn measuring 10 m by 8 m.
    She moves the sprinkler slowly around the edge of the lawn. Draw a scale diagram to
    show the region of the garden which will be watered.

Chapter 24

Rearranging formulae

▶ This chapter is about

- Rearranging formulae

▶ You should already know

- How to simplify and solve linear equations
- How to simplify an expression by, for example, collecting together 'like' terms
- How to substitute numbers into a formula

Rearranging formulae

Sometimes you need to find the value of a letter which is not on the left-hand side of the formula.

To find the value of the letter, you first need to **rearrange** the formula.

For example, the formula $d = st$ links distance (d), speed (s) and time (t). If you know the distance covered during a journey and the time it took, and want to find the average speed, you need to get the s by itself.

The method you use to rearrange a formula is similar to the method you use to solve an equation.

You learned how to solve equations in Chapters 6 and 17.

In this case, to get the s on its own, you need to divide by t. As with equations, you must do each operation to the whole of both sides of the formula.

$d = st$

$\dfrac{d}{t} = \dfrac{st}{t}$ Divide both sides by t.

$\dfrac{d}{t} = s$

$s = \dfrac{d}{t}$ A formula is usually written with the single term (in this case, s) on the left-hand side.

s is now the **subject** of the formula.

The formula gives s in terms of d and t.

Example 24.1

$y = mx + c$

Make x the subject.

Solution

$$y = mx + c$$
$$y - c = mx + c - c \quad \text{Subtract } c \text{ from both sides.}$$
$$y - c = mx$$
$$\frac{y - c}{m} = \frac{mx}{m} \qquad \text{Divide both sides by } m.$$
$$\frac{y - c}{m} = x$$
$$x = \frac{y - c}{m} \qquad \text{Swap sides so that } x \text{ is on the left-hand side.}$$

Example 24.2

The formula for the volume, v, of a square-based pyramid of side a is $v = \frac{1}{3}a^2h.$

Rearrange the formula to make h the subject.

Solution

$$v = \frac{1}{3}a^2h$$
$$3v = a^2h \qquad \text{Get rid of the fraction first by multiplying both sides by 3.}$$
$$\frac{3v}{a^2} = h \qquad \text{Divide both sides by } a^2.$$
$$h = \frac{3v}{a^2} \qquad \text{Swap sides so that } h \text{ is on the left-hand side.}$$

Example 24.3

Rearrange the formula $A = \pi r^2$ to make r the subject.

Solution

$$A = \pi r^2$$
$$\frac{A}{\pi} = r^2 \qquad \text{Divide both sides by } \pi.$$
$$r^2 = \frac{A}{\pi} \qquad \text{Swap sides so that } r^2 \text{ is on the left-hand side.}$$
$$r = \sqrt{\frac{A}{\pi}} \qquad \text{Take the square root of both sides.}$$

Exercise 24.1

1   Rearrange each of these formulae to make the letter in brackets the subject.

(a) $a = b - c$ (b)

(b) $4a = wx + y$ (x)

(c) $v = u + at$ (t)

(d) $c = p - 3t$ (t)

(e) $a = p(q + r)$ (q)

(f) $p = 2g - 2f$ (g)

(g) $F = \dfrac{m + 4n}{t}$ (n)

2   Make u the subject of the formula $s = \dfrac{3uv}{bn}$.

3   Rearrange the formula $a = \dfrac{bh}{2}$ to give h in terms of a and b.

4   The formula for calculating simple interest is $I = \dfrac{PRT}{100}$.

Make R the subject of this formula.

5   The volume of a cone is given by the formula $V = \dfrac{\pi r^2 h}{3}$ where V is the volume in cm^3, r is the radius of the base in centimetres and h is the height in centimetres.

(a) Rearrange the formula to make h the subject.

(b) Calculate the height of a cone with radius 5 cm and volume 435 cm³.
Use $\pi = 3.14$ and give your answer correct to 1 decimal place.

6   To change from degrees Celsius (°C) to degrees Fahrenheit (°F), you can use the formula

$$F = \tfrac{9}{5}(C + 40) - 40.$$

(a) The temperature is 60°C.
What is this in degrees Fahrenheit?

(b) Rearrange the formula to find C in terms of F.

7   Rearrange the formula $V = \dfrac{\pi r^2 h}{3}$ to make r the subject.

8   (a) Make a the subject of the formula $v^2 = u^2 + 2as$.

(b) Make u the subject of the formula $v^2 = u^2 + 2as$.

What you have learned

- To rearrange a formula, do each operation to the whole of both sides of the formula until you get the required term on its own, on the left-hand side of the formula

Mixed exercise 24

1 Rearrange each of these formulae to make the letter in brackets the subject.

(a) $p = q + 2r$ (q)

(b) $x = s + 5r$ (r)

(c) $m = \dfrac{pqr}{s}$ (r)

(d) $A = t(x - 2y)$ (y)

2 The cooking time, T minutes, for w kilograms of meat is given by the formula $T = 45w + 40$.

(a) Make w the subject of the formula.

(b) What is the value of w when the cooking time is 2 hours 28 minutes?

3 The area of a triangle is given by the formula $A = b \times h \div 2$, where b is the base and h is the height.

(a) Find the length of the base when $A = 12\,\text{cm}^2$ and $h = 6\,\text{cm}$.

(b) Find the height when $A = 22\,\text{cm}^2$ and $b = 5.5\,\text{cm}$.

4 The cost in £ of an advert in a local paper is given by the formula $C = 12 + \dfrac{w}{5}$, where w is the number of words in the advert.

How many words can you have if you are willing to pay

(a) £18?

(b) £24?

Unit B Contents

Chapter 1

Working with numbers

▶ **This chapter is about**

- Ordering integers
- Integer complements
- Recalling multiplication facts to 10×10 and deriving the corresponding division facts
- Adding and subtracting integers
- Adding and subtracting decimals
- Multiplying and dividing decimals

▶ **You should already know**

- How to add, subtract, multiply and divide integers
- How to multiply and divide numbers by 10, 100, 1000, …
- How to multiply and divide an integer by a decimal

Ordering integers

Integer is another word for a whole number. When you put integers into order you have to think about the place value of each digit. The digit on the left of each number has the largest value.

Example 1.1

Put these numbers in order of size, smallest first.

3421   412   3146   598   94

Solution

The smallest number is the number that has only tens and units: 94.

Then come the numbers that also have hundreds: 412 and 598.
Order these by looking at the hundreds digit.
Start with the smaller hundreds digit: 412, 598.

Then come the numbers that have thousands: 3421 and 3146.
They both have the same thousands digit so look at the hundreds digit.
Start with the smaller hundreds digit: 3146, 3421.

So the order is: 94   412   598   3146   3421

● Challenge 1.1

Write down six whole numbers.

Swap them with a friend.

Put your friend's numbers in order.

See who can do it first!

What makes it difficult to order whole numbers?

Exercise 1.1

1 Put these whole numbers in order of size, smallest first.

(a) 412	2179	57	361	21	4215
(b) 462	321	197	358	426	411
(c) 10 425	5427	3704	5821	6146	1256
(d) 89 125	39 171	4621	59 042	6317	9981
(e) 120 000	102 020	134 050	104 210	152 104	102 002
(f) 124	1792	75	631	12	415
(g) 9425	4257	7034	5218	6641	1611
(h) 86 525	34 771	54 621	9042	317	91 981
(i) 1 030 504	1 020 504	1 040 501	1 060 504	1 010 701	1 050 403

2 For each group of numbers write down the smallest and largest numbers that can be made with all the digits using each digit once only.

(a) 6, 1, 3, 4, 2   (b) 8, 4, 1, 2, 9   (c) 2, 5, 0, 4, 6, 9

3 Each of these groups of numbers are in order of size.
   Write each group out with an extra number between each one.

(a) 25	78	123	132	156	201	2001	5000
(b) 523	525	529	531	555	565	585	588
(c) 9	99	999	9999	99 999	999 999		

Integer complements

How many do you need to add to 4 to get 10?
The answer is 6 as $4 + 6 = 10$. The number 6 is the **complement to 10** of 4. It is useful in mental arithmetic. What is the complement to 10 of 2? The answer is 8.

Example 1.2

How many do you need to add to 61 to make 100?

Solution

$61 + 9 = 70$ First make 61 up to the next multiple of ten.
$70 + 30 = 100$ Then make 70 up to 100.
$9 + 30 = 39$ You need to add 39 to 61 to make 100.

Alternatively, this is the same as asking $100 - 61$.
$100 - 61 = 100 - 60 - 1$ Split the second number into tens and units.
$\qquad\quad = 40 - 1 = 39$

---

39 is the **complement to 100** of 61.

🔍 Check up 1.1

What is the complement to 100 of 43?

Exercise 1.2

1 How many do you need to add to these numbers to make 10?
   (a) 1   (b) 7   (c) 5

2 How many do you need to add to these numbers to make 100?
   (a) 44   (b) 58   (c) 27

3 Write down the complements to 10 of these numbers.
   (a) 2   (b) 4   (c) 9

4 Write down the complement to 100 of these numbers.
   (a) 84   (b) 93   (c) 23   (d) 54
   (e) 35   (f) 68   (g) 4

5 (a) What is the complement to 10 of 6?
   (b) What is $32 - 6$?
   (c) Add the complement to 10 of 6 to 22.
   (d) What can you say about your answers to parts (b) and (c)?

6 (a) What is the complement to 100 of 79?

(b) What is 131 − 79?

(c) Add the complement to 100 of 79 to 31.

(d) What can you say about your answers to parts (b) and (c)?

7 Use complements to work out these.

(a) 41 − 8   (b) 50 − 4   (c) 83 − 49   (d) 143 − 88   (e) 256 − 97

● Challenge 1.2

(a) Write down the complement to 100 of 94.

(b) Add 100 to 268 and subtract your answer to part (a).

(c) What other calculation have you done?

(d) Use complements to do these additions.

(i)   56 + 88   (ii) 154 + 96   (iii) 23 + 99   (iv) 341 + 199

● Challenge 1.3

Produce a poster to show how complements can be used in addition and subtraction.

Multiplication facts

◎ Check up 1.2

You need to know your times tables.

Copy and complete this multiplication grid.

From your table, check that 7 × 8 and 8 × 7 give the same answer.

TIP

The order of the numbers does not matter when multiplying.

For example, 7 × 8 = 8 × 7 = 56.

×	1	2	3	4	5	6	7	8	9	10
1										
2	2	4	6	8	10	12	14	16	18	20
3										
4										
5										
6										
7										
8										
9										
10										

Exercise 1.3

Work out these.

1 4 × 3	2 3 × 4	3 9 × 5	4 5 × 9
5 7 × 5	6 6 × 2	7 8 × 6	8 3 × 9
9 6 × 6	10 5 × 3	11 7 × 4	12 8 × 5
13 6 × 3	14 9 × 4	15 5 × 10	16 8 × 7
17 7 × 7	18 3 × 6	19 6 × 9	20 7 × 9

● Challenge 1.4

At an indoor football tournament there are eight teams of seven players.
How many players is this altogether?

Division facts

● Discovery 1.1

Explain how to use the grid in Check up 1.2 to work out these.
(a) 56 ÷ 8   (b) 63 ÷ 7   (c) 28 ÷ 4

Exercise 1.4

Work out these.

1 18 ÷ 3	2 3)‾12‾	3 15 ÷ 5	4 16 ÷ 4
5 25 ÷ 5	6 2)‾14‾	7 40 ÷ 8	8 36 ÷ 4
9 27 ÷ 9	10 6)‾54‾	11 48 ÷ 6	12 8)‾56‾
13 42 ÷ 6	14 7)‾63‾	15 36 ÷ 9	16 2)‾16‾
17 28 ÷ 4	18 48 ÷ 8	19 49 ÷ 7	20 8)‾32‾

● Challenge 1.5

Jasmine shared 63 sweets equally between 7 people.
How many did they each receive?

Challenge 1.6

Like addition and subtraction, multiplication and division are connected.

The numbers 24, 6 and 4 can be connected by $4 \times 6 = 24$, $24 \div 4 = 6$ and $24 \div 6 = 4$.

For each of these sets of three numbers, write down three different calculations.

(a) $5, 7, 35$ (b) $63, 9, 7$ (c) $9, 45, 5$

(d) $36, 9, 4$ (e) $8, 72, 9$

Adding and subtracting integers mentally

There are various ways to do addition and subtraction calculations mentally.

Challenge 1.7

Work in pairs.

Write some methods to do addition and subtraction questions in your head.

Partitioning

Example 1.3

Work out these.

(a) $123 + 456$ (b) $358 + 234$

Solution

(a) 456 is $400 + 50 + 6$ Split the second number into hundreds, tens and units.

$123 + 400 = 523$ Add the three parts on separately.
$523 + 50 = 573$
$573 + 6 = 579$

(b) 234 is $200 + 30 + 4$

$358 + 200 = 558$
$558 + 30 = 588$
$588 + 4 = 592$

Example 1.4

Work out 545 − 324.

Solution

324 is 300 + 20 + 4.

545 − 300 = 245	First take away 300.
245 − 20 = 225	Then take away 20.
225 − 4 = 221	Finally take away 4.

Using complements

Example 1.5

Work out 238 + 94.

Solution

The complement to 100 of 94 is 6.
So an easier way is to add 100 and subtract the complement, which is 6.

238 + 100 = 338
338 − 6 = 332
So 238 + 94 = 332.

Example 1.6

Work out 312 − 67.

Solution

The complement to 100 of 67 is 33.
So an easier way is to subtract 100 and add the complement, which is 33.

312 − 100 = 212
212 + 33 = 245
So 312 − 67 = 245.

Counting on

Example 1.7

Work out $132 - 59$.

Solution

The number line shows the route between 59 and 132.

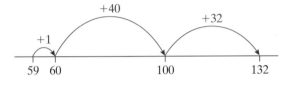

You count on $+1$, $+40$ and $+32$.
So add: $1 + 40 + 32 = 73$

Alternatively you could just add the complement to 100 of 59, which is 41, to 32 to get 73.

Remember when you use the complement you must reduce the original number.

You add 41 to 32 not 132.

Exercise 1.5

1   Use the strategies shown in the previous section to do these calculations mentally.

   (a) $326 + 77$ (b) $124 + 531$ (c) $438 + 253$ (d) $352 - 88$

   (e) $549 + 224$ (f) $678 + 69$ (g) $463 - 329$ (h) $543 + 99$

   (i) $746 + 136$ (j) $386 - 26$ (k) $386 - 215$ (l) $457 + 314$

   (m) $862 - 78$ (n) $538 - 246$

2   Add 416 and 222.

3   Add 361 and 525.

4   What is 925 take away 247?

5   What is 239 plus 424?

6   What is 365 take away 89?

7   What is 174 plus 68?
                                                 Exercise continues …

8 Find 868 minus 543.

9 What is the sum of 469 and 531?

10 What is 893 subtract 645?

11 What is the sum of 354 and 417?

12 What is 483 subtract 97?

13 Find the total of 657 and 412.

14 Meera has 469 buttons, Tom has 254.
How many buttons do they have in total?

15 It is 632 miles from Brisbane to Sydney and then 598 miles from Sydney to Melbourne.
How far is it from Brisbane to Melbourne via Sydney?

16 A pack of paper contains 500 sheets.
Kami uses 274 sheets from the pack.
How many sheets are left in the pack?

17 The road distance from Melbourne to Adelaide is 728 kilometres.
Carla has driven 493 kilometres already.
How many more kilometres does she have to drive?

18 An article costs £3.95.
How much change do I get from £10?

19 Mika has these numbers written on a piece of paper.

123   456   246   369   175   248

He claims that he can make exactly 1000 by adding just three of the numbers together.
Show whether this is possible.

20 Josie has these numbers written on a piece of paper.

147   345   248   139   247   642

She claims that, if she adds the two smallest numbers together and the two largest numbers together, the difference between her two answers is exactly 700.
Explain whether this is true.

Adding and subtracting decimals mentally

When you are adding decimals it is usually easier to deal with the decimal parts first and then add the integer parts.

Example 1.8

Work out $12.67 + 4.55$.

Solution

Step 1:   Add 0.67 and 0.55.
You can do this by adding 0.7 to 0.55 to get 1.25 and then compensating by subtracting 0.03 to get 1.22.
(There are other ways of doing this.)
Step 2:   Add the integers.
              $12 + 4 = 16$
Step 3:   Add the two results.
              $16 + 1.22 = 17.22$

○ Discovery 1.2

Use your ruler to draw a line which is 5.6 cm long.
Mark on the line a point which is 2.8 cm from one end.
Measure the distance from this point to the other end of your line.
Do an appropriate addition or subtraction to find out if your measurement is accurate.

Work in pairs. Repeat the instructions above using different measurements.
One person finds the distance by drawing, the other by doing a subtraction.

Example 1.9

Find the difference between 17.3 and 8.9.

Solution

Method 1

The complement to 10 of 8.9 is 1.1.
Subtract 10. $17.3 - 10 = 7.3$
Then add back 1.1. $7.3 + 1.1 = 8.4$

Method 2

Subtract the integer part first. $17.3 - 8 = 9.3$

Then you need to subtract 0.9.

You can break 0.9 down as $0.3 + 0.6$ and subtract it in two parts.
$9.3 - 0.3 = 9$
$9 - 0.6 = 8.4$

Method 3

You can count on from 8.9.

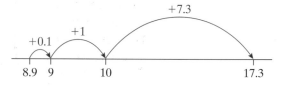

You count on $+0.1, +1$ and $+7.3$.
So add: $0.1 + 1 + 7.3 = 8.4$

Method 4

Subtract 9 from 17.3 to get 8.3.
You have taken away 0.1 too many.
You compensate by adding 0.1 to 8.3 and the answer is 8.4.

Example 1.10

Julie buys a kilogram of apples costing £1.21 and a punnet of strawberries costing 99p.

How much has she spent?

Solution

First, make the units the same. $99p = £0.99$

Then do the addition.
$1.21 + 0.99$

It is easier to add 1 and subtract 0.01.
$1.21 + 1 = 2.21$
$2.21 - 0.01 = 2.2$

So the answer is £2.20.

Alternatively you could have solved this problem by working in pence and then changing your answer into pounds.

TIP

When working with money you should write the answer with two decimal places: £2.20 not £2.2.

Example 1.11

A piece of wood is 2.3 m long.
75 cm is cut off.
How much remains?

Solution

First, make the units the same.   75 cm = 0.75 m
Then do the subtraction.   2.30 − 0.75

The complement of 0.75 to 1 is 0.25.
Subtract 1.   2.30 − 1 = 1.30
Then add back 0.25.   1.30 + 0.25 = 1.55

So the answer is 1.55 m.

Alternatively you could have solved this problem by working in
centimetres and then changing your answer into metres.

> **TIP**
>
> When you do a
> calculation the units
> of measurements
> must be the same.

Exercise 1.6

1 Work out these.
   (a) 6.7 + 7.2   (b) 18.9 + 23.4   (c) 2.7 + 3.6   (d) 5.9 + 8.7
   (e) 6.74 + 3.97   (f) 33.55 + 7.85   (g) 6.82 + 2.49   (h) 26.92 + 18.54

2 Work out these.
   (a) 9.7 − 7.1   (b) 8.7 − 3.4   (c) 28.4 − 16.5   (d) 18.4 − 7.8
   (e) 7.54 − 2.74   (f) 39.87 − 18.03   (g) 21.74 − 8.83   (h) 36.86 − 12.78

3 Work out these.
   Give your answers in the larger unit.
   (a) 6.1 m + 90 cm + 9.3 m   (b) 3.2 m + 28 cm + 6.74 m + 93 cm
   (c) 7.2 m − 165 cm   (d) 8.54 m − 66 cm
   (e) 7.6 cm − 8 mm

4 Work out these.
   (a) £6.84 + 37p + £9.41
   (b) £16.89 + 94p + £7.07
   (c) £61.84 + 76p + £41.32 + £10.55
   (d) £3.89 + 73p + 68p + £91.80

Exercise continues …

5 Two pieces of wood are put end to end.
   Their lengths are 2.5 m and 60 cm.
   Find the total length of the wood, in metres.

6 In the high jump Angela jumps 1.62 m and Sarah jumps 1.47 m.
   Find the difference between the heights of their jumps.

7 The times for the first and last places in a 100-metre race were 11.73 seconds and
   14.38 seconds.
   Find the difference between these times.

8 Sam buys a magazine for £2.25, a newspaper for 70p and a pack of sweets for 58p.
   How much does he spend altogether?

9 Mary has £7.19 in her purse.
   She spends 70p on parking.
   How much does she have left?

10 Winston is buying his lunch.
   He buys a sandwich for £1.89, a packet of crisps for 37p, an apple for 40p and a drink
   for 45p.
   He pays with a £5 note.
   How much change does he receive?

11 Here is a price list from a café.

   > Sandwich .... £1.49 + 30p for salad
   > Crisps 55p
   > Drinks 60p
   > Fruit 35p

   (a) Parool wants to buy a cheese sandwich with salad, a drink and a packet of crisps.
       How much does this cost?

   (b) Parool pays with three £1 coins.
       How much change should he receive?

   (c) Mika has £2.50.
       He wants a sandwich and a drink.
       What else could he buy?

Written methods for adding and subtracting decimals

When adding or subtracting decimals without using a calculator, take care to line up the decimal points.

Example 1.12

Work out these.
(a) 16.45 + 2.62
(b) 13.78 − 1.24

Solution

(a) 16.45 + 2.62 =
```
  16.45
+  2.62
-------
  19.07
```

(b) 13.78 − 1.24 =
```
  13.78
−  1.24
-------
  12.54
```

Example 1.13

Work out these.
(a) 124.5 + 79.87
(b) 23.4 − 9.58

Solution

(a)
```
  1 2 4 . 5
+   7 9 . 8 7
-----------
  2 0 4 . 3 7
    1 1   1
```

(b)
```
  ¹2 ¹²3̶ . ¹³4̶ ¹0
−      9 .  5 8
--------------
     1 3 .  8 2
```
Include an extra zero so that the first number has the same number of decimal places as the second number.

Remember, every time you add or subtract amounts of money you are dealing with decimals. Don't forget that you should write, for example, three pounds and twenty pence as £3.20 and not £3.2.

Exercise 1.7

1 Work out these.
   (a) £2.10 + £3.45   (b) £5.78 + £2.82   (c) £7.15 − £6.13
   (d) £34.02 + £14.89   (e) £123.67 − £65.77   (f) £4.37 + £6.53
   (g) £15.78 + £9.89   (h) £6.18 + £6.32   (i) £15.42 − £9.34
   (j) £19.21 − £17.04

2 Work out these.
   (a) 1.6 + 3.4   (b) 4.9 + 5.21   (c) 17.77 + 19.54
   (d) 10.78 − 4.9   (e) 21.99 − 11.9   (f) 5.53 − 3.09
   (g) 4.9 + 5.6   (h) 34.94 + 7.62   (i) 24.24 + 16.16
   (j) 8.6 − 3.4   (k) 9.42 − 8.57   (l) 24.2 − 4.63

3 Anna buys a T-shirt costing £8.99 and a jumper costing £18.99.
   How much change does she get from £30?

4 Helen buys a packet of sandwiches costing £1.60, a bar of chocolate costing 45p and a
   drink costing 65p. How much change does she get from £5?

5 Class 7 are collecting money each week for a charity.
   The totals collected in the first four weeks are £12.56, £14.66, £18.13 and £11.82.
   How much more do they need to raise to reach their target of £100?

6 Kate buys a 4-metre length of material. She makes two curtains, each of which is 1.8 m
   long. How much material does she have left?

Multiplying and dividing decimals

Multiplying decimals

⊙ Discovery 1.3

(a) 39 × 8 = 312

   Without using your calculator, write down the answers to these multiplications.
   (i)   3.9 × 8   (ii) 39 × 0.8   (iii) 0.39 × 8   (iv) 0.39 × 0.8

(b) 37 × 56 = 2072

   Without using your calculator, write down the answers to these multiplications.
   (i)   3.7 × 56   (ii) 37 × 5.6   (iii) 3.7 × 5.6
   (iv) 0.37 × 56   (v)   0.37 × 5.6   (vi) 0.37 × 0.56

Now check your answers with your calculator.

These are the steps you take to multiply decimals.

1 Carry out the multiplication, ignoring the decimal points. The digits in this answer will be the same as the digits in the final answer.

2 Count the total number of decimal places in the two numbers to be multiplied.

3 Place the decimal point into your answer from step 1 so that the final answer has the same number of decimal places as you found in step 2.

Remember you can always do a rough estimate to check that your answer is of the correct size. In the next example, 8×0.7 will be a bit less than $8 \times 1 = 8$.

Example 1.14

Work out 8×0.7.

Solution

1 First do $8 \times 7 = 56$.

2 The total number of decimal places in 8 and 0.7 is $0 + 1 = 1$.

3 The answer is 5.6.

TIP

Notice that when you multiply by a number between 0 and 1, such as 0.7, the answer is smaller than the original number (5.6 is smaller than 8).

Example 1.15

Work out 8.3×3.4.

Solution

1 First do 83×34.

$$
\begin{array}{r}
83 \\
\times\ 34 \\
\hline
2490 \\
332 \\
\hline
2822
\end{array}
$$

The method used here is the traditional 'long multiplication'. You may prefer another method.

2 The total number of decimal places in 8.3 and 3.4 is $1 + 1 = 2$.

3 The answer is 28.22.

The next example shows some other methods for multiplying.

Example 1.16

Work out 47.6 × 4.6.

Solution

Step 1: Work out 476 × 46.

Method 1

```
        476
  ×      46
    19 040    (476 × 40)
     2 856    (476 × 6)
    21 896
```

Method 2 Using brackets

$(400 + 70 + 6) \times 40 + (400 + 70 + 6) \times 6$

$400 \times 40 + 70 \times 40 + 6 \times 40 = 16\,000 + 2800 + 240$
$$= 19\,040$$

$400 \times 6 + 70 \times 6 + 6 \times 6 = 2400 + 420 + 36$
$$= 2856$$

$19\,040 + 2856 = 21\,896$

Method 3 Using a grid

×	400	70	6		
40	16 000	2800	240	=	19 040
6	2400	420	36	=	+ 2 856
					21 896

Method 4 By doubling

$476 \times 1\ \ = 476$
$476 \times 2\ \ = 952$
$476 \times 4\ \ = 1904$
$476 \times 8\ \ = 3808$
$476 \times 16 = 7616$
$476 \times 32 = 15\,232$

Now $46 = 32 + 8 + 4 + 2$

So $476 \times 46 = 476 \times 32 + 476 \times 8 + 476 \times 4 + 476 \times 2$
$$= 15\,232 + 3808 + 1904 + 952$$
$$= 21\,896$$

Method 5 The lattice method

Add along the diagonal lines and carry the 'tens' digit to the next column as shown.

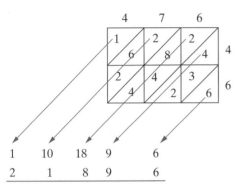

Once you have done the multiplication, continue with the method for dealing with decimals.

Step 2: Count the number of decimal places in the numbers 47.6 and 4.6.
$1 + 1 = 2$

Step 3: The answer to 47.6×4.6 will have two decimal places. So the answer is 218.96.

Sometimes you can use the rules you know for adding and multiplying numbers to simplify calculations, as shown in the next example.

Example 1.17

Work out $78.5 \times 6 + 78.5 \times 4$.

Solution

Notice the same number, 78.5, is used in both parts of the calculation.

You can rewrite the calculation as
$78.5 \times (6 + 4) = 78.5 \times 10 = 785$.

Dividing decimals

Dividing a decimal by a whole number is very like dividing two whole numbers. Line up your work carefully, keeping the decimal points in line.

As with multiplication, there are different methods for doing divisions. The next example shows some different methods.

Example 1.18

Work out $126.4 \div 8$.

Solution

Method 1 Long division

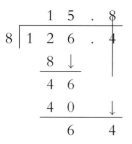

```
        1  5  .  8
    8 │ 1  2  6  .  4
        8  ↓
        4  6
        4  0     ↓
           6     4
```

Method 2 Short division

```
      1 5. 8
  8)12⁴6.⁶4
```

Method 3 Chunking

```
        1  2  6  .  4
    −      8  0          (10 × 8)
        4  6  .  4
    −      4  0          (5 × 8)
           6  .  4
    −      6  .  4       (0.8 × 8)
           0  .  0
```

$10 + 5 + 0.8 = 15.8$
So $126.4 \div 8 = 15.8$.

○ Discovery 1.4

(a) Do these calculations on your calculator.
   (i) $26 \div 1.3$ (ii) $260 \div 13$

(b) What do you notice?

(c) Now do these calculations on your calculator.
   (i) $5.92 \div 3.7$ (ii) $59.2 \div 37$
   (iii) $3.995 \div 2.35$ (iv) $399.5 \div 235$

(d) Can you explain your results?

The result of a division is unchanged when you multiply both numbers by 10 (i.e. move the digits one place to the left in both numbers).

The result is also unchanged when you multiply both numbers by 100 (i.e. move the digits two places to the left in both numbers).

You will see that this rule is also true when you are writing equivalent fractions.

For example, $\frac{3}{5} = \frac{30}{50} = \frac{300}{500}$.

You use this rule when you are dividing by a decimal.

Example 1.19

Work out $6 \div 0.3$.

Solution

First multiply both numbers by 10 by moving the digits one place to the left.
The number you are dividing by is then a whole number.

The calculation becomes $60 \div 3$.

$60 \div 3 = 20$.
$6 \div 0.3$ is also 20.

TIP

Notice that when you divide by a number between 0 and 1, such as 0.3, the answer is larger than the original number (20 is larger than 6).

Example 1.20

Work out $4.68 \div 0.4$.

Solution

First multiply both numbers by 10 by moving the digits one place to the left.

The calculation becomes $46.8 \div 4$.

$$\begin{array}{r} 11.7 \\ 4\overline{)46.^28} \end{array}$$ The decimal point in the answer goes above the decimal point in 46.8.

$4.68 \div 0.4$ is also 11.7.

Example 1.21

Work out $3.64 \div 1.3$.

Solution

First multiply both numbers by 10 by moving the digits one place to the left.

The calculation becomes $36.4 \div 13$.

$$\begin{array}{r} 2.\ 8 \\ 13\overline{)36.^{10}4} \end{array}$$ You may have been taught to do this by long division rather than by short division.

$3.64 \div 1.3$ is also 2.8.

Exercise 1.8

1 Given that $63 \times 231 = 14\,553$ and $124 \times 85 = 10\,540$, write the answers to these calculations.

   (a) 6.3×231 (b) 6.3×23.1 (c) 23.1×63

   (d) 12.4×85 (e) 124×8.5 (f) 12.4×8.5

   (g) $1455.3 \div 63$ (h) $105.4 \div 85$

2 Work out these.

   (a) 4.8×4 (b) 0.3×0.2 (c) 2.4×0.6

   (d) 24×2.3 (e) 2.1×1.8 (f) 9.3×1.5

   (g) 8.2×1.6 (h) 36×4.2 (i) 83×5.4

   (j) 104×0.4 (k) 6.2×36 (l) 0.8×14.8

   (m) 35.2×18 (n) 0.6×1.7

3 Work out these.

   (a) $13.6 \div 4$ (b) $11.4 \div 3$ (c) $27.2 \div 0.8$

   (d) $40.5 \div 0.9$ (e) $5.28 \div 0.03$ (f) $7.25 \div 0.5$

   (g) $8.54 \div 0.7$ (h) $6.36 \div 0.06$ (i) $0.85 \div 0.05$

   (j) $2.52 \div 0.09$ (k) $19.2 \div 1.2$ (l) $71.4 \div 0.21$

   (m) $1.95 \div 0.15$ (n) $3.6 \div 2.4$ (o) $2.88 \div 0.018$

Exercise continues …

4 (a) Calculate the area of this rectangle.

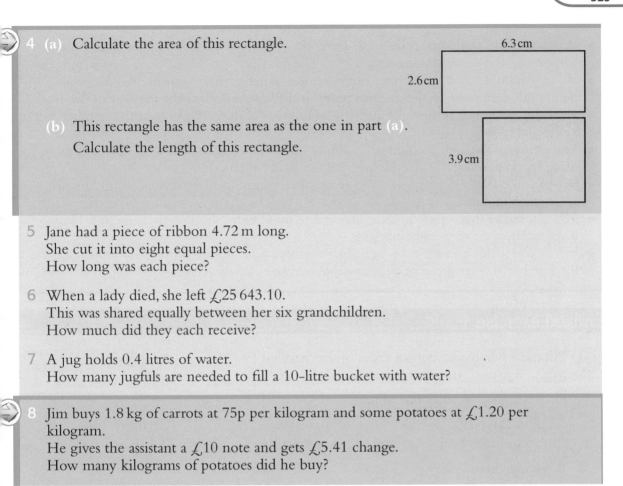

6.3 cm

2.6 cm

(b) This rectangle has the same area as the one in part (a).
Calculate the length of this rectangle.

3.9 cm

5 Jane had a piece of ribbon 4.72 m long.
She cut it into eight equal pieces.
How long was each piece?

6 When a lady died, she left £25 643.10.
This was shared equally between her six grandchildren.
How much did they each receive?

7 A jug holds 0.4 litres of water.
How many jugfuls are needed to fill a 10-litre bucket with water?

8 Jim buys 1.8 kg of carrots at 75p per kilogram and some potatoes at £1.20 per
kilogram.
He gives the assistant a £10 note and gets £5.41 change.
How many kilograms of potatoes did he buy?

What you have learned

- You order integers by looking at the place value of the digits in the numbers
- The complement to 10 of a number is the number you have to add to it to make 10; for
  example, the complement to 10 of 2 is 8 because 2 + 8 = 10
- The complement to 100 of a number is the number you have to add to it to make 100;
  for example, the complement to 100 of 32 is 68 because 32 + 68 = 100
- You can derive division facts from the corresponding multiplication facts; for example,
  4 × 7 = 28 so 28 ÷ 4 = 7 and 28 ÷ 7 = 4
- You can use a range of methods to add and subtract numbers mentally; these include
  partitioning, using complements, counting on and compensation
- You can use these methods with both integers and decimals
- When you add and subtract measurements you must use the same units for all the
  measurements

- When you add and subtract decimals using the column method, you must make sure that you line up the decimal points
- When you multiply decimals first carry out the multiplication ignoring the decimal points and then position the decimal point in the answer so that the number of decimal places in the answer is the same as the total number of decimal places in the numbers being multiplied
- You can use a range of methods to multiply numbers; these include long multiplication, using brackets, using a grid, doubling and the lattice method
- When you divide a decimal by a whole number, you must make sure that you line up the decimal points
- You can use a range of methods to divide one number by another; these include long division, short division and chunking
- When you divide a number by a decimal multiply both numbers by 10 or 100 so that the number you are dividing by is a whole number

Mixed exercise 1

1 Put these whole numbers in order of size, smallest first.

612   231   917   583   646   111

2 How many do you need to add to
- (a) 3 to make 10?
- (b) 8 to make 10?
- (c) 6 to make 10?
- (d) 55 to make 100?
- (e) 32 to make 100?
- (f) 74 to make 100?

3 Work out these.
- (a) £4.64 + £5.92
- (b) £16.34 + £8.26
- (c) £5.96 − £1.48
- (d) £24.33 − £13.74

4 Work out these.
- (a) 2.46 + 1.70
- (b) 19.83 + 16.42
- (c) 36.95 − 14.43
- (d) 134.2 − 99.9
- (e) 12.8 + 28.57
- (f) 108.64 + 69.8
- (g) 73.6 − 38.78
- (h) 152.46 − 85.7

5 Work out these.
- (a) 5 × 0.7
- (b) 0.3 × 6
- (c) 4 × 0.6
- (d) 0.7 × 9
- (e) 50 × 0.3
- (f) 0.6 × 70
- (g) 8 × 7.6
- (h) 7 × 4.9
- (i) 3 × 0.04
- (j) 8 × 0.9
- (k) 5000 × 2.4
- (l) 6.99 × 400
- (m) 1.88 × 3000

6 Work out these.
- (a) 0.3 × 0.1
- (b) 0.9 × 0.6
- (c) 0.4 × 0.2
- (d) 0.5 × 0.3
- (e) $(0.5)^2$
- (f) $(0.1)^2$
- (g) $(0.3)^2$

Mixed exercise 1 continues …

7 Work out these.

   (a) 60.5 ÷ 5   (b) 330.3 ÷ 3   (c) 85.5 ÷ 5

   (d) 4 ÷ 0.8   (e) 3.2 ÷ 0.4   (f) 3.9 ÷ 0.6

8 Harry is collecting stickers.
There are 276 in a complete set and he has collected 193 so far.
How many more stickers does Harry need to complete his set?

9 There are 845 tickets for a concert.
379 tickets have been sold so far.
How many tickets are still available?

10 Niko is selling raffle tickets.
He had 750 tickets to begin with and has already sold 387.
How many tickets does he have left?

11 In a 4 by 400 m relay race, the four members of a team ran these times.

44.5 seconds   45.6 seconds   45.8 seconds   43.9 seconds

What was their average time?

12 Alice buys a magazine costing £1.95 each week for five weeks.
How much does she spend?

13 John buys four bags of these sweets.

   (a) How much do the sweets he buys weigh in total?

   (b) How much does he spend on sweets?

14 The special–offer DVDs in a shop cost £8.35 each.
Emily buys five.
How much does she spend?

15 Nasreen has a fifty-pound gift card for a clothes shop.
She buys a pair of jeans costing £28.99 and two T-shirts costing £4.95 each.
How much does she have left on her gift card?

16 Tom's grandmother buys two cabbages costing 74p each and three grapefruits costing 35p each.
She pays with a ten-pound note.
How much change does she receive?

Angles, triangles and quadrilaterals

▶ This chapter is about

- The fact that angles in a triangle add up to 180°
- The properties of equilateral, isosceles and right-angled triangles
- The properties of quadrilaterals

▶ You should already know

- A right angle is 90°
- A straight line is 180°
- What acute, obtuse and reflex angles are
- How to measure and draw angles

Angles

You learned about angles in Unit A Chapter 11.

🔍 Check up 2.1

(a) Match these angles with the types of angle in the box.

right angle   obtuse angle   reflex angle   acute angle

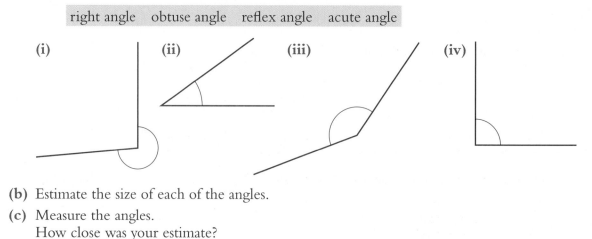

(i)   (ii)   (iii)   (iv)

(b) Estimate the size of each of the angles.

(c) Measure the angles.
How close was your estimate?

Triangles

A **triangle** has three sides.

The angles in a triangle

● Discovery 2.1

(a) Make a rough copy of this triangle
on a piece of paper.
Tear off the three corners.
Arrange the three corners
next to each other.
What do you notice?

(b) Repeat this with a different triangle.
Does the same thing happen?

(c) Copy and complete this sentence.
The three angles inside a triangle add up to _____°.

The angles in a triangle add up to 180°.

In this triangle, there is one right angle (90°) and
two angles of 45°.

$$90° + 45° + 45° = 180°$$

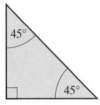

You can use the fact that the angles in a triangle add up to 180° to
find the size of a missing angle in a triangle.

Example 2.1

Find the size of angle a in this triangle.

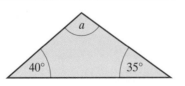

Solution

The sum of the angles in a triangle is 180° therefore

$$a + 40° + 35° = 180°$$
$$a + 75° = 180°$$
$$a = 105°$$

Properties of triangles

Triangle	Properties
Right-angled	• One right angle • Can be isosceles
Equilateral	• All three sides the same length • All three angles the same size and equal 60° (Notice how small arcs can be used in the angles to show equal angles.)
Isosceles	• Two sides equal in length • The angles opposite the equal sides are equal (The equal sides are often shown by small lines crossing the sides, as in the diagram.)
Scalene	• All three sides of different length • All three angles of different size

You can use the properties of special triangles together with the fact that angles in a triangle add up to 180°.

Example 2.2

Find the size of angle m in this isosceles triangle.

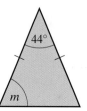

Solution

Since the triangle is isosceles the unmarked angle $= m$.

The sum of the angles in a triangle is 180° therefore

$$2m + 44 = 180°$$
$$2m = 136°$$
$$m = 68°$$

Exercise 2.1

1 Calculate the third angle of each of these triangles.

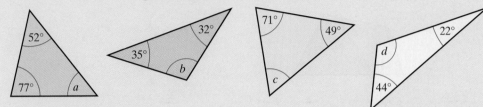

2 Calculate the angles marked with letters in these isosceles triangles.

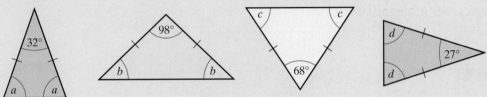

3 (a) Calculate the angles marked with letters in these triangles.

(i)   (ii)   (iii)

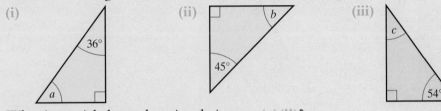

(b) What is special about the triangle in part (a)(ii)?

4 (a) Lily says that she has measured the angles of a triangle and they are 94°, 64° and 40°.
        Is she correct?

   (b) Henry says that he has measured the angles of a triangle and they are 78°, 46° and 57°.
        Is he correct?

5 Explain why you cannot have two obtuse angles in a triangle.

● Challenge 2.1

(a) Draw a triangle and extend the
     bottom line like this.
     Measure the size of angles *a*, *b* and *c*.
     What do you notice about
     the three angles?

(b) Try again with triangles with different angles.
     Can you find a rule that always works?

Quadrilaterals

A **quadrilateral** has four sides.

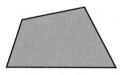

◯ Discovery 2.2

Here are seven types of quadrilateral.

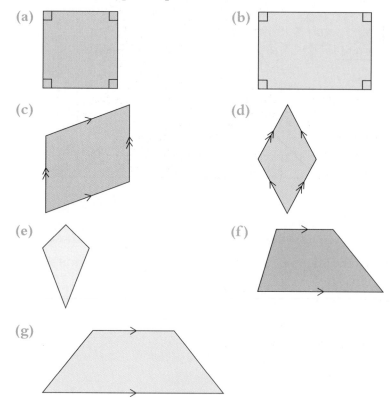

In a small group or on your own

- name each one.
- write down its properties.
- draw a diagonal and write down the properties involving diagonals.

The names and geometrical properties of the seven types of quadrilaterals shown on the previous page are given in the table below.

| Name | Angles | Sides | | Diagonals |
		Lengths	Parallel	
(a) Square	All 90°	All equal	Opposite sides parallel	Equal length Bisect at 90°
(b) Rectangle	All 90°	Opposite sides equal	Opposite sides parallel	Equal length Bisect, but not at 90°
(c) Parallelogram	Opposite angles equal	Opposite sides equal	Opposite sides parallel	Not equal Bisect, but not at 90°
(d) Rhombus	Opposite angles equal	All equal	Opposite sides parallel	Not equal Bisect at 90°
(e) Kite	One pair of opposite angles equal	Two pairs of adjacent sides equal	None parallel	Not equal Only one bisected by the other They cross at 90°
(f) Trapezium	Can be different	Can be different	One pair of sides parallel	Nothing special
(g) Isosceles trapezium	Two pairs of adjacent angles equal	One pair of opposite sides equal	Other pair of opposite sides parallel	Equal length Do not bisect or cross at 90°

TIP

Many people think that all trapezia are isosceles, but *any* quadrilateral that has one pair of opposite sides parallel is a trapezium.

Example 2.3

Which quadrilaterals have
(a) both pairs of opposite sides equal?
(b) just one pair of opposite angles equal?

Solution

(a) Square, rectangle, parallelogram, rhombus
(b) Kite

○ Discovery 2.3

(a) Draw a large rectangle on squared paper.
Draw lines to split it into as many of the different quadrilaterals as
you can (you may be left with some triangles as well).
Label each shape with its name.

(b) Look around you at school and at home and find as many
examples of the different quadrilaterals as you can.
Make a list of your examples.
What type of quadrilateral is each of them?

Exercise 2.2

1 Name each of the shapes A to E in the diagram as fully as you can.

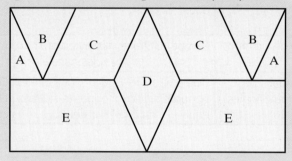

2 (a) Which four triangles in the diagram below form a kite?

(b) Write down a pair of triangles which make a parallelogram.

(c) Which triangles are not isosceles?

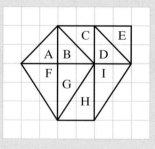

3 Which quadrilaterals have all four angles the same?

4 Which quadrilaterals have opposite sides the same, but not all four sides the same?

5 A rectangle can be described as a special type of parallelogram.
What else is true about a rectangle?

Exercise continues …

6   Which quadrilaterals have all four sides the same?

7   Which quadrilaterals have opposite angles the same, but not 90°?

8   A rhombus can be described as a special type of parallelogram.
    What else is true about a rhombus?

9   (a) Draw a square and clearly mark any angles that are the same and any sides that are
         the same.
    (b) Draw in the diagonals and check what is true about them.

10 (a) Draw a kite and clearly mark any angles that are the same and any sides that are
         the same.
    (b) Draw in the diagonals and check what is true about them.

11 (a) Draw a rectangle and clearly mark any angles that are the same and any sides that
         are the same.
    (b) Draw in the diagonals and check what is true about them.

12 A quadrilateral has angles 80°, 100°, 80° and 100° as you go round the quadrilateral.
    The sides are not all the same length.
    (a) Sketch the quadrilateral.
    (b) What special type of quadrilateral is this?

13 A quadrilateral has angles 80°, 80°, 100° and 100° as you go round the quadrilateral.
    (a) Sketch the quadrilateral.
    (b) What special type of quadrilateral is this?

14 (a) Draw a trapezium that is not isosceles.
    (b) What can you say about the angles?

15 Which quadrilaterals have diagonals that cross at 90°?

16 (a) Draw a parallelogram and mark any sides that are the same or parallel, and any
         angles that are the same.
    (b) Draw in the diagonals and check what is true about them.

17 A quadrilateral has the diagonals the same length.
    What types of quadrilateral could it be?

18 List all the quadrilaterals that have both pairs of opposite sides parallel and equal.

Exercise continues …

19 Name each of these shapes.

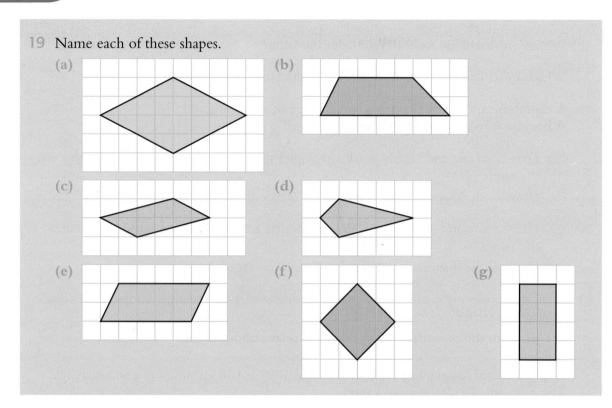

(a)

(b)

(c)

(d)

(e)

(f)

(g)

● Discovery 2.4

Some quadrilaterals are special types of other quadrilaterals. For instance, a square is a rectangle with all sides equal. There are also connections from the square to other quadrilaterals.

Go through all seven types of quadrilaterals named in this chapter and list which ones are special types of other quadrilaterals.

What you have learned

- The angles in a triangle add up to 180° and you can use this fact to calculate a missing angle in a triangle
- The properties of special types of triangles and how to classify triangles by their properties
- The properties of special types of quadrilaterals and how to classify quadrilaterals by their properties

Mixed exercise 2

1 Find the missing angles in these triangles.

2 Work out angles x and y in this triangle.
Give reasons for your answers.

3 Copy and complete this table showing the properties that are always true for the special quadrilaterals.

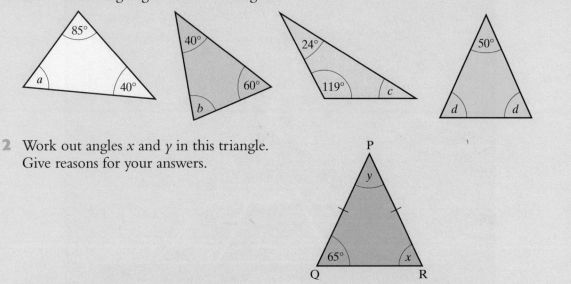

Quadrilateral	Sides		Diagonals	Parallel sides	
	Four equal sides	**Two different pairs of equal sides**	**Equal diagonals**	**Two pairs of parallel sides**	**Only one pair of parallel sides**
Square	✓		✓	✓	
Rectangle		✓	✓	✓	
Rhombus					
Parallelogram					
Trapezium					
Kite					

4 What quadrilaterals are these? There may be more than one answer.
  (a) Only one pair of parallel sides
  (b) Diagonals at right angles
  (c) Two pairs of sides equal and parallel and at least one right angle
  (d) Two pairs of adjacent sides equal

Mixed exercise 2 continues …

5 This is the Eiffel Tower in Paris, France.
When it was built in 1889 it was the tallest building in the world.

(a) What quadrilaterals can you see in the building?

(b) The criss-cross pattern of the metal bars gives the tower strength.
Describe the triangles you can see in one of the squares of the middle section as fully as you can.

(c) There is a tall isosceles triangle in the base of the top section.
The top angle of the triangle is 10°.
What size are the other two angles?

Chapter 3

Fractions

▶ This chapter is about

- Equivalent fractions
- Expressing a fraction in its lowest terms
- Ordering fractions
- Finding a fraction of a quantity
- Expressing one number as a fraction of another
- Adding and subtracting fractions
- Converting a fraction to a decimal

▶ You should already know

- How to use fraction notation

Equivalent fractions

🔍 Check up 3.1

Copy the diagrams and copy and complete the statements below.

(a)

(b)

(c)

$\dfrac{\Box}{12}$ of the shape is purple.

$\dfrac{\Box}{6}$ of the shape is purple.

$\dfrac{\Box}{\Box}$ of the shape is purple.

What do you notice?

To make **equivalent fractions** you multiply or divide the **numerator** and **denominator** by the same number.

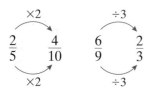

$$\frac{2}{5} \xrightarrow{\times 2} \frac{4}{10} \qquad \frac{6}{9} \xrightarrow{\div 3} \frac{2}{3}$$

> **TIP**
>
> Remember: the numerator is the top of the fraction and the denominator is the bottom of the fraction.

Exercise 3.1

1 Fill in the missing numbers in each set of equivalent fractions.

(a) $\dfrac{1}{4} = \dfrac{\square}{8} = \dfrac{\square}{12} = \dfrac{5}{\square}$

(b) $\dfrac{1}{5} = \dfrac{\square}{10} = \dfrac{\square}{20} = \dfrac{7}{\square}$

(c) $\dfrac{2}{5} = \dfrac{4}{\square} = \dfrac{\square}{25} = \dfrac{12}{\square}$

(d) $\dfrac{2}{9} = \dfrac{4}{\square} = \dfrac{\square}{36} = \dfrac{6}{\square}$

(e) $\dfrac{1}{7} = \dfrac{2}{\square} = \dfrac{\square}{35}$

(f) $\dfrac{4}{9} = \dfrac{16}{\square} = \dfrac{\square}{72}$

(g) $\dfrac{1}{6} = \dfrac{4}{\square} = \dfrac{\square}{12}$

(h) $\dfrac{2}{3} = \dfrac{\square}{6} = \dfrac{12}{\square} = \dfrac{\square}{24}$

2 Fill in the missing numbers in each set of equivalent fractions.

(a) $\dfrac{3}{4} = \dfrac{\square}{12}$

(b) $\dfrac{10}{16} = \dfrac{5}{\square}$

(c) $\dfrac{1}{2} = \dfrac{\square}{18}$

(d) $\dfrac{30}{50} = \dfrac{3}{\square}$

(e) $\dfrac{12}{18} = \dfrac{\square}{9}$

(f) $\dfrac{2}{7} = \dfrac{10}{\square}$

(g) $\dfrac{4}{5} = \dfrac{\square}{30}$

(h) $\dfrac{3}{21} = \dfrac{1}{\square}$

(i) $\dfrac{2}{9} = \dfrac{\square}{27}$

(j) $\dfrac{3}{11} = \dfrac{\square}{44}$

(k) $\dfrac{15}{35} = \dfrac{3}{\square}$

(l) $\dfrac{28}{70} = \dfrac{\square}{10}$

Expressing a fraction in its lowest terms

Look at these equivalent fractions.

$$\frac{4}{12} \qquad \frac{2}{6} \qquad \frac{1}{3}$$

$\frac{1}{3}$ is the simplest of these fractions. It cannot be simplified any further because there is no number, except 1, that will divide into both 1 and 3.

Changing $\frac{4}{12}$ to $\frac{1}{3}$ is called expressing $\frac{4}{12}$ in its lowest terms.

It can also be called expressing $\frac{4}{12}$ in its **simplest form** or **lowest terms**.

You may also say you are **cancelling** a fraction.

Notice that you can change $\frac{4}{12}$ to $\frac{1}{3}$

- in two steps by dividing both the numerator and the denominator by 2 and then by 2 again or
- in one step by dividing both the numerator and the denominator by 4.

TIP

Always try to spot as large a number as possible that will divide into both the numerator and denominator.

Example 3.1

Express these fractions in their lowest terms.

(a) $\frac{18}{20}$ **(b)** $\frac{35}{40}$ **(c)** $\frac{60}{80}$ **(d)** $\frac{45}{60}$

Solution

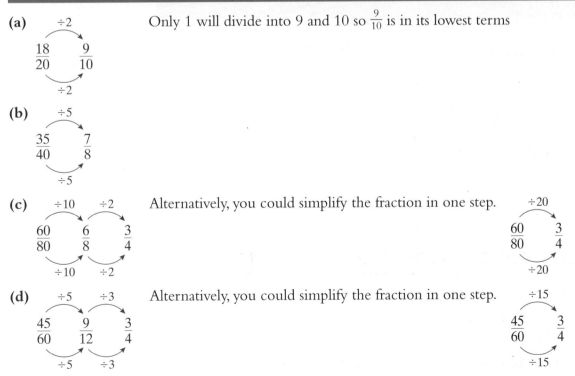

(a)

$\overset{\div 2}{\frac{18}{20}} \quad \frac{9}{10}$

$\div 2$

Only 1 will divide into 9 and 10 so $\frac{9}{10}$ is in its lowest terms

(b)

$\overset{\div 5}{\frac{35}{40}} \quad \frac{7}{8}$

$\div 5$

(c)

$\overset{\div 10}{\frac{60}{80}} \quad \overset{\div 2}{\frac{6}{8}} \quad \frac{3}{4}$

$\div 10 \qquad \div 2$

Alternatively, you could simplify the fraction in one step.

$\overset{\div 20}{\frac{60}{80}} \quad \frac{3}{4}$

$\div 20$

(d)

$\overset{\div 5}{\frac{45}{60}} \quad \overset{\div 3}{\frac{9}{12}} \quad \frac{3}{4}$

$\div 5 \qquad \div 3$

Alternatively, you could simplify the fraction in one step.

$\overset{\div 15}{\frac{45}{60}} \quad \frac{3}{4}$

$\div 15$

Exercise 3.2

1 Express these fractions in their lowest terms.

(a) $\frac{8}{10}$ (b) $\frac{2}{12}$ (c) $\frac{15}{21}$ (d) $\frac{12}{16}$

(e) $\frac{14}{21}$ (f) $\frac{25}{30}$ (g) $\frac{20}{40}$ (h) $\frac{18}{30}$

(i) $\frac{16}{24}$ (j) $\frac{150}{300}$ (k) $\frac{20}{120}$ (l) $\frac{500}{1000}$

(m) $\frac{56}{70}$ (n) $\frac{64}{72}$ (o) $\frac{60}{84}$ (p) $\frac{120}{180}$

Ordering fractions

To put fractions in order, change them to equivalent fractions all with the same denominator and order them by the numerator.

Example 3.2

Which is the bigger, $\frac{3}{4}$ or $\frac{5}{6}$?

Solution

First find a common denominator. 24 is an obvious one, as $4 \times 6 = 24$, but a smaller one is 12.

So convert both fractions into twelfths.

$$\frac{3 \times 3}{4 \times 3} = \frac{9}{12} \qquad \frac{5 \times 2}{6 \times 2} = \frac{10}{12}$$

$\frac{10}{12}$ is bigger than $\frac{9}{12}$, so $\frac{5}{6}$ is bigger than $\frac{3}{4}$.

> **TIP**
>
> Multiplying the two denominators together will always work to find a common denominator but the lowest common multiple is sometimes smaller.

Exercise 3.3

1 For each pair of fractions
  - find the lowest common denominator.
  - state which is the bigger fraction.

(a) $\frac{2}{3}$ or $\frac{7}{9}$ (b) $\frac{5}{6}$ or $\frac{7}{8}$ (c) $\frac{3}{8}$ or $\frac{7}{20}$

2 Which of these fractions is the bigger?

(a) $\frac{3}{4}$ or $\frac{5}{8}$ (b) $\frac{7}{9}$ or $\frac{5}{6}$ (c) $\frac{3}{10}$ or $\frac{4}{15}$

Exercise continues ...

3 Write each of these sets of fractions in order, smallest first.

(a) $\dfrac{7}{10}$ $\dfrac{3}{4}$ $\dfrac{11}{20}$ $\dfrac{3}{5}$

(b) $\dfrac{7}{12}$ $\dfrac{3}{4}$ $\dfrac{7}{8}$ $\dfrac{5}{6}$

(c) $\dfrac{13}{15}$ $\dfrac{2}{3}$ $\dfrac{3}{10}$ $\dfrac{2}{5}$ $\dfrac{1}{2}$

(d) $\dfrac{13}{16}$ $\dfrac{5}{8}$ $\dfrac{3}{4}$ $\dfrac{7}{16}$ $\dfrac{1}{2}$

(e) $\dfrac{2}{5}$ $\dfrac{1}{2}$ $\dfrac{9}{20}$ $\dfrac{17}{40}$ $\dfrac{3}{8}$

(f) $\dfrac{7}{8}$ $\dfrac{8}{9}$ $\dfrac{5}{12}$

(g) $\dfrac{1}{12}$ $\dfrac{1}{3}$ $\dfrac{3}{4}$ $\dfrac{1}{6}$ $\dfrac{1}{2}$

(h) $\dfrac{11}{16}$ $\dfrac{7}{8}$ $\dfrac{3}{4}$ $\dfrac{17}{32}$

(i) $\dfrac{1}{2}$ $\dfrac{1}{8}$ $\dfrac{3}{8}$ $\dfrac{1}{4}$

(j) $\dfrac{11}{12}$ $\dfrac{5}{8}$ $\dfrac{3}{4}$ $\dfrac{7}{24}$ $\dfrac{1}{2}$

Finding a fraction of a quantity

Fractions with 1 as the numerator

A fraction such as $\frac{1}{2}$ means a whole divided by two.

If the whole is 20 then $\frac{1}{2}$ of 20 = 10.
This is the same as saying 20 ÷ 2 = 10.

Similarly finding $\frac{1}{3}$ of something is the same as dividing by 3, finding $\frac{1}{4}$ of something is the same as dividing by 4 and so on.

Example 3.3

(a) Find $\frac{1}{4}$ of £34.

(b) Find $\frac{1}{5}$ of 24 metres.

Solution

(a) $\dfrac{8.5}{4\overline{)34.^20}}$

Answer: £8.50

(b) $\dfrac{4.8}{5\overline{)24.^40}}$

Answer: 4.8 metres

> **TIP**
>
> Remember: with money you must always give the answer as £8.50 rather than £8.5.

Fractions with other numbers as the numerator

If you want to find a fraction such as $\frac{3}{5}$, simply find $\frac{1}{5}$ and then multiply by 3.

Example 3.4

(a) Find $\frac{3}{5}$ of 40. **(b)** Find $\frac{2}{7}$ of 28.

Solution

(a) $40 \div 5 = 8$ First divide by 5 to find $\frac{1}{5}$.

$8 \times 3 = 24$ Then multiply by 3 to find $\frac{3}{5}$.

Answer: 24

(b) $28 \div 7 = 4$ First divide by 7 to find $\frac{1}{7}$.

$4 \times 2 = 8$ Then multiply by 2 to find $\frac{2}{7}$.

Answer: 8

Exercise 3.4

1 Find $\frac{1}{4}$ of these quantities.
   (a) 20   (b) 36   (c) 68   (d) £100   (e) £10

2 Find $\frac{1}{5}$ of these quantities.
   (a) 30   (b) 45   (c) 80   (d) £120   (e) 26 m

3 Find $\frac{3}{4}$ of these quantities.
   (a) 24   (b) 48   (c) 200   (d) £56   (e) £140

4 Find $\frac{5}{6}$ of these quantities.
   (a) 30   (b) 48   (c) 120   (d) 42 cm   (e) £90

5 Emma receives £8 pocket money. She saves $\frac{1}{5}$ of it. How much does she save?

6 Adam had a part-time job. He earned £24.
   He spent $\frac{1}{8}$ of it on sweets and $\frac{3}{8}$ on books and magazines.
   (a) How much did he spend on sweets?
   (b) How much did he spend on books and magazines?

7 Mr Green has 72 metres of hose. He cuts off $\frac{2}{9}$ of it. How much is left?

Exercise continues …

8 There were 180 students on a trip. $\frac{2}{5}$ of them were boys.
   How many boys were there?

9 A school with 560 students had a mock election.
   $\frac{3}{8}$ of the students voted for the Green Party.
   How many votes did the Green Party receive?

10 Which is the larger, a $\frac{5}{8}$ share of £120 or a $\frac{3}{4}$ share of £96.
   Show your working.

11 John wants to buy the latest mobile phone. It costs £220.
   His dad offers him $\frac{2}{5}$ of £540.
   His mum offers him $\frac{5}{12}$ of £510.
   His grandad offers him $\frac{3}{8}$ of £590.
   His aunt offers him $\frac{9}{10}$ of £240.
   Which relative would give John enough money to buy the phone?

> **TIP**
>
> You must show how you worked out your answer.

Expressing one number as a fraction of another

To write one number as a fraction of another, write the first number as the numerator of the fraction and the second number as the denominator. Write the fraction in its lowest terms if necessary.

Example 3.5

Express 14 as a fraction of 60.

Solution

$\frac{14}{60} = \frac{7}{30}$ Write the fraction in its lowest terms.

Exercise 3.5

1 What fraction is
   (a) 7 of 14?   (b) 5 of 15?   (c) 8 of 18?   (d) 12 of 30?
   (e) 16 of 24?   (f) 11 of 55?   (g) 6 of 54?   (h) 12 of 64?
   Write each fraction in its lowest terms.

Adding and subtracting fractions

Adding fractions

In this diagram, each square is divided into thirds.

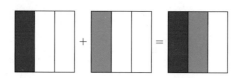

The diagram shows $\frac{1}{3} + \frac{1}{3} = \frac{2}{3}$.

Counting columns, 1 column + 1 column = 2 columns.

To add fractions with the same denominator, just add the numerators.

In this diagram there are 12 small squares in each rectangle.

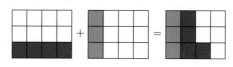

The diagram shows $\frac{1}{3} + \frac{1}{4}$.

Counting squares, 4 squares + 3 squares = 7 squares or $\frac{7}{12}$.

To add fractions with different denominators, change them to equivalent fractions, all with the same denominator.

$\frac{1}{3} = \frac{4}{12}$ $\frac{1}{4} = \frac{3}{12}$

$\frac{4}{12} + \frac{3}{12} = \frac{7}{12}$

TIP

Remember that you add the numerators but not the denominators.

Example 3.6

Add the fractions.

(a) $\frac{2}{5} + \frac{1}{5}$ **(b)** $\frac{1}{5} + \frac{3}{10}$ **(c)** $\frac{1}{6} + \frac{3}{4}$

Solution

(a) $\frac{2}{5} + \frac{1}{5} = \frac{3}{5}$ They have the same denominator so just add the numerators.

(b) $\frac{1}{5} + \frac{3}{10}$ They need to be changed so that they both have the same denominator.

     $= \frac{2}{10} + \frac{3}{10}$ Change $\frac{1}{5}$ to $\frac{2}{10}$ so they both have 10 as the denominator.

     $= \frac{5}{10} = \frac{1}{2}$ Write the answer in the lowest terms.

(c) $\frac{1}{6} + \frac{3}{4}$ This time they both need to be changed to have the same denominator.

     $= \frac{2}{12} + \frac{9}{12} = \frac{11}{12}$ The smallest possible common denominator is 12 so convert both fractions to twelfths.

Subtracting fractions

In this diagram there are 20 small squares in each rectangle.

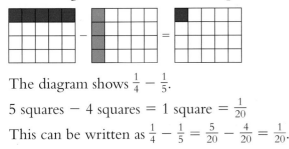

The diagram shows $\frac{1}{4} - \frac{1}{5}$.

5 squares $-$ 4 squares $= 1$ square $= \frac{1}{20}$

This can be written as $\frac{1}{4} - \frac{1}{5} = \frac{5}{20} - \frac{4}{20} = \frac{1}{20}$.

> **TIP**
>
> When adding or subtracting fractions, change to equivalent fractions all with the same denominator and add or subtract the numerators. The most common error is to add the denominators and add the numerators.

Change both fractions so that they have the same denominator, as with adding, but then subtract the numerators.

Example 3.7

Work out these.

(a) $\frac{3}{5} - \frac{2}{5}$ **(b)** $\frac{3}{4} - \frac{2}{3}$ **(c)** $\frac{5}{6} + \frac{1}{4} - \frac{1}{3}$

Solution

(a) $\frac{3}{5} - \frac{2}{5} = \frac{1}{5}$

They have the same denominator so just subtract the numerators.

(b) $\frac{3}{4} - \frac{2}{3} = \frac{9}{12} - \frac{8}{12}$

$= \frac{1}{12}$

4 and 3 both divide into 12, so make 12 the denominator for both. Multiply $\frac{3}{4}$ by 3 top and bottom and $\frac{2}{3}$ by 4 top and bottom.

(c) $\frac{5}{6} + \frac{1}{4} - \frac{1}{3} = \frac{10}{12} + \frac{3}{12} - \frac{4}{12}$

$= \frac{9}{12}$

$= \frac{3}{4}$

All the denominators divide into 12 so make the common denominator 12.

Write the answer in its lowest terms.

Discovery 3.1

(a) Use shading in a rectangle with 12 squares to show the addition $\frac{1}{6} + \frac{3}{4}$.
Then write the working using common denominators.

(b) Use shading in a rectangle with 24 squares to show the subtraction $\frac{5}{8} - \frac{1}{3}$.
Then write the working using common denominators.

Exercise 3.6

1 Add these fractions.

(a) $\frac{1}{8} + \frac{1}{8}$ (b) $\frac{1}{8} + \frac{3}{8}$ (c) $\frac{1}{4} + \frac{1}{8}$ (d) $\frac{1}{2} + \frac{3}{8}$ (e) $\frac{5}{8} + \frac{1}{4}$

2 Subtract these fractions.

(a) $\frac{3}{8} - \frac{1}{8}$ (b) $\frac{5}{8} - \frac{3}{8}$ (c) $\frac{1}{4} - \frac{1}{8}$ (d) $\frac{1}{2} - \frac{3}{8}$ (e) $\frac{5}{8} - \frac{1}{4}$

3 Add these fractions.

(a) $\frac{1}{10} + \frac{1}{10}$ (b) $\frac{1}{10} + \frac{3}{10}$ (c) $\frac{1}{5} + \frac{1}{10}$ (d) $\frac{1}{2} + \frac{3}{10}$ (e) $\frac{3}{10} + \frac{2}{5}$

4 Subtract these fractions.

(a) $\frac{3}{10} - \frac{1}{10}$ (b) $\frac{5}{10} - \frac{3}{10}$ (c) $\frac{1}{5} - \frac{1}{10}$ (d) $\frac{1}{2} - \frac{3}{10}$ (e) $\frac{7}{10} - \frac{2}{5}$

5 Add these fractions.

(a) $\frac{2}{3} + \frac{1}{3}$ (b) $\frac{1}{3} + \frac{1}{2}$ (c) $\frac{3}{5} + \frac{1}{4}$

(d) $\frac{1}{6} + \frac{2}{3}$ (e) $\frac{2}{5} + \frac{3}{8}$ (f) $\frac{3}{4} + \frac{1}{6}$

6 Add these fractions.

(a) $\frac{2}{7} + \frac{4}{7}$ (b) $\frac{1}{3} + \frac{1}{6}$ c) $\frac{2}{3} + \frac{1}{4}$

(d) $\frac{1}{5} + \frac{3}{4}$ (e) $\frac{3}{8} + \frac{1}{5}$ (f) $\frac{1}{3} + \frac{2}{5}$

7 Subtract these fractions.

(a) $\frac{2}{7} - \frac{1}{7}$ (b) $\frac{5}{6} - \frac{1}{3}$ (c) $\frac{2}{3} - \frac{1}{4}$

(d) $\frac{11}{12} - \frac{2}{3}$ (e) $\frac{5}{8} - \frac{1}{3}$ (f) $\frac{7}{9} - \frac{5}{12}$

8 Subtract these fractions.

(a) $\frac{3}{4} - \frac{1}{4}$ (b) $\frac{1}{2} - \frac{1}{3}$ (c) $\frac{3}{4} - \frac{3}{5}$

(d) $\frac{3}{4} - \frac{1}{6}$ (e) $\frac{3}{5} - \frac{1}{2}$ (f) $\frac{7}{8} - \frac{2}{3}$

9 Add these fractions.

(a) $\frac{1}{8} + \frac{1}{8} + \frac{1}{8}$ (b) $\frac{1}{8} + \frac{3}{8} + \frac{1}{4}$ (c) $\frac{1}{4} + \frac{1}{8} + \frac{1}{8}$

(d) $\frac{1}{2} + \frac{1}{8} + \frac{3}{8}$ (e) $\frac{1}{8} + \frac{1}{4} + \frac{1}{8}$

10 Add these fractions.

(a) $\frac{1}{12} + \frac{1}{12} + \frac{3}{12}$ (b) $\frac{1}{12} + \frac{1}{4} + \frac{1}{3}$ (c) $\frac{1}{4} + \frac{5}{12} + \frac{1}{6}$

(d) $\frac{1}{10} + \frac{1}{5} + \frac{3}{10}$ (e) $\frac{1}{5} + \frac{1}{20} + \frac{3}{10}$

Exercise continues ...

11 Work out these.

(a) $\frac{1}{2} + \frac{3}{8}$　　(b) $\frac{4}{9} + \frac{1}{3}$　　(c) $\frac{5}{6} - \frac{1}{4}$　　(d) $\frac{11}{12} - \frac{2}{3}$

(e) $\frac{8}{9} - \frac{1}{6}$　　(f) $\frac{7}{15} + \frac{3}{10}$　　(g) $\frac{4}{9} - \frac{1}{12}$　　(h) $\frac{7}{20} + \frac{5}{8}$

12 Work out these

(a) $\frac{4}{5} + \frac{7}{10} - \frac{3}{5}$　　(b) $\frac{3}{5} + \frac{5}{6} - \frac{2}{3}$　　(c) $\frac{2}{3} + \frac{3}{4} - \frac{1}{2}$

(d) $\frac{2}{5} + \frac{5}{8} - \frac{3}{4}$　　(e) $\frac{1}{5} + \frac{3}{10} - \frac{1}{2}$　　(f) $\frac{3}{7} + \frac{5}{14} - \frac{1}{2}$

13 Work out these.

(a) $\frac{3}{5} + \frac{2}{5} - \frac{7}{10}$　　(b) $\frac{1}{4} + \frac{3}{8} - \frac{1}{6}$　　(c) $\frac{1}{6} + \frac{2}{3} - \frac{1}{4}$

(d) $\frac{5}{8} + \frac{3}{5} - \frac{3}{4}$　　(e) $\frac{3}{4} - \frac{5}{6} + \frac{2}{3}$　　(f) $\frac{3}{20} - \frac{2}{5} + \frac{3}{4}$

14 In an exam Sally answered $\frac{9}{10}$ of the questions correctly and Val answered $\frac{11}{12}$ of the questions correctly.
Who answered more correctly and by what fraction?

15 Three friends shared the money they won on the lottery.
David received one third, Malcolm received one quarter and Steve received the rest.
What fraction did Steve receive?

16 Mark went on a three-day walk.
He walked $\frac{1}{2}$ the distance on the first day and $\frac{3}{10}$ of the distance on the second day.
What fraction was left to walk on the third day?

17 Dalia bought a bag of sweets.
She said she gave $\frac{1}{4}$ of them to Emily, $\frac{2}{5}$ of them to Beccy and ate $\frac{1}{2}$ of them herself.
What is wrong with what she said?

● Challenge 3.1

Make up a story problem similar to the ones in questions 14 to 17 of Exercise 3.6.

Solve it yourself and then swap stories with a friend.

Solve your friend's problem.

Check you get the same answer.

Converting a fraction to a decimal

Since a fraction like $\frac{5}{8}$ means the same as $5 \div 8$, you can use division to change a fraction into a decimal.

Example 3.8

Convert $\frac{5}{8}$ to a decimal.

Solution

First write 5 as 5.000.

You may need more or fewer zeros depending on the fraction.

Now work out $5.000 \div 8$

$$\begin{array}{r} 0.6\ 2\ 5 \\ 8\overline{)5.0^20^40} \end{array}$$

If the division is not exact, you may need to round your answer to a given number of decimal places.

Exercise 3.7

1 Change each of these fractions to a decimal.
   If necessary, give your answer to 3 decimal places.

   (a) $\frac{4}{5}$ (b) $\frac{3}{8}$ (c) $\frac{9}{20}$ (d) $\frac{1}{5}$ (e) $\frac{3}{4}$ (f) $\frac{3}{15}$

   (g) $\frac{3}{7}$ (h) $\frac{7}{10}$ (i) $\frac{1}{2}$ (j) $\frac{2}{5}$ (k) $\frac{4}{9}$

What you have learned

- To change a fraction to an equivalent fraction you multiply or divide both the numerator and the denominator by the same number
- To express a fraction in its lowest terms or simplest form you write it as an equivalent fraction using the smallest possible numbers
- To order fractions you write them as equivalent fractions all with the same denominator and then order them by the numerator
- To find a fraction of a quantity first divide the quantity by the denominator of the fraction and then multiply the result by the numerator
- To express one number as a fraction of another, write the first number as the numerator of the fraction and the second number as the denominator then express the fraction in its lowest terms
- To add or subtract fractions write them as equivalent fractions all with the same denominator and then add or subtract the numerators
- To convert a fraction to a decimal divide the numerator by the denominator

Mixed exercise 3

1 Copy and complete the following.

(a) $\dfrac{3}{4} = \dfrac{\square}{20}$ (b) $\dfrac{15}{21} = \dfrac{5}{\square}$ (c) $\dfrac{1}{2} = \dfrac{\square}{22}$ (d) $\dfrac{18}{60} = \dfrac{3}{\square}$

(e) $\dfrac{16}{18} = \dfrac{\square}{9}$ (f) $\dfrac{3}{7} = \dfrac{12}{\square}$ (g) $\dfrac{4}{9} = \dfrac{\square}{90}$ (h) $\dfrac{8}{24} = \dfrac{1}{\square}$

2 Cancel these fractions to their lowest terms.

(a) $\dfrac{16}{24}$ (b) $\dfrac{80}{100}$ (c) $\dfrac{20}{55}$ (d) $\dfrac{36}{60}$ (e) $\dfrac{18}{45}$

(f) $\dfrac{21}{77}$ (g) $\dfrac{66}{88}$ (h) $\dfrac{75}{90}$ (i) $\dfrac{120}{150}$ (j) $\dfrac{26}{52}$

3 For each pair of fractions
- find the common denominator.
- state which is the bigger fraction.

(a) $\dfrac{4}{5}$ or $\dfrac{5}{6}$ (b) $\dfrac{1}{3}$ or $\dfrac{2}{7}$ (c) $\dfrac{13}{20}$ or $\dfrac{5}{8}$

4 Work out these.

(a) $\dfrac{1}{4}$ of 32   (b) $\dfrac{1}{5}$ of 55   (c) $\dfrac{3}{4}$ of 60   (d) $\dfrac{4}{5}$ of 200

5 A rope was 48 metres long. Charlotte cut off $\dfrac{1}{6}$ of it.
How much was left?

6 There were 12 000 spectators at a football match. $\dfrac{9}{10}$ of them were adults.
How many adults were there?

7 Imran cycled 42 km. He stopped for a rest after $\dfrac{2}{3}$ of the journey.
How far had he travelled before he stopped?

8 Which is the larger, $\dfrac{3}{4}$ of 180 or $\dfrac{7}{10}$ of 200?
Show your working.

9 18 of the students in a class are girls. There are 30 students in the class.
What fraction of the students are girls?
Give your answer in its lowest terms.

10 Work out these.

(a) $\dfrac{3}{5} + \dfrac{1}{5}$ (b) $\dfrac{1}{7} + \dfrac{2}{3}$ (c) $\dfrac{5}{8} - \dfrac{1}{6}$ (d) $\dfrac{7}{10} + \dfrac{2}{15}$ (e) $\dfrac{11}{12} - \dfrac{3}{8}$

11 Change each of these fractions to a decimal.
Where necessary, give your answer correct to 3 decimal places.

(a) $\dfrac{1}{8}$ (b) $\dfrac{2}{9}$ (c) $\dfrac{5}{7}$ (d) $\dfrac{3}{11}$

Solving problems

- Solving problems involving time without using a calculator
- Converting between metric units
- Converting between metric and imperial units
- Reading tables

- How to tell and write the time in words and figures
- How to convert between metric units
- How to convert between metric and imperial units
- How to add, subtract, multiply and divide decimals

Problems involving time

🔍 Check up 4.1

One scale you are used to is the clock face.
It has equal divisions around a circle.

(a) How many hours does it take for the little hand to go once around the clock face?

(b) How many minutes does it take for the big hand to go once around the clock face?

(c) The time shown on this clock is 3:35 or twenty-five to four.

Write the times shown on these clocks using numbers and using words.

(i)

(ii)

(iii)

(iv)

(v)

(vi)

(vii)

(viii)

(ix)

(x)

(xi)

(xii)

When you are working with times, you have to remember how the different units relate to each other.

- 1 day = 24 hours
- 1 hour = 60 minutes
- 1 minute = 60 seconds

Example 4.1

How many minutes are there between these times?

(a) 8:18 and 8:45

(b) $\frac{1}{4}$ past 3 and 10 past 4

(c) 13:20 and 14:19

(d) 14:25 and 16:09

Solution

(a) A simple method to use is to count on from the start time.
From 18 minutes to 30 minutes is 12 minutes.
From 30 minutes to 45 minutes is 15 minutes.
12 + 15 = 27
There are 27 minutes between 8:18 and 8:45.

(b) From $\frac{1}{4}$ past 3 to 4 o'clock is 45 minutes.
From 4 o'clock to 10 past 4 is 10 minutes.
There are 55 minutes between $\frac{1}{4}$ past 3 and 10 past 4.

(c) Sometimes you need to combine counting on with compensating.
From 13:20 to 14:20 is 60 minutes.
From 13:20 to 14:19 is 1 minute less.
There are 59 minutes between 13:20 and 14:19.

(d) From 14:25 to 16:25 is 2 hours = 120 minutes.
From 16:09 to 16:25 is 16 minutes so you need to subtract
16 minutes from 120 minutes.
120 − 16 = 104
There are 104 minutes between 14:25 and 16:09.

Example 4.2

Work out the finishing time for each of these TV programmes.

(a) The programme starts at 1:25 and lasts for 25 minutes.

(b) The programme starts at $\frac{1}{2}$ past 10 and lasts for 45 minutes.

Solution

(a) You add 25 on to 25 to give a finishing time of 1:50.

(b) You add on 30 of the 45 minutes to take the time to 11 o'clock.
Then you add on the remaining 15 minutes to give a finishing
time of 11:15 or $\frac{1}{4}$ past 11.

Exercise 4.1

1 At Queen's School lessons last for 55 minutes.
The last lesson finishes at 15:20.
What time does it start?

2 (a) A train leaves Derby for London at 07:55.
The journey lasts 1 hour and 40 minutes.
At what time does the train arrive in London?
(b) A train leaves London for Derby at 22:38.
It takes 1 hour and 35 minutes.
At what time does the train arrive in Derby?

3 Jane watches these TV programmes.
  • Eastenders, which runs from 7:30 p.m. to 8 p.m.
  • Friends, which lasts from 9 p.m. to 9:35 p.m.
  • Will and Grace, which follows Friends and finishes at 10:10 p.m.

How long does she spend watching TV?

4 The time by Simon's watch is 2 minutes past 6.
Simon's watch is 7 minutes fast.
What is the correct time?

5 How many minutes are there between the following times?
  (a) 9:00 and 9:33   (b) 5:33 and 5:53
  (c) 4:10 and 4:49   (d) 12:33 and 13:13
  (e) $\frac{1}{4}$ to 10 and $\frac{1}{2}$ past 10   (f) 10 minutes to 5 and 23 minutes past 5

6 At what time do these programmes finish?
  (a) Starts at 3:05 and lasts 30 minutes.
  (b) Starts at 2:15 and lasts 50 minutes.
  (c) Starts at 12:25 and lasts 45 minutes.
  (d) Starts at 12:33 and lasts 50 minutes.
  (e) Starts at $\frac{1}{4}$ to 10 and lasts 20 minutes.
  (f) Starts at 10 to 5 and lasts 35 minutes

7 A train leaves Derby at 07:46.
It arrives in Leicester 35 minutes later.
At what time does the train reach Leicester?

Exercise continues ...

8 A train is due in Birmingham at 15:40.
It is 55 minutes late.
At what time does it arrive?

9 Melissa leaves Nottingham at 8:40.
The journey to Skegness takes 2 hours and 35 minutes.
At what time does she arrive in Skegness?

10 Jason sets a security light to switch on at 21:35 and switch off at 23:10.
For how long is the light on?

Problems involving units

Some problems require you to know how to convert between one
metric unit and another. You need to know the conversion rates. Other
problems require you to change between metric and imperial units.
For these you will be told the conversion rate to use.

Example 4.3

(a) A farmer supplies a dairy with 500 litres of milk.
1 litre is roughly 1.75 pints.
How many pints of milk does the farmer supply?

(b) A fence is 20 metres long.
1 metre is about 3.3 feet.
How long is the fence in feet?

Solution

(a) 1 litre = 1.75 pints
500 litres = 500 × 1.75 = 875 pints

(b) 1 metre = 3.3 feet
20 metres = 20 × 3.3 = 66 feet

TIP

You can check your answers. A litre is bigger than a pint
so the number of pints must be more than the number of
litres. Similarly, a metre is bigger than a foot so the number
of feet must be more than the number of metres.

Exercise 4.2

Here are some approximate conversions.

8 km is about 5 miles.
1 m is about 40 inches.
1 foot is about 30 cm.
1 inch is about 2.5 cm.
1 kg is about 2 pounds.
25 g is about 1 ounce.
4 litres is about 7 pints.

1   A recipe for a fish pie uses 350 g of fish.
    About how much is this in ounces?

2   To make leek soup for five people you need 750 g of leeks.
    About how much is this in pounds?

3   (a)   Copy the table.
          Use the conversions above to give these quantities in metric units.

Imperial	Metric
5 feet	
12 pounds	
5 pints	

   (b)   Copy the table.
         Use the conversions above to give these quantities in imperial units.

Metric	Imperial
20 km	
3 m	
10 litres	
15 kg	

Reading tables

You will often need to extract information from tables. You need to take care that you look in the correct part of the table.

You will find distance charts in many road atlases and it is worth knowing how to use these.

Example 4.4

Use the distance chart to find how far it is from Carlisle to Fort William.

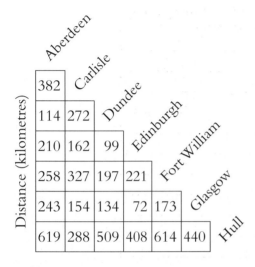

Solution

You look down the Carlisle column and across the Fort William row.

Where the column and the row meet you will find the entry you need.

The entry you need is shaded in the chart below.

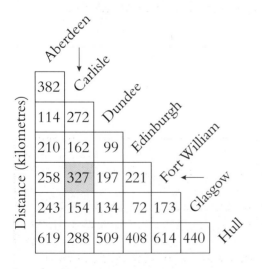

The distance from Carlisle to Fort William is 327 km.

Exercise 4.3

1   The table shows the cost per minute, in pence, of phone calls to various parts of the world.

Country	Price per minute	Country	Price per minute
Australia	3.5	New Zealand	4.16
Canada	4.16	Pakistan	16.66
China	3	Philippines	14.18
France	3.5	Poland	3.5
Germany	3.5	Russia	5
Hungary	4.5	South Africa	6
India	12.5	Spain	3.5
Ireland	3.5	Thailand	7.5
Italy	3.5	UK	3.5
Jamaica	7.5	United States	3.5
Lithuania	9	Zimbabwe	6.5

(a) How much does a 5-minute call to Poland cost?

(b) Which is more expensive and by how much:
    a 7-minute call to Jamaica or an 8-minute call to Zimbabwe?

(c) For how many countries is the cost per minute greater than 10p?

2   Look at the distance chart in Example 4.4.

(a) How far is it from Hull to Dundee?

(b) How much further is it from Aberdeen to Fort William than from Aberdeen to Glasgow?

What you have learned

- Time does not use metric units; there are 60 seconds in a minute, 60 minutes in an hour and 24 hours in a day
- You can work out time intervals, start times and finish times by counting on or back
- You can convert between different metric units or between metric and imperial units by multiplying or dividing by the conversion rates
- When you are reading a table you have to take care that you are looking at the correct part of the table

Mixed exercise 4

1   Amy wants to record a 2–hour film which starts at ten minutes to midnight.
   What are the start and stop times? Give your answers using the 24-hour clock.

2   Work out the costs of these.
   Give your answers in pounds (£).

   (a) Four books at 225p each

   (b) Eight bars of chocolate at 75p each

   (c) 3.5 kg of potatoes at £1.46 per kilogram

3   Work out the costs of these.
   Give your answers in pounds (£).

   (a) Seven blu-ray discs at 995p each

   (b) Five bottles of lemonade at 79p each

   (c) Four tubs of butter at 63p each

4   To work out the number of rolls of wallpaper needed to paper a room a DIY shop
   shows this table.

Height of room (feet)	Distance around room (feet)						
	40	45	50	55	60	65	70
8	7	8	8	9	10	11	12
10	8	9	10	11	12	13	14
12	10	11	12	14	15	16	17

   (a) The distance around Tom's room is 65 feet and the height is 10 feet.
       How many rolls does he need?

   (b) The distance around Phoebe's room is 50 feet and the height is 8 feet.
       How many rolls does she need?

5   1 mile is approximately 1.6 kilometres.

   (a) On a British road the speed limit is 30 mph.
       Convert this to kilometres per hour.

   (b) On an Italian road the speed limit is 80 km/h.
       Convert this to miles per hour.

6   The table shows an approximate relationship between gallons and litres.

Gallons	1	2	5	10	15	20	100	200
Litres	4.5	9	22.5	45	67.5	90	450	900

   Use the table to calculate how many litres there are in these volumes.

   (a) 7 gallons   (b) 50 gallons   (c) 300 gallons

Angles

▶ **This chapter is about**

- Angles on a straight line, angles around a point and vertically opposite angles
- Recognising perpendicular and parallel lines
- The angles associated with parallel lines

▶ **You should already know**

- What an angle is
- That angles are measured in degrees
- That a right angle is 90°, a straight line is 180° and a full turn is 360°
- That an acute angle is less than 90°, an obtuse angle is between 90° and 180° and a reflex angle is greater than 180°
- That the angles in a triangle add up to 180°

Angles on a straight line

You learned in Unit A Chapter 11 that a straight line is an angle of 180°.

So, in these two diagrams

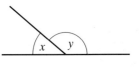   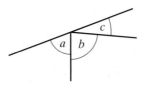

$$x + y = 180° \qquad \text{and} \qquad a + b + c = 180°.$$

Angle fact 1: Angles on a straight line add up to 180°.

Example 5.1

Work out the size of angle x in this diagram.

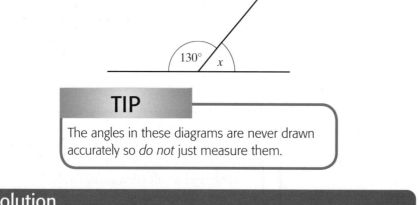

TIP

The angles in these diagrams are never drawn accurately so *do not* just measure them.

Solution

You use the fact that angles on a straight line add up to $180°$.

$x = 180 - 130$
$x = 50°$

TIP

There is no need to put the degree signs in your working but you must put the degree sign in your answer.

Example 5.2

Work out the size of angle y in this diagram.

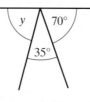

Solution

$y = 180 - (35 + 70)$ You use the fact that angles on a straight line add up to $180°$.

$y = 180 - 105$ Add together the angles given and then subtract from 180.

$y = 75°$

Exercise 5.1

Work out the size of the unknown angle in each of these diagrams.

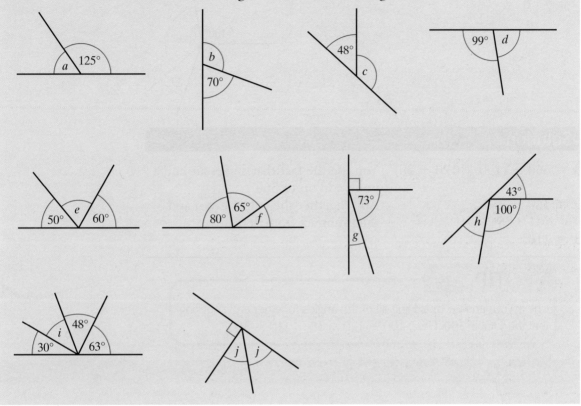

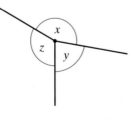

Angles around a point

You learned in Unit A Chapter 11 that a full turn is an angle of 360°.

So, in this diagram

$$x + y + z = 360°.$$

Angle fact 2: Angles around a point add up to 360°.

Example 5.3

Work out the size of angle x in this diagram.

Solution

$x = 360 - (100 + 120 + 30)$ You use the fact that angles around a point add up to 360°.

$x = 360 - 250$ Add together the angles given and then subtract from 360.

$x = 110°$

> **TIP**
>
> Check your answer by adding all of the angles together and checking that the total is 360. Here, $100 + 120 + 30 + 110 = 360$.

Exercise 5.2

Work out the size of the unknown angle in each of these diagrams.

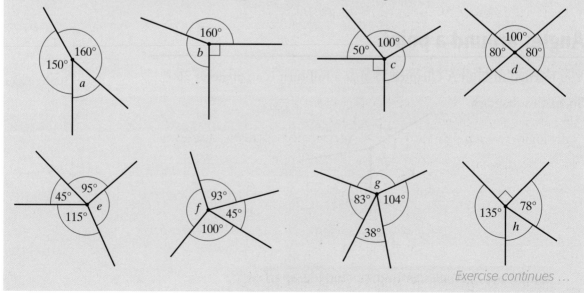

Exercise continues …

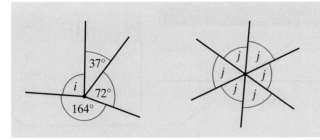

● Challenge 5.1

How many different ways can you fit these angles
(a) on a straight line?
(b) around a point?

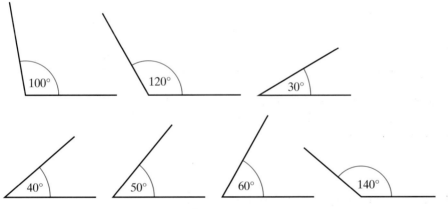

Check with a friend. Who found more ways?

Vertically opposite angles

When two straight lines cross they form four angles.
The angles across from each other are equal.
The angles in each pair are known as **vertically opposite angles**.

So, in this diagram

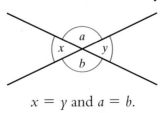

$$x = y \text{ and } a = b.$$

Angle fact 3: Vertically opposite angles are equal.

Example 5.4

Work out the size of angles b, c and d in this diagram.
Give a reason for each of your answers.

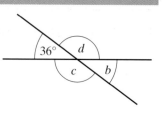

Solution

$b = 36°$ Vertically opposite angles are equal.
$c = 180 - 36$ Angles on a straight line add up to 180°.
   $= 144°$
$d = 144°$ Vertically opposite angles are equal.

Exercise 5.3

For each question   ● work out the size of each unknown angle.
                   ● give a reason for each answer.

1

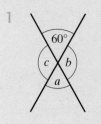

2

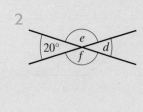

3

4

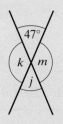

5

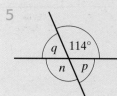

6

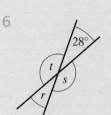

7

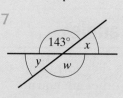

8

9

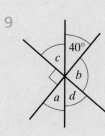

10

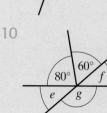

Angles made with parallel lines

Lines that meet or cross at right angles are **perpendicular**.

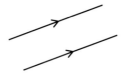

Lines that never cross and are always the same distance apart are **parallel**.

You show that lines are parallel by marking them with arrow heads.

This is a map of part of New York.

(a) Find Broadway and W 32nd Street on the map.
Find some more angles equal to the angle between Broadway and W 32nd Street.

Two angles that add up to 180° are called **supplementary** angles.

(b) Find an angle that is supplementary to the angle between Broadway and W 32nd Street.

(c) Explain you results.

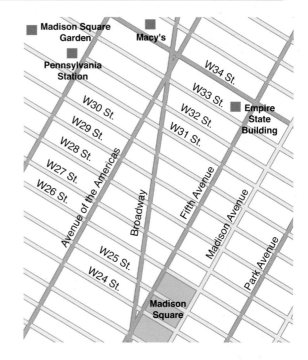

In Discovery 5.1 you should have found three sorts of angles made with parallel lines.

Corresponding angles

The diagrams show equal angles made by a line cutting across a pair of parallel lines.

These equal angles are called **corresponding** angles.

Corresponding angles occur in an F–shape.

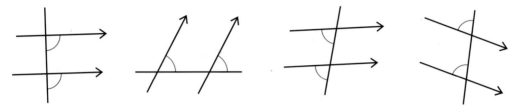

Alternate angles

These diagrams also show equal angles made by a line cutting across a pair of parallel lines.

These equal angles are called **alternate** angles.

Alternate angles occur in a Z–shape.

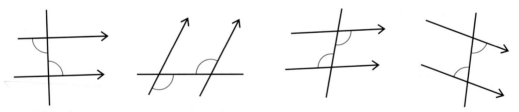

Allied angles

You can see that the two angles marked in these diagrams are not equal. Instead, they are supplementary. (Remember that supplementary angles add up to 180°.)

These angles are called **allied** angles or **co-interior** angles and occur in a C–shape.

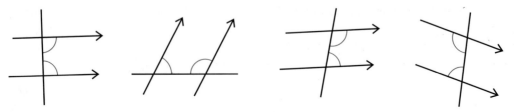

You can use the properties associated with parallel lines to prove some of the properties associated with special quadrilaterals.

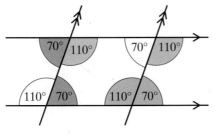

Alternate angles are equal.

Corresponding angles are equal.

Angles on a straight line add up to 180°.

Therefore the opposite angles of a parallelogram are equal.

Questions about finding the size of angles often ask you to give reasons for your answers. This means that you must say why, for example, angles are equal. Stating that the angles are alternate angles or corresponding angles would be possible reasons.

Example 5.5

Work out the size of the lettered angles.

Give a reason for each answer.

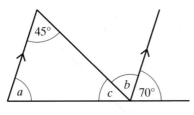

Solution

$a = 70°$ Corresponding angles
$b = 45°$ Alternate angles
$c = 65°$ Angles on a straight line add up to 180°
            or Allied angles add up to 180°
            or Angles in a triangle add up to 180°

Exercise 5.4

Find the size of the lettered angles. Give a reason for each answer.

1
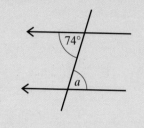

2
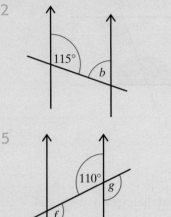

3
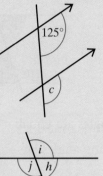

4
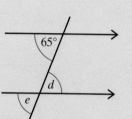

5

6
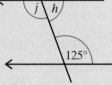

7
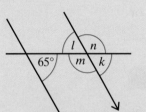

8
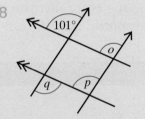

9
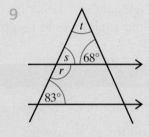

10

11
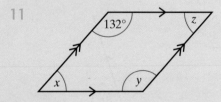

What you have learned

- The angles on a straight line add up to 180°
- The angles around a point add up to 360°
- Vertically opposite angles are equal
- Perpendicular lines cross at right angles
- Parallel lines never cross and are always the same distance apart
- When a line crosses a pair of parallel lines, corresponding angles are equal
- When a line crosses a pair of parallel lines, alternate angles are equal
- When a line crosses a pair of parallel lines, allied angles add up to 180°

Mixed exercise 5

1 For each diagram, work out the size of the unknown angle.

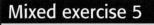
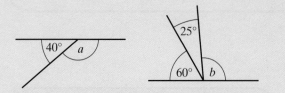

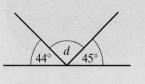

2 For each diagram, work out the size of the unknown angle.

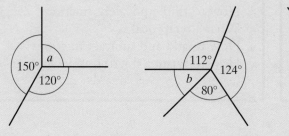
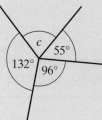
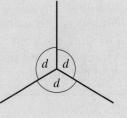

3 Work out the size of the unknown angles.
Give a reason for each of your answers.

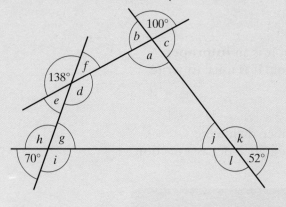

4 Find the size of the lettered angles. Give a reason for each answer.

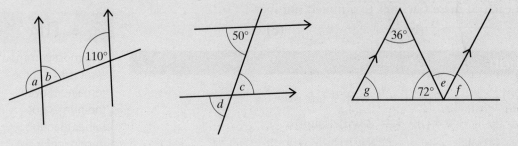

Fractions and mixed numbers

▶ You should already know

- How to find a fraction of a quantity
- How to find equivalent fractions
- What a reciprocal is
- How to add and subtract fractions
- How to write a fraction in its lowest terms

Mixed numbers

Look at this fraction.

$$\frac{4}{3}$$

The numerator is larger than the denominator. It is an **improper fraction**, sometimes called a 'top-heavy' fraction. It is usual to write fractions like this as **mixed numbers**.

$$\frac{4}{3} = 1\frac{1}{3}$$

To write an improper fraction as a mixed number, divide the denominator into the numerator and write the remainder as a fraction over the denominator.

Example 6.1

Change each of these fractions to a mixed number.

(a) $\frac{7}{4}$ **(b)** $\frac{11}{5}$ **(c)** $\frac{24}{7}$

Solution

(a) $\frac{7}{4} = 1\frac{3}{4}$ $7 \div 4 = 1$ with 3 left over.

(b) $\frac{11}{5} = 2\frac{1}{5}$ $11 \div 5 = 2$ with 1 left over.

(c) $\frac{24}{7} = 3\frac{3}{7}$ $24 \div 7 = 3$ with 3 left over.

TIP

The most common error is to put the remainder over the numerator rather than the denominator.

Mixed numbers can be changed into improper fractions. Just reverse the process.

Example 6.2

Change each of these mixed numbers to an improper fraction.

(a) $3\frac{1}{4}$ **(b)** $2\frac{3}{5}$ **(c)** $3\frac{5}{6}$

Solution

(a) $3\frac{1}{4} = 3 + \frac{1}{4}$

$\qquad = \frac{12}{4} + \frac{1}{4}$ Change the whole number to quarters and then add.

$\qquad = \frac{13}{4}$

Another way to think of it is to multiply the whole number by the denominator and add on the numerator. $(3 \times 4 + 1 = 13)$

(b) $2\frac{3}{5} = \frac{13}{5}$ $2 \times 5 + 3 = 13$

(c) $3\frac{5}{6} = \frac{23}{6}$ $3 \times 6 + 5 = 23$

Exercise 6.1

1   Change each of these improper fractions to a mixed number.

   (a) $\frac{11}{8}$ (b) $\frac{15}{8}$ (c) $\frac{9}{4}$ (d) $\frac{7}{2}$ (e) $\frac{15}{7}$

   (f) $\frac{11}{4}$ (g) $\frac{5}{2}$ (h) $\frac{9}{5}$ (i) $\frac{7}{3}$ (j) $\frac{37}{9}$

2   Change each of these improper fractions to a mixed number.

   (a) $\frac{11}{5}$ (b) $\frac{10}{7}$ (c) $\frac{10}{3}$ (d) $\frac{13}{2}$ (e) $\frac{14}{3}$

   (f) $\frac{11}{6}$ (g) $\frac{19}{8}$ (h) $\frac{20}{7}$ (i) $\frac{23}{4}$ (j) $\frac{33}{10}$

3   Change each of these mixed numbers to an improper fraction.

   (a) $1\frac{1}{8}$ (b) $2\frac{5}{8}$ (c) $3\frac{1}{4}$ (d) $5\frac{1}{2}$ (e) $5\frac{1}{5}$

   (f) $1\frac{1}{2}$ (g) $2\frac{2}{5}$ (h) $3\frac{1}{3}$ (i) $5\frac{1}{4}$ (j) $2\frac{3}{8}$

Multiplying fractions

Multiplying a fraction by a whole number

In this diagram $\frac{1}{8}$ is red.

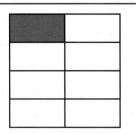

In this diagram five times as much is red, so $\frac{1}{8} \times 5 = \frac{5}{8}$.

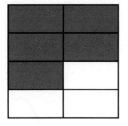

This shows that, to multiply a fraction by an integer, you just multiply the numerator by the integer. You then simplify by cancelling and changing to a mixed number if possible.

Example 6.3

Work out these.

(a) $\frac{3}{10} \times 3$ **(b)** $\frac{2}{3} \times 9$ **(c)** $\frac{3}{8} \times 2$ **(d)** $5 \times \frac{1}{2}$

Solution

(a) $\frac{3}{10} \times 3 = \frac{9}{10}$ Multiply the numerator by 3.

(b) $\frac{2}{3} \times 9 = \frac{18}{3} = \frac{6}{1} = 6$ $\frac{6}{1}$ means $6 \div 1 = 6$.

(c) $\frac{3}{8} \times 2 = \frac{6}{8} = \frac{3}{4}$ Give $\frac{6}{8}$ in its lowest terms.

(d) $5 \times \frac{1}{2} = \frac{5}{2} = 2\frac{1}{2}$ Change $\frac{5}{2}$ to a mixed number.

Multiplying a fraction by a fraction

In this diagram, $\frac{1}{4}$ is red.

In this diagram, $\frac{1}{3}$ as much is red, so $\frac{1}{4} \times \frac{1}{3} = \frac{1}{12}$.

To multiply fractions, multiply the numerators and multiply the denominators, then simplify if possible.

Example 6.4

Work out these.

(a) $\frac{1}{3} \times \frac{1}{2}$ **(b)** $\frac{3}{5} \times \frac{1}{2}$ **(c)** $\frac{3}{4}$ of $\frac{6}{7}$

Solution

(a) $\frac{1}{3} \times \frac{1}{2} = \frac{1}{6}$

Multiply the numerators and multiply the denominators.

(b) $\frac{3}{5} \times \frac{1}{2} = \frac{3}{10}$

(c) $\frac{3}{4} \times \frac{6}{7} = \frac{18}{28}$

'of' means the same as '×'

$\frac{18}{28} = \frac{9}{14}$

Simplify by dividing the numerator and the denominator by 2.

$\frac{3}{\overset{2}{\cancel{4}}} \times \frac{\overset{3}{\cancel{6}}}{7} = \frac{9}{14}$

Another way is to cancel a term in the numerator with one in the denominator. This can make the arithmetic simpler. Divide both 4 and 6 by 2, then multiply the numerators and denominators.

> **TIP**
>
> Note: $\frac{1}{2} \times \frac{1}{3} = \frac{1}{6}$
> A common error is to multiply 1×1 and get 2.

> **TIP**
>
> When cancelling, cancel a term in the numerator with one in the denominator.

Exercise 6.2

Work out these. Where necessary, write your answers as proper fractions or mixed numbers in their lowest terms.

1 $\frac{1}{2} \times 4$

2 $7 \times \frac{1}{2}$

3 $9 \times \frac{1}{3}$

4 $\frac{3}{4} \times 12$

5 $\frac{2}{5} \times 5$

6 $24 \times \frac{5}{12}$

7 $\frac{2}{3} \times 4$

8 $\frac{4}{9} \times 2$

9 $\frac{4}{5} \times 3$

10 $\frac{1}{5} \times 3$

11 $\frac{1}{4} \times \frac{2}{3}$

12 $\frac{2}{3} \times \frac{3}{5}$

13 $\frac{4}{9} \times \frac{1}{2}$

14 $\frac{1}{3} \times \frac{2}{3}$

15 $\frac{5}{6} \times \frac{3}{5}$

16 $\frac{3}{7} \times \frac{7}{9}$

17 $\frac{1}{2} \times \frac{5}{6}$

18 $\frac{3}{10} \times \frac{5}{11}$

19 $\frac{2}{3} \times \frac{5}{8}$

20 $\frac{3}{5} \times \frac{5}{12}$

21 $\frac{4}{9} \times \frac{1}{3}$

22 $\frac{5}{6} \times \frac{1}{4}$

23 $\frac{5}{7} \times \frac{3}{4}$

24 $\frac{8}{9} \times \frac{1}{6}$

25 $\frac{8}{9} \times \frac{5}{6}$

26 $\frac{7}{20} \times \frac{5}{8}$

Exercise continues …

27 Jane makes 120 cakes for a car boot sale.
She sells $\frac{5}{6}$ of them.
How many cakes does she sell?

28 After everyone has helped themselves to part of a pizza, $\frac{2}{5}$ of the pizza is left over.
For a second helping, Dom has $\frac{1}{4}$ of what is left.
What fraction of the whole pizza does Dom have for a second helping?
Give your answer in its lowest terms.

● Challenge 6.1

Stephen and Vicky bought a house together. It cost £245 000.

Vicky paid three-fifths of the cost. How much did she pay?

● Challenge 6.2

Iain gave one-tenth of his income to charity.

Two-thirds of the donation went to overseas aid.

What fraction of his income did he give to overseas aid?

Dividing fractions

When you work out $6 \div 3$, you are finding how many 3s there are in 6.

Finding $6 \div \frac{1}{3}$ is the same as finding how many $\frac{1}{3}$s there are in 6, which is $6 \times 3 = 18$.

So dividing by $\frac{1}{3}$ is the same as multiplying by 3.

This can be extended, for example, $4 \div \frac{2}{3} = 4 \times \frac{3}{2} = \frac{12}{2} = 6$.

To divide by a fraction you multiply by its **reciprocal**.

TIP

The reciprocal of a fraction is a fraction with the numerator and denominator swapped round. You can think of this as 'turning the fraction upside-down'.
You met reciprocals in Unit A Chapter 14.

Any non-zero number multiplied by its reciprocal is 1.

$\frac{1}{2} \times \frac{2}{1} = \frac{2}{2} = \frac{1}{1}$ which is 1. $\frac{5}{9} \times \frac{9}{5} = \frac{45}{45} = 1$

Zero does not have a reciprocal. You cannot divide by zero.

Example 6.5

Work out these.

(a) $\frac{8}{9} \div \frac{1}{3}$ (b) $\frac{8}{9} \div 2$ (c) $\frac{3}{4} \div \frac{2}{7}$ (d) $\frac{5}{8} \div \frac{3}{4}$

Solution

(a) $\frac{8}{9} \div \frac{1}{3} = \frac{8}{9} \times \frac{3}{1}$ The reciprocal of $\frac{1}{3}$ is $\frac{3}{1}$.

$\qquad = \frac{8}{{}_3\cancel{9}} \times \frac{\cancel{3}^1}{1}$ Divide 9 and 3 by 3.

$\qquad = \frac{8}{3} \times 1$

$\qquad = \frac{8}{3}$

$\qquad = 2\frac{2}{3}$ Change to a mixed number.

(b) $\frac{8}{9} \div 2 = \frac{8}{9} \div \frac{2}{1}$ Write 2 as $\frac{2}{1}$.

$\qquad = \frac{8}{9} \times \frac{1}{2}$ The reciprocal of 2 or $\frac{2}{1}$ is $\frac{1}{2}$.

$\qquad = \frac{\cancel{8}^4}{9} \times \frac{1}{\cancel{2}_1}$ Divide 8 and 2 by 2.

$\qquad = \frac{4}{9} \times 1$

$\qquad = \frac{4}{9}$

(c) $\frac{3}{4} \div \frac{2}{7} = \frac{3}{4} \times \frac{7}{2}$ The reciprocal of $\frac{2}{7}$ is $\frac{7}{2}$.

$\qquad = \frac{21}{8}$ Multiply the numerators and denominators.

$\qquad = 2\frac{5}{8}$ Change the improper fraction into a mixed number.

(d) $\frac{5}{8} \div \frac{3}{4} = \frac{5}{8} \times \frac{4}{3}$ The reciprocal of $\frac{3}{4}$ is $\frac{4}{3}$.

$\qquad = \frac{5}{{}_2\cancel{8}} \times \frac{\cancel{4}^1}{3}$ Divide 4 and 8 by 4.

$\qquad = \frac{5}{6}$

TIP

Never cancel fractions at the divide stage. Wait until it has changed to a multiplication.

Exercise 6.3

Work out these. Where necessary, write your answers as proper fractions or mixed numbers in their lowest terms.

1 $9 \div \frac{1}{3}$ 2 $12 \div \frac{3}{4}$ 3 $\frac{3}{4} \div \frac{1}{2}$ 4 $\frac{2}{3} \div 3$ 5 $\frac{2}{3} \div 5$

Exercise continues ...

6 $\frac{1}{3} \div \frac{3}{4}$ **7** $\frac{4}{9} \div 2$ **8** $\frac{3}{8} \div \frac{1}{4}$ **9** $\frac{4}{5} \div 4$ **10** $\frac{2}{3} \div \frac{1}{3}$

11 $\frac{2}{3} \div \frac{1}{6}$ **12** $\frac{5}{6} \div 10$ **13** $\frac{7}{9} \div \frac{1}{9}$ **14** $\frac{3}{4} \div \frac{1}{8}$ **15** $\frac{4}{5} \div \frac{3}{10}$

16 $\frac{1}{4} \div \frac{3}{8}$ **17** $\frac{3}{4} \div \frac{5}{6}$ **18** $\frac{2}{5} \div \frac{1}{15}$ **19** $\frac{2}{5} \div \frac{7}{10}$ **20** $\frac{1}{6} \div \frac{3}{4}$

21 $\frac{3}{8} \div \frac{1}{4}$ **22** $\frac{2}{3} \div \frac{5}{8}$ **23** $\frac{3}{4} \div \frac{3}{8}$ **24** $\frac{2}{3} \div \frac{4}{15}$ **25** $\frac{3}{5} \div \frac{9}{10}$

26 $\frac{1}{8} \div \frac{5}{12}$ **27** $\frac{5}{7} \div 3$ **28** $6 \div \frac{2}{3}$ **29** $\frac{2}{3} \div 5$ **30** $\frac{11}{12} \div \frac{2}{3}$

31 $\frac{4}{5} \div \frac{1}{2}$ **32** $\frac{7}{10} \div \frac{4}{5}$ **33** $\frac{7}{15} \div \frac{3}{10}$ **34** $\frac{4}{9} \div \frac{1}{12}$

35 Out of Peter's take-home pay, he gives his Mum £50 for food and rent.
He spends $\frac{3}{8}$ of the rest and saves what is left over.
How much is his take-home pay in a week when he saves £60?

36 In an election, the Green candidate got 240 votes.
This was $\frac{3}{8}$ of the votes.
The Independent candidate got $\frac{1}{5}$ of the votes.
How many votes did the Independent candidate get?

Multiplying and dividing mixed numbers

When multiplying and dividing mixed numbers, you first have to
change the mixed numbers into improper fractions.

Example 6.6

Work out these.
(a) $2\frac{1}{2} \times 4\frac{3}{5}$ **(b)** $2\frac{3}{4} \div 1\frac{5}{8}$

Solution

(a) $2\frac{1}{2} \times 4\frac{3}{5} = \frac{5}{2} \times \frac{23}{5}$ First change the mixed numbers into improper fractions.

$= \frac{\cancel{5}^{1}}{2} \times \frac{23}{\cancel{5}_{1}}$ The arithmetic is much easier if you cancel the 5s.

$= \frac{23}{2}$ Multiply the numerators and multiply the denominators.

$= 11\frac{1}{2}$ Change the result back to a mixed number.

TIP

A common error
is to multiply the
whole numbers first.
You must change
mixed numbers into
improper fractions to
multiply or divide.

(b) $2\frac{3}{4} \div 1\frac{5}{8} = \frac{11}{4} \div \frac{13}{8}$ Change the mixed numbers to improper fractions.

$= \frac{11}{\underset{1}{\cancel{4}}} \times \frac{\cancel{8}^2}{13}$ Turn the second fraction upside-down and multiply.
The arithmetic is easier if you cancel the 4s.

$= \frac{22}{13}$ Multiply the numerators and multiply the denominators.

$= 1\frac{9}{13}$ Change back to a mixed number.

> ### TIP
>
> When cancelling, divide a term in the numerator and a term in the denominator by the same number.
> Only cancel a division calculation when it is at the multiplication stage.

Note that if you are multiplying or dividing by a whole number like 6, you can write it as $\frac{6}{1}$.

Exercise 6.4

Work out these.

1. (a) $4\frac{1}{2} \times 2\frac{1}{6}$ (b) $1\frac{1}{2} \times 3\frac{2}{3}$ (c) $4\frac{1}{5} \times 1\frac{2}{3}$

   (d) $3\frac{1}{3} \times 2\frac{2}{5}$ (e) $2\frac{2}{5} \times \frac{3}{4}$ (f) $3\frac{1}{5} \times 1\frac{2}{3}$

2. (a) $2\frac{1}{3} \div 1\frac{1}{3}$ (b) $2\frac{2}{5} \div 1\frac{1}{2}$ (c) $3\frac{1}{5} \div \frac{4}{15}$

   (d) $3\frac{1}{8} \div 1\frac{1}{4}$ (e) $2\frac{1}{4} \div 3\frac{1}{2}$ (f) $1\frac{1}{5} \div \frac{4}{15}$

3. (a) $3\frac{1}{2} \times 2\frac{1}{5}$ (b) $4\frac{2}{7} \times \frac{1}{2}$ (c) $2\frac{3}{4} \div 1\frac{3}{4}$

   (d) $1\frac{5}{12} \div 3\frac{1}{3}$ (e) $3\frac{1}{5} \times 2\frac{5}{8}$ (f) $2\frac{7}{8} \div 1\frac{3}{4}$

   (g) $2\frac{7}{9} \times 3\frac{3}{5}$ (h) $5\frac{5}{6} \div 1\frac{3}{4}$ (i) $3\frac{5}{7} \times 2\frac{1}{13}$

   (j) $5\frac{2}{5} \div 2\frac{1}{4}$ (k) $5\frac{2}{7} \times 3\frac{1}{2}$ (l) $4\frac{1}{12} \div 3\frac{1}{4}$

4. (a) $2\frac{1}{2} \times 1\frac{1}{3} \times 1\frac{3}{8}$ (b) $1\frac{1}{2} \times 2\frac{2}{3} \div 1\frac{3}{5}$ (c) $1\frac{1}{4} \times 3\frac{1}{5} \div 1\frac{1}{2}$

   (d) $3\frac{1}{2} \times 1\frac{2}{3} \times \frac{5}{7}$ (e) $1\frac{1}{4} \times 1\frac{2}{3} \div 1\frac{1}{9}$ (f) $3\frac{1}{3} \times 1\frac{1}{4} \div 2\frac{1}{2}$

Challenge 6.3

(a) Find the perimeter of this rectangle.
(b) Find the area of this rectangle.

$5\frac{1}{4}$ cm

$3\frac{2}{3}$ cm

Adding and subtracting mixed numbers

There are two methods you can use to add mixed numbers. You can
either add the whole number parts first and then add the fraction
parts. Alternatively you can convert the mixed numbers into improper
fractions and then add.

Example 6.7

Work out $1\frac{1}{4} + 2\frac{1}{2}$.

Solution

Method 1

$$1\frac{1}{4} + 2\frac{1}{2} = 1 + 2 + \frac{1}{4} + \frac{1}{2}$$

$$= 3 + \frac{1}{4} + \frac{1}{2} \qquad \text{Add the whole numbers first.}$$

$$= 3 + \frac{1}{4} + \frac{2}{4} \qquad \begin{array}{l}\text{Change the fractions into equivalent} \\ \text{fractions with a common denominator.}\end{array}$$

$$= 3\frac{3}{4} \qquad \text{Add the fractions.}$$

Method 2

$$1\frac{1}{4} + 2\frac{1}{2} = \frac{5}{4} + \frac{5}{2} \qquad \begin{array}{l}\text{Change both mixed numbers into} \\ \text{improper fractions.}\end{array}$$

$$= \frac{5}{4} + \frac{10}{4} \qquad \begin{array}{l}\text{Change the fractions into equivalent} \\ \text{fractions with a common denominator.}\end{array}$$

$$= \frac{15}{4} \qquad \text{Add the fractions.}$$

$$= 3\frac{3}{4} \qquad \begin{array}{l}\text{Change the improper fraction into a} \\ \text{mixed number.}\end{array}$$

Example 6.8

Work out $2\frac{3}{5} + 4\frac{2}{3}$.

Solution

Method 1

$2\frac{3}{5} + 4\frac{2}{3} = 6 + \frac{3}{5} + \frac{2}{3}$ Add the whole numbers first.

$= 6 + \frac{9}{15} + \frac{10}{15}$ Change the fractions into equivalent fractions with a common denominator.

$= 6 + \frac{19}{15}$ Add the fractions. $\frac{19}{15}$ is an improper fraction. You need to change it into a mixed number and add the whole number to the 6 you already have.

$= 7\frac{4}{15}$ $\frac{19}{15} = 1\frac{4}{15}$ and $6 + 1 = 7$.

Method 2

$2\frac{3}{5} + 4\frac{2}{3} = \frac{13}{5} + \frac{14}{3}$ Change both mixed numbers into improper fractions.

$= \frac{39}{15} + \frac{70}{15}$ Change the fractions into equivalent fractions with a common denominator.

$= \frac{109}{15}$ Add the fractions.

$= 7\frac{4}{15}$ Change the improper fraction into a mixed number.

You subtract mixed numbers in a similar way.

Example 6.9

Work out $3\frac{3}{4} - 1\frac{1}{3}$.

Solution

Method 1

$3\frac{3}{4} - 1\frac{1}{3} = 3 - 1 + \frac{3}{4} - \frac{1}{3}$ Split the calculation into two parts.

$= 2 + \frac{3}{4} - \frac{1}{3}$ Subtract the whole numbers first.

$= 2 + \frac{9}{12} - \frac{4}{12}$ Change the fractions into equivalent fractions with a common denominator.

$= 2\frac{5}{12}$ Subtract the fractions.

Method 2

$3\frac{3}{4} - 1\frac{1}{3} = \frac{15}{4} - \frac{4}{3}$ Change both mixed numbers into improper fractions.

$= \frac{45}{12} - \frac{16}{12}$ Change the fractions into equivalent fractions with a common denominator.

$= \frac{29}{12}$ Subtract the fractions.

$= 2\frac{5}{12}$ Change the improper fraction into a mixed number.

Example 6.10

Work out $5\frac{3}{10} - 2\frac{3}{4}$.

Solution

Method 1

$$5\frac{3}{10} - 2\frac{3}{4} = 5 - 2 + \frac{3}{10} - \frac{3}{4}$$ Split the calculation into two parts.

$$= 3 + \frac{3}{10} - \frac{3}{4}$$ Subtract the whole numbers first.

$$= 3 + \frac{6}{20} - \frac{15}{20}$$ Change the fractions into equivalent fractions with a common denominator.

$$= 2 + \frac{20}{20} + \frac{6}{20} - \frac{15}{20}$$ $\frac{6}{20}$ is smaller than $\frac{15}{20}$ and would give a negative answer. Take 1 of the whole units and change it to $\frac{20}{20}$.

$$= 2 + \frac{26}{20} - \frac{15}{20}$$ Add it to $\frac{6}{20}$.

$$= 2\frac{11}{20}$$ Subtract the fractions.

Method 2

$$5\frac{3}{10} - 2\frac{3}{4} = \frac{53}{10} - \frac{11}{4}$$ Change both mixed numbers into improper fractions.

$$= \frac{106}{20} - \frac{55}{20}$$ Change the fractions into equivalent fractions with a common denominator.

$$= \frac{51}{20}$$ Subtract the fractions.

$$= 2\frac{11}{20}$$ Convert the improper fraction into a mixed number.

Exercise 6.5

1 Add each of these. Write your answers as simply as possible.

(a) $1\frac{1}{3} + 3\frac{1}{4}$ (b) $1\frac{1}{2} + 2\frac{1}{6}$ (c) $3\frac{1}{5} + \frac{7}{10}$ (d) $1\frac{4}{5} + 2\frac{1}{10}$

(e) $1\frac{3}{4} + 4\frac{2}{5}$ (f) $6\frac{1}{6} + 1\frac{4}{9}$ (g) $2\frac{5}{6} + 7\frac{4}{9}$ (h) $2\frac{4}{7} + 1\frac{2}{3}$

(i) $\frac{2}{7} + \frac{1}{2} + \frac{5}{14}$ (j) $\frac{4}{5} + 1\frac{3}{4} + 2\frac{1}{2}$ (k) $1\frac{1}{2} + \frac{3}{4} + 2\frac{3}{8}$ (l) $6\frac{1}{3} + 1\frac{4}{9} + 1\frac{2}{9}$

2 Subtract each of these. Write your answers as simply as possible.

(a) $2\frac{4}{5} - 1\frac{3}{5}$ (b) $2\frac{2}{3} - 1\frac{1}{6}$ (c) $5\frac{3}{8} - 2\frac{1}{4}$ (d) $3\frac{5}{8} - 1\frac{1}{4}$

(e) $3\frac{2}{3} - \frac{1}{2}$ (f) $2\frac{4}{5} - \frac{1}{2}$ (g) $3\frac{2}{5} - 1\frac{3}{4}$ (h) $4\frac{2}{5} - 1\frac{1}{4}$

(i) $5\frac{1}{6} - 3\frac{2}{3}$ (j) $8\frac{1}{6} - 5\frac{3}{8}$ (k) $5\frac{1}{5} - \frac{2}{3}$ (l) $1\frac{1}{4} - \frac{5}{8}$

Exercise continues …

3 There were three books in a pile on Faisal's desk.
   The first book was $2\frac{3}{4}$ inches high, the second $\frac{7}{8}$ inch high and the third $1\frac{5}{6}$ inches high.
   What was the total height of the pile?

4 The blade of a knife was $5\frac{3}{4}$ inches long.
   The handle was $4\frac{2}{5}$ inches long.
   What was the total length of the knife?

5 Caroline bought a piece of ribbon
   24 inches long. She cut off two pieces, each $5\frac{5}{8}$ inches long.
   How long was the piece she had left?

6 Sam had a piece of wood $28\frac{1}{2}$ inches long.
   After using some, $9\frac{5}{8}$ inches were left.
   What length did he use?

What you have learned

- To change an improper fraction into a mixed number, divide the denominator into the numerator and write the remainder as a fraction over the denominator
- To change a mixed number into an improper fraction, multiply the whole number by the denominator, add on the numerator and write the result as a fraction over the denominator
- To multiply a fraction by an integer, multiply just the numerator by the integer then simplify the answer if possible
- To multiply a fraction by a fraction, multiply the numerators and multiply the denominators then simplify the answer if possible
- To divide by a fraction, you multiply by its reciprocal
- The reciprocal of a fraction is a fraction with the numerator and the denominator swapped around; you can think of it as turning the fraction upside-down
- To multiply or divide mixed numbers, you must change the mixed numbers into improper fractions first
- To add or subtract fractions you change the fractions into equivalent fractions with a common denominator; you then add or subtract the numerators only
- There are two methods you can use to add mixed numbers: you can add the whole number parts first and then add the fraction parts; alternatively, you can convert the mixed numbers into improper fractions and then add.
- Similarly, there are two methods you can use to subtract mixed numbers; either dealing with the whole numbers and fractions separately or converting to improper fractions

Mixed exercise 6

1 (a) Write each of these as a mixed number.
  (i) $\frac{4}{3}$ (ii) $\frac{12}{7}$ (iii) $\frac{15}{4}$

(b) Write each of these as an improper fraction.
  (i) $2\frac{1}{3}$ (ii) $1\frac{5}{8}$ (iii) $3\frac{1}{4}$

2 Work out these. Where possible, cancel the fractions to their lowest terms.
(a) $\frac{1}{9} \times 5$ (b) $\frac{1}{12} \times 4$ (c) $4 \times \frac{3}{20}$ (d) $6 \times \frac{2}{17}$ (e) $\frac{2}{9} \times 3$

3 Work out these. Write your answers as mixed numbers.
(a) $\frac{3}{4} \times 5$ (b) $\frac{2}{5} \times 7$ (c) $6 \times \frac{3}{7}$ (d) $10 \times \frac{2}{3}$ (e) $\frac{8}{15} \times 3$

4 Georgina cut eight pieces of string, each $\frac{3}{5}$ metres long.
How much string did she use?
Write your answer as a mixed number.

5 Work out these.
(a) $\frac{3}{5} \times \frac{2}{3}$ (b) $\frac{4}{7} \times \frac{5}{6}$ (c) $\frac{5}{8} \div \frac{2}{3}$ (d) $\frac{9}{10} \div \frac{3}{7}$ (e) $\frac{15}{16} \times \frac{12}{25}$

6 Work out these.
(a) $1\frac{2}{3} \times 2\frac{1}{5}$ (b) $2\frac{5}{6} \div 1\frac{3}{4}$ (c) $2\frac{5}{8} \times 1\frac{3}{7}$ (d) $1\frac{7}{10} \div 4\frac{2}{5}$ (e) $2\frac{3}{4} \times 3\frac{3}{7}$

7 Work out these.
(a) $\frac{3}{5} + \frac{4}{5}$ (b) $\frac{3}{7} + \frac{2}{3}$ (c) $\frac{5}{8} - \frac{1}{6}$ (d) $\frac{7}{10} + \frac{2}{15}$ (e) $\frac{11}{12} - \frac{3}{8}$

8 Work out these.
(a) $3\frac{1}{4} + 2\frac{1}{6}$ (b) $4\frac{3}{4} - 1\frac{2}{5}$ (c) $5\frac{1}{2} + 2\frac{7}{8}$ (d) $3\frac{5}{6} + 2\frac{2}{9}$ (e) $4\frac{1}{4} - 2\frac{3}{5}$

9 Work out these.
(a) $\frac{2}{11} + \frac{5}{6}$ (b) $\frac{7}{8} - \frac{3}{5}$ (c) $2\frac{2}{7} \times 1\frac{3}{8}$ (d) $8\frac{2}{5} \div 2\frac{7}{10}$

Circles and polygons

▶ This chapter is about

- The language associated with circles
- Types of polygon
- Constructing regular polygons in a circle
- The angles in a triangle
- The angles in a quadrilateral
- The angles in a polygon
- Regular polygons

▶ You should already know

- How to measure and draw angles
- How to measure and draw lines
- About special triangles and quadrilaterals
- That the angles in a triangle add up to 180°

Circles

You learned in Unit A Chapter 23 that a **circle** is the locus of points that are the same distance from a given point and that point is the **centre** of the circle.

You need to be able to identify the parts of a circle.

The **circumference** of a circle is the distance all the way round – the perimeter of the circle.

Part of the circumference is called an **arc**.

The **radius** of a circle is the distance from the centre of the circle to its edge.

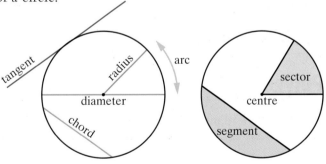

A line all the way across the circle and passing through its centre is a **diameter** of the circle.

So the length of the diameter = 2 × the length of the radius.

A **tangent** to a circle is a straight line which touches the circumference at one point only.

A **chord** is a straight line joining any two points on the circumference. A chord divides a circle into two **segments**.

A **sector** is a part of a circle made by two radii and an arc.

● Challenge 7.1

(a) Use a pair of compasses to draw a circle.

(b) Keeping the compasses open the same amount, put the compass point on the circumference of the circle. Draw an arc of a circle starting from the circumference, going through the centre of the circle to meet the circumference again.

(c) Put your compass point where the arc meets the circumference and repeat part **(b)**.

(d) Continue until you have completed the pattern.

If you have drawn this accurately, you will have a petal pattern like this.

You can colour your pattern.

◎ Discovery 7.1

Look around your classroom.

How many circles can you see?

Which object has the largest diameter?

Which object has the smallest radius?

Polygons

Polygons are many-sided shapes.

You have already met triangles (three-sided polygons) and quadrilaterals (four-sided polygons). Here are the names of some more polygons.

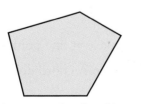

A pentagon has five sides.

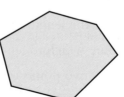

A hexagon has six sides.

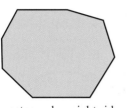

An octagon has eight sides.

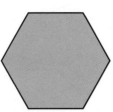

A decagon has ten sides.

When the sides and angles of a polygon are all the same, it is called a **regular** polygon.

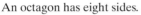

A regular pentagon

A regular hexagon

A regular polygon can be **constructed** by using a circle as shown in the next example. When you construct a shape it means you draw it accurately using compasses, ruler and protractor.

Example 7.1

Use a circle of radius 5 cm to construct a regular pentagon.

Solution

Step 1: Open the compasses to 5 cm and draw a circle.

Step 2: Draw a radius as a starting line for measuring.

Step 3: Work out the angle to measure. You learned in Chapter 5 that angles round a point add up to 360°. For a pentagon you need five angles so each one is 360 ÷ 5 = 72°.

Step 4: Measure an angle of 72° and draw the radius.

Step 5: Continue round the circle, measuring 72° angles and drawing the radius for each angle.

Step 6: Now join the points on the circumference to form a regular pentagon.

Exercise 7.1

1 Name these parts of a circle.

(a)   (b)   (c)

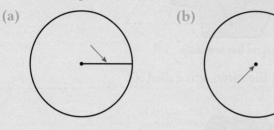

2 Name these polygons.

(a)   (b)   (c)

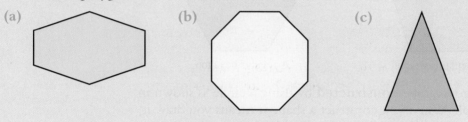

3 Draw any pentagon.
   Draw diagonals across the pentagon from each vertex (corner) of the pentagon to
   another vertex.
   How many diagonals can you draw in the pentagon altogether?

4 Draw any octagon.
   Draw diagonals across the octagon from each vertex of the octagon to another vertex.
   How many diagonals can you draw in the octagon altogether?

5 A regular decagon is constructed in a circle.
   How many degrees are measured at the centre to draw each radius required?

6 A nine-sided regular polygon is constructed in a circle.
   How many degrees are measured at the centre to draw each radius required?

7 Draw a circle of radius 5 cm and use it to construct a regular hexagon.
   Measure the length of a side of your hexagon.

8 Draw a circle of radius 6 cm and use it to construct a regular octagon.
   Measure the length of a side of your octagon.

● Challenge 7.2

- Draw a circle.
- Use angles of 72° at the centre to mark five evenly-spaced points on the circumference.
- Join a point to the next-but-one point and continue joining points like this to make a star shape in the circle.

Draw another circle and use more points on the edge of the circle to make different star shapes, such as a six-pointed star. You will need to work out the angle to use.

The angles in a triangle

In Chapter 2 you learned that the angles in a triangle add up to 180°. You can use the properties of angles associated with parallel lines to prove this fact.

● Proof 7.1

You draw a line parallel to the base of a triangle.

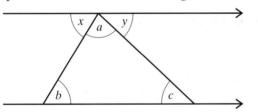

$x + a + y = 180°$ Angles on a straight line add up to 180°
$b = x$ Alternate angles
$c = y$ Alternate angles
So $b + a + c = 180°$ Since $b = x$ and $c = y$

This proves that the three angles in any triangle add up to 180°.

The angles inside a triangle (or any polygon) are called **interior** angles. If you extend a side of the triangle, there is an angle between the extended side and the next side. This angle is called an **exterior** angle.

> The exterior angle of a triangle is equal to the sum of the opposite, interior angles.

Here is a proof of this fact.

● Proof 7.2

One side is extended, as shown
in the diagram.

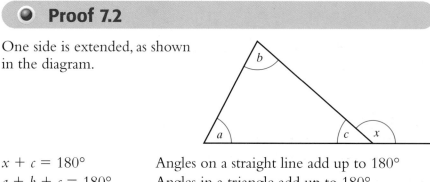

$x + c = 180°$	Angles on a straight line add up to 180°
$a + b + c = 180°$	Angles in a triangle add up to 180°
So $x = a + b$	Since $x = 180° - c$ and $a + b = 180° - c$

This proves that the exterior angle of a triangle is equal to the sum of
the opposite, interior angles.

● Challenge 7.3

There is another way to prove that the exterior angle of a triangle is
equal to the sum of the opposite, interior angles. It uses angle facts
associated with parallel lines.

Complete a proof for this diagram. Remember to give a reason for
each step.

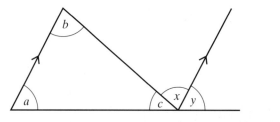

You can use the angle facts associated with triangles to work out
missing angles. This is shown in the next example.

Example 7.2

Work out the size of the lettered angles.
Give a reason for each answer.

Solution

$d = 180° - (51° + 90°)$ Angles in a triangle add up to 180°
$d = 39°$
$e = 51° + 90°$ Exterior angle of a triangle equals the
$e = 141°$ sum of the opposite interior angles

The angles in a quadrilateral

You learned in Chapter 2 that a **quadrilateral** is a four-sided shape.

> The angles in a quadrilateral add up to 360°.

You can divide a quadrilateral into two triangles. You can then use the
fact that angles in a triangle add up to 180° to prove this fact.

● Proof 7.3

A quadrilateral is divided into two triangles,
as shown in the diagram.

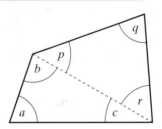

$a + b + c = 180°$ Angles in a triangle add up to 180°
$p + q + r = 180°$ Angles in a triangle add up to 180°
$a + b + c + p + q + r = 360°$, so
the interior angles of a quadrilateral add up to 360°.

Example 7.3

Work out the size of angle x

Give a reason for your answer.

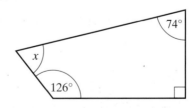

Solution

$x = 360° - (126° + 90° + 74°)$ Angles in a quadrilateral add up to 360°
$x = 70°$

Exercise 7.2

Find the size of the lettered angles. Give a reason for each answer.

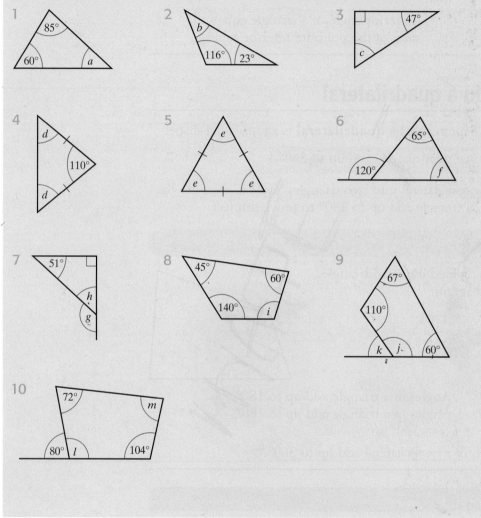

The angles in a polygon

In the previous section you saw that by dividing a quadrilateral into two triangles you could show that the angles in a quadrilateral add up to 360°.

In the same way you can divide any polygon into triangles to find the sum of its interior angles.

Discovery 7.2

Copy and complete this table.

Number of sides	Diagram	Name	Sum of interior angles
3		Triangle	1 × 180° = 180°
4		Quadrilateral	2 × 180° = 360°
5			3 × 180° =
6			

Discovery 7.3

Here is a pentagon showing all its exterior angles.

(a) Measure each of the exterior angles and find the total.
Check with your neighbour.
Do you both have the same total?

(b) Draw another polygon.
Extend its sides and measure each of the exterior angles.
Is the total of these angles the same as for the pentagon?

At each **vertex**, or corner, of a polygon there is an interior angle and an exterior angle.

Since these form a straight line you know the sum of the angles.

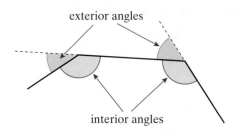

exterior angles

interior angles

Interior angle + Exterior angle = 180°.

You may have noticed that the five exterior angles of the pentagon go round in a full circle. This gives you another fact about the angles of a polygon.

The sum of the exterior angles of a polygon is 360°.

Example 7.4

Two of the exterior angles of this pentagon are equal.

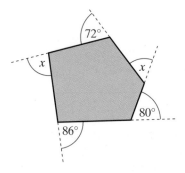

Find their size.

Solution

$x + x + 72° + 80° + 86° = 360°$ Exterior angles of a
$2x = 360° - 238°$ polygon add up to 360°
$2x = 122°$
$x = 61°$

Regular polygons

Earlier in this chapter you learned that, for a regular polygon, the sides are all the same length and the interior angles are all the same size. You can see now that the exterior angles are also all the same size.

You also know that

$$\text{The angle at the centre} = \frac{360°}{\text{number of sides}}$$

Note that the size of the angle at the centre of a polygon is the same as the exterior angle of the polygon. Can you see why?

🔍 Check up 7.1

Lines are drawn from the centre of a regular pentagon to each of its vertices.

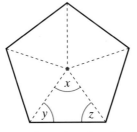

(a) What can you say about the triangles formed?
(b) What sort of triangles are they?
(c) Calculate the size of the lettered angles.
    Give a reason for each of your answers.

Example 7.5

Find the size of the exterior and interior angles of a regular octagon.

Solution

An octagon has eight sides.

Exterior angle $= \frac{360°}{8}$ Since the sum of the exterior angles of any polygon is 360°

Exterior angle $= 45°$

Interior angle $= 180° - 45°$ Since the interior angle and the exterior angle add up to 180°

Interior angle $= 135°$

Exercise 7.3

1 A polygon has 15 sides.
Work out the sum of the interior angles of this polygon.

2 A polygon has 20 sides.
Work out the sum of the interior angles of this polygon.

3 Three of the exterior angles of a quadrilateral are 94°, 50° and 85°.

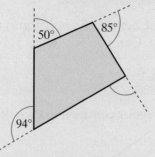

(a) Work out the size of the fourth exterior angle.
(b) Work out the size of the interior angles of the quadrilateral.

4 Four of the exterior angles of a pentagon are 90°, 80°, 57° and 75°.

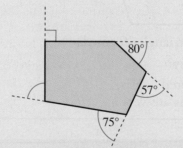

(a) Work out the size of the other exterior angle.
(b) Work out the size of the interior angles of the pentagon.

5 A regular polygon has 12 sides.
Find the size of the exterior and interior angles of this polygon.

6 A regular polygon has 100 sides.
Find the size of the exterior and interior angles of this polygon.

7 A regular polygon has an exterior angle of 24°.
Work out the number of sides that the polygon has.

8 A regular polygon has an interior angle of 162°.
Work out the number of sides that the polygon has.

What you have learned

- A circle is the locus of points that are the same distance from a given point and that point is the centre of the circle
- The distance all the way round a circle is its circumference
- The language used to describe parts of a circle

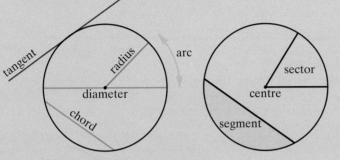

- A polygon is a many-sided shape
- A regular polygon has all its angles equal and all its sides equal
- A regular polygon can be constructed in a circle
- The angle at the centre of a regular polygon $= \dfrac{360°}{\text{number of sides}}$
- The sum of the interior angles of a triangle is 180°
- The exterior angle of a triangle is equal to the sum of the opposite, interior angles
- A quadrilateral has four sides and the sum of its interior angles is 360°
- A pentagon has five sides and the sum of its interior angles is 540°
- A hexagon has six sides and the sum of its interior angles is 720°
- The interior angle of a polygon and the exterior angle next to it add up to 180°

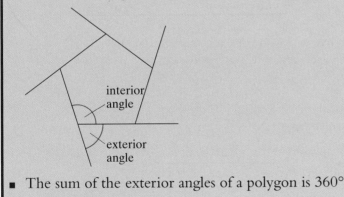

- The sum of the exterior angles of a polygon is 360°

Mixed exercise 7

1 Copy and complete these sentences.

(a) The distance from the centre of a circle to its edge is the ....

(b) The .... is the distance round a circle.

(c) A .... is a line across a circle, passing though the centre.

2 Draw a circle of radius 5 cm.
Draw a diameter of the circle. What is its length?

3 Draw a circle with a diameter of 12 cm.
Construct a regular octagon with vertices on the circumference of the circle.

4 Find the size of the lettered angles. Give a reason for each answer.

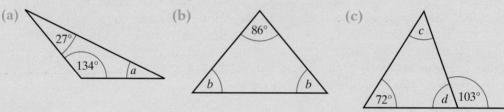

(a)   (b)   (c)

5 Find the size of the lettered angles. Give a reason for each answer.

(a)   (b)

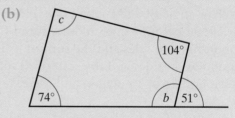

6 Find the size of the lettered angles. Give a reason for each answer.

(a)   (b)

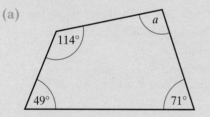

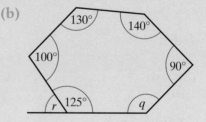

7 (a) A regular polygon has 15 sides.
Find the size of the exterior and interior angles of this polygon.

(b) A regular polygon has an exterior angle of 10°.
Work out the number of sides that the polygon has.

Powers and indices

▶ **This chapter is about**

- Squares and square roots
- Cubes and cube roots
- Using index notation
- The rules of indices

▶ **You should already know**

- That an integer is a whole number
- What square numbers and square roots are
- What cubes and cube roots are
- How to use index notation
- How to multiply whole numbers
- How to find the area of a square

Squares and square roots

As you can see, the square with side 3 has an area of $3 \times 3 = 9$ squares and the square with side 4 has an area of $4 \times 4 = 16$ squares. You can say that the **square** of 3 is 9, and write it as $3^2 = 9$.

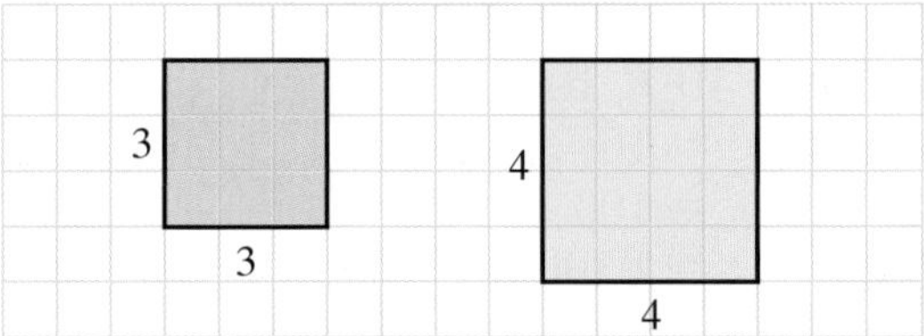

The set of whole numbers, $1, 4, 9, 16, 25, 36, \ldots$ are the squares of the counting numbers.

Because $16 = 4^2$ the positive **square root** of 16 is 4, written as $\sqrt{16} = 4$.

Similarly $\sqrt{36} = 6$ and $\sqrt{81} = 9$.

Remember though, if two negative numbers are multiplied together the answer is positive.

$(-4)^2 = 16$

So each positive number has two square roots, one positive and one negative.

The positive square root of 16 is 4 and the **negative square root** of 16 is -4.

This is often written $\sqrt{16} = \pm 4$.

> ### TIP
>
> In many practical problems where the answers are square roots, the negative answer would not have a meaning and so it should be left out.

In Unit A you learned how to square and square root numbers using your calculator but you also need to know the squares of numbers 1 to 15 without using a calculator. These are shown in the table. When you know these square numbers you also know the corresponding square roots.

Number	1	2	3	4	5	6	7	8	9	10	11	12	13	14	15
Square	1	4	9	16	25	36	49	64	81	100	121	144	169	196	225

Other square numbers can be worked out using those above.

Example 8.1

Work out these.

(a) 50^2 **(b)** 100^2

Solution

(a) $50 = 5 \times 10$
so $50^2 = 5^2 \times 10^2$
$ = 25 \times 100$
$ = 2500$

(b) $100 = 10 \times 10$
so $100^2 = 10^2 \times 10^2$
$ = 100 \times 100$
$ = 10\,000$

Cubes and cube roots

The **cube** of a number is the number multiplied by itself and then by itself again. For example, 2 cubed is written 2^3 and equals $2 \times 2 \times 2 = 8$.

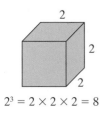

$2^3 = 2 \times 2 \times 2 = 8$

In Unit A you learned how to cube and cube root numbers using your calculator but you also need to know the cubes of the numbers 1 to 5 and the cube of 10 without using a calculator. When you know these cube numbers you also know the corresponding cube roots.

Number	1	2	3	4	5	...	10	...
Cube	1	8	27	64	125	...	1000	...

The numbers 1, 8, 27, 64, 125, ..., 1000, ... are **cube numbers** because they can be written as the cube of a whole number.

Exercise 8.1

1 Write down the square of each of these numbers.
   (a) 7   (b) 12   (c) 5   (d) 10   (e) 9
   (f) 8   (g) 11   (h) 3   (i) 6   (j) 4

2 Write down the positive square root of each of these numbers.
   (a) 49   (b) 81   (c) 25   (d) 144   (e) 100

3 Write down the positive and negative square roots of each of these numbers.
   (a) 121   (b) 36   (c) 81   (d) 16   (e) 64

4 Work out these.
   (a) 8^2 (b) 11^2 (c) 3^2 (d) 6^2 (e) 9^2

5 Work out these.
   (a) $\sqrt{81}$ (b) $\sqrt{144}$ (c) $\sqrt{16}$ (d) $\sqrt{100}$ (e) $\sqrt{64}$

6 Work out these.
   (a) 30^2 (b) 80^2

7 Work out these.
   (a) $6^2 - 5^2$ (b) $2^2 + 3^2$ (c) $7^2 - 4^2$
   (d) $3^2 - 2^2$ (e) $4^2 + 5^2$ (f) $6^2 - 3^2$

8 Write down the cube of each number
   (a) 4   (b) 5   (c) 3   (d) 10   (e) $\sqrt[3]{18}$

9 Write down the cube root of each number.
   (a) 8   (b) 1   (c) 64   (d) 1000   (e) 20^3

Challenge 8.1

Differences between squares

(a) Write down two consecutive numbers, square both these numbers and find the difference between the squares.

Try several pairs of consecutive numbers.

What do you notice?

(b) Write down three consecutive numbers, square the middle one, multiply the first and the third numbers together and find the difference between the answers.

Try several sets of consecutive numbers.

What do you notice?

Challenge 8.2

I think of a number.

I find its cube root and then square that.

The answer is 9.

What is the number I thought of?

Challenge 8.3

Find two numbers less than 200 which are both square numbers and cube numbers.

Using index notation

Indices (or **powers**) are a form of mathematical shorthand. You have already met squares and cubes in this chapter and you met higher powers in Unit A.

Discovery 8.1

Copy and complete these.

$5 \times 5 = 5^2$
$5 \times 5 \times 5 = 5^3$
$5 \times 5 \times 5 \times 5 =$
$5 \times 5 \times 5 \times 5 \times 5 =$
$5 \times 5 \times 5 \times 5 \times 5 \times 5 =$

In Discovery 8.1 the products are written as powers of 5.
5^4 means '5 to the power 4'.

The power is often called the **index** and you say that the number is written using **index notation**.

○ Discovery 8.2

Use the pattern to copy and complete these.

$2^2 = 2 \times 2 = 4$
$2^3 = 2 \times 2 \times 2 = 8$
$2^4 = 2 \times 2 \times 2 \times 2 =$
$2^5 = \qquad\qquad\qquad =$
$2^6 = \qquad\qquad\qquad\quad =$
$2^7 = \qquad\qquad\qquad\qquad =$
$2^8 = \qquad\qquad\qquad\qquad\quad =$
$2^9 = \qquad\qquad\qquad\qquad\qquad =$
$2^{10} = \qquad\qquad\qquad\qquad\qquad\quad =$

You should have found that $2^n = 2 \times 2 \times 2 \times 2 \dots n$ times

Example 8.2

Write each of these using index notation.
(a) $4 \times 4 \times 4 \times 4 \times 4$ **(b)** $6 \times 6 \times 6 \times 6$ **(c)** $25 \times 25 \times 25$

Solution

(a) 4^5 **(b)** 6^4 **(c)** 25^3

Example 8.3

Work out the value of each of these.
(a) 3^5 **(b)** 7^3 **(c)** 10^6

Solution

(a) $3 \times 3 \times 3 \times 3 \times 3 = 243$
(b) $7 \times 7 \times 7 = 343$
(c) $10 \times 10 \times 10 \times 10 \times 10 \times 10 = 1\,000\,000$

Exercise 8.2

1 Write each of these using index notation.
   (a) $2 \times 2 \times 2 \times 2$ (b) 7×7
   (c) $8 \times 8 \times 8 \times 8 \times 8 \times 8 \times 8$ (d) $10 \times 10 \times 10 \times 10 \times 10$
   (e) $9 \times 9 \times 9 \times 9 \times 9$

2 Find the value of each of these.
   (a) 6^3 (b) 4^3 (c) 8^3 (d) 1^3 (e) 9^3
   (f) 1^4 (g) 3^4 (h) 4^4 (i) 5^4 (j) 6^4
   (k) 1^5 (l) 3^5 (m) 2^5 (n) 5^5 (o) 4^5

The rules of indices

● Discovery 8.3

$$2^2 \times 2^5 = (2 \times 2) \times (2 \times 2 \times 2 \times 2 \times 2)$$
$$= (2 \times 2 \times 2 \times 2 \times 2 \times 2 \times 2) = 2^7$$

$$3^5 \div 3^2 = (3 \times 3 \times 3 \times 3 \times 3) \div (3 \times 3)$$
$$= (3 \times 3 \times 3) = 3^3$$

Copy and complete the following.
(a) $5^2 \times 5^3 = (5 \times 5) \times (\dots\dots\dots)$
$$= (\dots\dots\dots) = \dots\dots\dots$$
(b) $2^4 \times 2^2 =$ (c) $6^5 \times 6^3 =$ (d) $5^5 \div 5^3 =$
(e) $3^6 \div 3^3 =$ (f) $7^5 \div 7^2 =$
What do you notice?

Your answers to Discovery 8.3 were examples of these two rules.

$$n^a \times n^b = n^{a+b} \quad \text{and} \quad n^a \div n^b = n^{a-b}$$

You have already met a number with an index of 1.
A number to the power 1 equals the number itself.

$$n^1 = n \qquad \text{For example: } 3^1 = 3.$$

Any number with an index 0 is 1.

$$n^0 = 1 \qquad \text{For example: } 3^0 = 1.$$

TIP

To confirm this, put $a = b$ in $n^a \div n^b = n^{a-b}$.
$n^a \div n^a = 1$ and $n^{a-a} = n^0$.

Example 8.4

Write each of these as a single power of 3.

(a) $3^4 \times 3^2$ (b) $3^7 \div 3^2$ (c) $\dfrac{3^5 \times 3}{3^6}$

Solution

(a) When you multiply powers you add the indices.
$$3^4 \times 3^2 = 3^{4+2}$$
$$= 3^6$$

(b) When you divide powers you subtract the indices.
$$3^7 \div 3^2 = 3^{7-2}$$
$$= 3^5$$

(c) You can also combine operations.
$$\frac{3^5 \times 3}{3^6} = 3^{5+1-6}$$
$$= 3^0$$

> **TIP**
>
> $3^0 = 1$ but you have been asked to write your answer as a power of 3 so you leave your answer as 3^0. If you are asked to simplify the expression, you write $3^0 = 1$.

Exercise 8.3

1 Write these in simpler form using indices.

(a) $3 \times 3 \times 3 \times 3 \times 3$ (b) $7 \times 7 \times 7$

(c) $3 \times 3 \times 3 \times 3 \times 5 \times 5$ Hint: Write the 3s separately from the 5s.

2 Work out these, giving your answers in index form.

(a) $5^2 \times 5^3$ (b) $10^5 \times 10^2$ (c) 8×8^3 (d) $3^6 \times 3^4$ (e) $2^5 \times 2$

3 Work out these, giving your answers in index form.

(a) $5^4 \div 5^2$ (b) $10^5 \div 10^2$ (c) $8^6 \div 8^3$ (d) $3^6 \div 3^4$ (e) $2^3 \div 2^3$

4 Work out these, giving your answers in index form.

(a) $5^4 \times 5^2 \div 5^2$ (b) $10^7 \times 10^6 \div 10^2$ (c) $8^4 \times 8 \div 8^3$ (d) $3^5 \times 3^3 \div 3^4$

5 Work out these, giving your answers in index form.

(a) $\dfrac{2^6 \times 2^3}{2^4}$ (b) $\dfrac{3^6}{3^2 \times 3^2}$ (c) $\dfrac{5^3 \times 5^4}{5 \times 5^2}$ (d) $\dfrac{7^4 \times 7^4}{7^2 \times 7^3}$

What you have learned

- A square number is the result of multiplying a number by itself
- When you square root a number you get two numbers, one positive and one negative, for example, $\sqrt{16} = \pm 4$
- 5^2 means '5 squared' or 5×5
- The squares from 1^2 to 15^2 are 1, 4, 9, 16, 25, 36, 49, 64, 81, 100, 121, 144, 169, 196 and 225
- 4^3 means '4 cubed' or $4 \times 4 \times 4$
- The cubes from 1^3 to 5^3 are 1, 8, 27, 64 and 125 and 10^3 is 1000
- 3^4 means '3 to the power 4' or $3 \times 3 \times 3 \times 3$
- You can multiply numbers in index form using the rule $n^a \times n^b = n^{a+b}$
- You can divide numbers in index form using the rule $n^a \div n^b = n^{a-b}$

Mixed exercise 8

1 Write down these.
   (a) 8^2 (b) 10^2

2 Write down these.
   (a) $\sqrt{49}$ (b) $\sqrt{225}$ (c) $\sqrt{36}$ (d) $\sqrt{169}$

3 Work out these.
   (a) 40^2 (b) 500^2

4 Write down these.
   (a) 10^3 (b) 1^3

5 Write down these.
   (a) $\sqrt[3]{64}$ (b) $\sqrt[3]{125}$

6 Work out 30^3

7 Write each of these using index notation.
   (a) $3 \times 3 \times 3 \times 3 \times 3$ (b) $7 \times 7 \times 7$ (c) $8 \times 8 \times 8 \times 8 \times 8$
   (d) $4 \times 4 \times 4 \times 4 \times 4$ (e) $8 \times 8 \times 8$ (f) $2 \times 2 \times 2 \times 2 \times 2$

8 Find the value of these.
   (a) 2^6 (b) 10^7

9 Write these in a simpler form, using indices.
   (a) $7^2 \times 7^3$ (b) $6^3 \times 6^6$ (c) $10^9 \times 10^3$ (d) $3^4 \times 3^8$ (e) $8^9 \div 8^3$
   (f) $6^9 \div 6^7$ (g) $4^3 \div 4^2$ (h) $9^6 \div 9^2$ (i) $\dfrac{3^7 \times 3^3}{3^4}$

Chapter 9

Decimals and fractions

⊙ This chapter is about

- Ordering decimals
- Changing decimals into fractions
- Recurring decimals
- Converting between fractions and recurring decimals

⊙ You should already know

- The place value of integers and decimals
- How to order integers
- The decimal equivalents of half, quarters and fifths
- How to change a fraction to a decimal by division
- How to find the prime factors of a number

Ordering decimals

You learned about place value in Unit A Chapter 4. The position of a digit in a number tells you its value.

🔎 Check up 9.1

What does the 2 represent in each of these numbers?
(a) 12   (b) 120   (c) 14.25   (d) 3.125   (e) 41.712

You learned how to order integers in Chapter 1 of this unit.

🔎 Check up 9.2

Put these numbers in order of size, smallest first.
1421   512   3146   598   94

You order decimals in a similar way to ordering integers. You look where the first digit is, the nearer to the decimal point the bigger the number.

Example 9.1

Put these decimals in order of size, smallest first.

0.412   0.0059   0.325   0.046   0.012   0.007

Solution

0.0059   0.007   0.012   0.046   0.325   0.412

The smallest are the ones starting with a digit in the third place after the decimal point, and these are in order of size of that digit, no matter how many other digits there are. The next smallest are those with a digit in the second place, and then come those with a digit in the first place.

When the numbers to order include a whole number, you deal with the whole number part first and then put the decimals in order as needed.

Example 9.2

Put these in order of size, smallest first.

3.612   4.15   0.273   0.046   4.051   4.105

Solution

0.046   0.273   3.612   4.051   4.105   4.15

The ones with no whole number are the smallest, ordered as before. Then comes the one with 3 units, followed by those with 4 units, ordered by the decimals.

Exercise 9.1

1   Put these decimals in order of size, smallest first.
    (a) 0.123   0.456   0.231   0.201   0.102   0.114
    (b) 0.871   0.561   0.271   0.914   0.832   0.9
    (c) 0.01   0.003   0.1   0.056   0.066   0.008
    (d) 0.213   0.0256   0.0026   0.000 141 0.031   0.07
    (e) 0.0404   0.404   0.004 04   0.044   0.0044   0.400 04
    (f) 0.213   0.614   0.317   0.401   0.502   0.514
    (g) 0.71   0.51   0.112   0.149   0.2   0.641
    (h) 0.05   0.301   0.8   0.516   0.686   0.083
    (i) 0.913   0.0946   0.009 16   0.090 11   0.091   0.097
    (j) 0.03   0.304   0.004 304 0.034   0.0043   0.300 04

Exercise continues …

2 Put these numbers in order of size, smallest first.

    (a) 7.6   3.42   1.63   0.84   8.9   4.2

    (b) 3.12   3.21   3.001   3.102   3.201   3.02

    (c) 1.21   2.12   12.1   121   0.12   0.21

    (d) 53.27   46.34   51.96   51.04   52   46.57

    (e) 7.023   7.69   7.015   7.105   7.41   7.14

    (f) 34.2   163   0.84   8.9   4.2   71

    (g) 5.321   5.001   5.0102   5.0201   5.02   5.21

    (h) $\frac{1}{100}$ 12.02   0.0121   1.201   0.0012 $\frac{21}{100}$

    (i) 546.27 514.3   15.96   514.04   512   516.57

    (j) 8.097   8.79   8.01 $8\frac{1}{10}$ $8\frac{4}{100}$ 8.104

> **TIP**
>
> Change fractions into decimals first.

Challenge 9.1

(a) Write a number between each of these pairs of numbers.

    (i)   5 and 6   (ii) 5.5 and 5.6   (iii) 5.05 and 5.06

    (iv) 5.999 and 6   (v) 5.0005 and 5.000 51

(b) How many numbers can you write that are between 0.001 and 0.002?

Changing a decimal into a fraction

Discovery 9.1

In Chapter 3 you learned that to change a fraction into a decimal you divide the numerator (top) by the denominator (bottom).

(a) Change these fractions into decimals.

    $\frac{3}{10}$, $\frac{7}{10}$, $\frac{7}{100}$, $\frac{37}{100}$, $\frac{37}{1000}$, $\frac{137}{1000}$, $\frac{7}{1000}$

(b) Explain how to change a decimal like the ones you got in part (a) into a fraction.

You can use place value to change decimals into fractions.

0.45 means no units, $\frac{4}{10}$ and $\frac{5}{100}$.

But $\frac{4}{10}$ is the same as $\frac{40}{100}$. So 0.45 can just be written as $\frac{45}{100}$.

In the same way 0.327 means $\frac{3}{10}$, $\frac{2}{100}$ and $\frac{7}{1000}$.

But $\frac{3}{10}$ is the same as $\frac{300}{1000}$, and $\frac{2}{100}$ is the same as $\frac{20}{1000}$.

So 0.327 can be written as $\frac{327}{1000}$.

Remember too that you already know the decimal equivalents of quarters and fifths. For example $\frac{1}{4} = 0.25$, $\frac{1}{5} = 0.2$, $\frac{2}{5} = 0.4$ and so on.

Example 9.3

Write these decimals as fractions.

(a) 0.3   (b) 0.75   (c) 0.317

(d) 0.041   (e) 2.3   (f) 4.6

Solution

(a) $\frac{3}{10}$ Only one decimal place, so tenths.

(b) $\frac{3}{4}$ You should recognise this.

(c) $\frac{317}{1000}$ Three decimal places, so thousandths.

(d) $\frac{41}{1000}$ Again three decimal places, but no tenths, so just $\frac{41}{1000}$.

(e) $2\frac{3}{10}$ 2 stays the same, and then there are only tenths.

(f) $4\frac{3}{5}$ 4 stays the same and you should recognise that 0.6 is $\frac{3}{5}$.

Exercise 9.2

1 Write these decimals as fractions.

(a) 0.9   (b) 0.87   (c) 0.654   (d) 0.121   (e) 0.301   (f) 1.5
(g) 0.03   (h) 3.004   (i) 0.108   (j) 0.1205   (k) 3.2   (l) 1.05
(m) 2.74   (n) 0.48   (o) 0.719   (p) 5.4

2 Write these decimals as fractions.

(a) 0.8   (b) 0.75   (c) 0.5   (d) 0.213   (e) 0.031   (f) 6.4
(g) 2.09   (h) 5.001   (i) 0.728   (j) 0.5905   (k) 0.81   (l) 0.37
(m) 0.316   (n) 1.49   (o) 23.564   (p) 10.548

Recurring decimals

◉ Discovery 9.2

Write each of these fractions as a decimal.

$$\frac{2}{3}, \quad \frac{1}{5}, \quad \frac{2}{11}, \quad \frac{3}{4}, \quad \frac{3}{7}, \quad \frac{5}{6}, \quad \frac{1}{18}, \quad \frac{1}{20}, \quad \frac{7}{8}$$

What do you notice about the decimal values?

Example 9.4

Convert $\frac{5}{8}$ to a decimal

Solution

$\frac{5}{8} = 5 \div 8 = 0.625$

Example 9.5

Convert $\frac{1}{6}$ to a decimal

Solution

$\frac{1}{6} = 1 \div 6 = 0.166\,66\ldots$

In Example 9.4, the decimal equivalent of $\frac{5}{8}$ is exactly 0.625.
This is an example of a **terminating decimal**.
It is called terminating because it finishes at the digit 5.

In Example 9.5, the decimal equivalent of $\frac{1}{6}$ goes on forever, with the digit 6 repeating itself over and over again.
This is an example of a **recurring decimal**.

The dot notation for recurring decimals

To save a lot of writing, there is a special notation for recurring decimals. You put a dot over the digit that recurs.

So, for example, $\frac{1}{3} = 0.333\,333\ldots$ is written as $0.\dot{3}$.

Similarly, $\frac{1}{6} = 0.166\,666\ldots$ is written as $0.1\dot{6}$.

Example 9.6

Write $\frac{7}{11}$ as a recurring decimal.

Solution

$7 \div 11 = 0.636\,363\ldots$

This time both the 6 and 3 recur. To show this you put a dot over both the 6 and the 3.

You write $0.\dot{6}\dot{3}$.

When three or more digits recur you put a dot over the first and last recurring digit.

⊙ Discovery 9.3

(a) Change these fractions into recurring decimals.

$$\frac{1}{9}, \quad \frac{1}{99}, \quad \frac{1}{999}$$

(b) Use your answers to part (a) to change these decimals into fractions.

(i)	$0.\dot{1}$	(ii)	$0.\dot{2}$
(iii)	$0.\dot{3}$	(iv)	$0.\dot{7}$
(v)	$0.\dot{0}\dot{7}$	(vi)	$0.\dot{4}\dot{5}$
(vii)	$0.\dot{0}0\dot{2}$	(viii)	$0.\dot{1}2\dot{3}$

● Challenge 9.2

Change $0.\dot{1}23\dot{4}$ and $0.\dot{1}234\dot{5}$ to fractions.

Hint: Using the result of Discovery 9.3 may help you.

⊙ Discovery 9.4

Some of the fractions in this list terminate and some recur.

$$\frac{1}{2}, \quad \frac{1}{3}, \quad \frac{1}{4}, \quad \frac{1}{5}, \quad \frac{1}{9}, \quad \frac{1}{10}, \quad \frac{1}{16}, \quad \frac{1}{20}, \quad \frac{1}{25}, \quad \frac{1}{33}, \quad \frac{1}{35}$$

Find which fractions terminate and which recur.

Explain what numbers in the denominators make fractions terminate.

Try some more fractions if you need to.

○ Discovery 9.5

(a) Work out $1 \div 3$.
   Write your answer as
   (i)   a decimal rounded to 2 decimal places
   (ii)   a decimal rounded to 5 decimal places
   (iii) a recurring decimal using dot notation

(b) How can you write the exact value of $1 \div 3$?

Exercise 9.3

For this exercise use the facts below to help you.

$$0.\dot{1} = \frac{1}{9} \qquad 0.\dot{0}\dot{1} = \frac{1}{99} \qquad 0.\dot{0}0\dot{1} = \frac{1}{999}$$

1 Use the facts above to show that $0.\dot{3}$ is equal to $\frac{1}{3}$.

2 Copy and complete these sentences.
   (a) $\frac{7}{9}$ is .... $\times \frac{1}{9}$ so it equals .... $\times 0.\dot{1} = $ ....
   (b) $\frac{17}{99}$ is .... $\times \frac{1}{99}$ so it equals .... $\times 0.\dot{0}\dot{1} = $ ....
   (c) $\frac{107}{999}$ is .... $\times \frac{1}{999}$ so it equals .... $\times 0.\dot{0}0\dot{1} = $ ....

3 Copy and complete these sentences.
   (a) $0.\dot{2}$ is .... $\times 0.\dot{1}$ so it equals .... $\times \frac{1}{9} = $ ....
   (b) $0.\dot{5}\dot{2}$ is .... $\times 0.\dot{0}\dot{1}$ so it equals .... $\times \frac{1}{99} = $ ....
   (c) $0.\dot{1}1\dot{2}$ is .... $\times 0.\dot{0}0\dot{1}$ so it equals .... $\times \frac{1}{999} = $ ....

4 Use the facts at the beginning of the exercise to change these fractions into decimals.
   (a) $\frac{8}{9}$ (b) $\frac{2}{99}$ (c) $\frac{8}{99}$ (d) $\frac{31}{99}$
   (e) $\frac{51}{99}$ (f) $\frac{128}{999}$ (g) $\frac{568}{999}$ (h) $\frac{5}{999}$

5 Change these recurring decimals to fractions.
   Simplify them where possible.
   (a) $0.\dot{4}$ (b) $0.\dot{6}$ (c) $0.\dot{2}\dot{3}$ (d) $0.\dot{0}\dot{8}$ (e) $0.\dot{2}\dot{1}$
   (f) $0.\dot{0}0\dot{3}$ (g) $0.\dot{3}6\dot{3}$ (h) $0.\dot{7}\dot{2}$ (i) $0.\dot{5}$ (j) $0.\dot{9}0\dot{9}$

6 Divide the fractions below into two groups, one group of fractions that will terminate and the other that will recur when changed into decimals.
   Explain how you made your choice.

   $$\frac{1}{40}, \quad \frac{1}{27}, \quad \frac{1}{45}, \quad \frac{1}{32}, \quad \frac{1}{128}, \quad \frac{3}{80}, \quad \frac{16}{63}, \quad \frac{2}{99}, \quad \frac{17}{50}, \quad \frac{23}{150}, \quad \frac{11}{160}$$

Challenge 9.3

Change $0.\dot{9}$ to a fraction and explain your answer.

Challenge 9.4

(a) Work out $\frac{1}{2} + \frac{1}{3}$.

(b) Write $\frac{1}{2}$ and $\frac{1}{3}$ as decimals.

(c) Hence show that $0.8\dot{3} = \frac{5}{6}$.

Challenge 9.5

Use the method of Challenge 9.4 to write $0.5\dot{6}\dot{0}$ as a fraction.

Challenge 9.6

Work in pairs or small groups.

(a) In one minute write down as many terminating fractions as you can. They must be in their lowest terms.

(b) In one minute write down as many recurring fractions as you can. They must be in their lowest terms.

Score two points for each correct fraction (check on your calculator).
Score one point for a correct fraction that is not in its lowest terms.
Lose one point for each incorrect fraction or repeated fraction.
Who scored the most points?

What you have learned

- You can order decimals using the place value of the digits
- To change a terminating decimal into a fraction you use the place value of the digits
- A terminating decimal is an exact number which stops after a certain number of digits
- A recurring decimal has one or more repeating digits and the number continues forever
- You can use facts you know to convert some recurring decimals into fractions, for example $\frac{1}{9} = 0.\dot{1}$ so $0.\dot{1} = \frac{1}{9}$ and $0.\dot{2} = \frac{2}{9}$
- A fraction will terminate when changed into a decimal if the denominator has only 2 and/or 5 as it prime factors
- The exact value of a recurring decimal can only be written as a fraction

Mixed exercise 9

1 Put these in order, smallest first.

(a) 0.23 $\frac{21}{100}$ $\frac{18}{100}$ 0.2   0.17

(b) 0.49 $\frac{37}{100}$ $\frac{56}{100}$ 0.50   0.51

(c) $\frac{91}{100}$ 0.81 $\frac{8}{10}$ 0.9   0.84

(d) $1\frac{21}{100}$ $1\frac{8}{100}$ 1.2   1.17   1.09

(e) 3.25 $3\frac{2}{10}$ $3\frac{18}{100}$ 3.204   3.17

(f) 0.64 $\frac{61}{100}$ $\frac{58}{100}$ 0.62   0.57

(g) 0.39 $\frac{37}{100}$ $\frac{41}{100}$ 0.47   0.42

(h) $\frac{61}{100}$ 0.64 $\frac{6}{10}$ 0.7   0.59

(i) $5\frac{21}{100}$ $5\frac{3}{10}$ 5.2   5.27   5.03

(j) 4.75 $4\frac{7}{10}$ $4\frac{9}{100}$ 4.704   4.71

2 Write these decimals as fractions.

(a) 0.7   (b) 0.35   (c) 0.105   (d) 0.007   (e) 5.4

3 Write these decimals as fractions.

(a) $0.\dot{3}$ (b) $0.\dot{1}$ (c) $0.0\dot{1}$ (d) $0.\dot{5}$ (e) $0.\dot{6}$

4 Divide the fractions below into two groups, one group of fractions that will terminate and the other that will recur when changed into decimals.

$$\frac{7}{8}, \quad \frac{6}{7}, \quad \frac{8}{9}, \quad \frac{11}{40}, \quad \frac{127}{128}, \quad \frac{7}{150}, \quad \frac{7}{240}, \quad \frac{11}{160}$$

Real-life graphs

▶ This chapter is about

- Drawing and interpreting conversion graphs
- Drawing and interpreting travel graphs

▶ You should already know

- How to plot and read coordinates of points
- The units of time, distance, temperature and money

Check up 10.1

You need to be able to plot and read coordinates correctly.
Write the coordinates of these points.

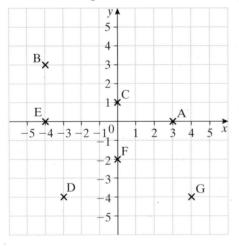

Conversion graphs

These graphs can be used to find equivalent quantities in different units.

Example 10.1

This graph converts between pounds (£) and euros (€).

(a) How many euros is £40?

(b) How many pounds is €30?

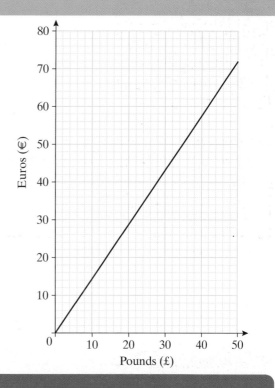

Solution

(a) You read from the horizontal axis at £40 to the graph line. Then from the graph line to the vertical axis. This is shown with red arrows on the graph. The value on the vertical axis is €58 so £40 is €58.

(b) Similarly, you read from €30 on the vertical axis to the graph line. Then from the graph line to the horizontal axis. This is shown with green arrows on the graph. The value on the horizontal axis is £21 so €30 is £21.

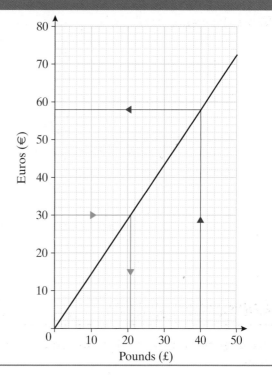

TIP

Take care when reading scales. In this case, each small square represents 2 units.

How many euros is each pound worth according to the graph?

Find a strategy which gives an accurate answer.

Find out what the conversion rate is today.

Exercise 10.1

1 This graph converts pounds (£)
to dollars ($).
Use the graph to find how many

(a) dollars is £35.

(b) pounds is $70.

(c) dollars is £1.

2 This graph converts between
gallons and litres.
Use the graph to find how many

(a) litres is 5 gallons.

(b) gallons is 40 litres.

(c) litres is 1 gallon.

Exercise continues …

3 This graph converts between kilometres and miles.

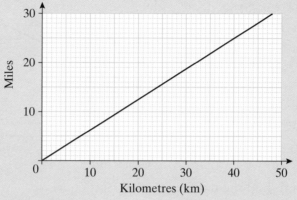

Use the graph to find how many

(a) miles is 40 km.

(b) kilometres is 15 miles.

(c) kilometres is 1 mile.

4 This conversion graph is for pounds (£) to New Zealand dollars (NZ$) for amounts up to £100.

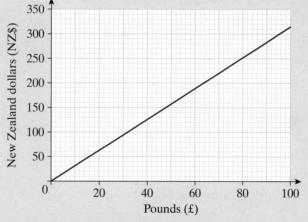

(a) Use the graph to find the number of New Zealand dollars equal to
   (i) £20.
   (ii) £85.

(b) Use the graph to find the number of pounds equal to
   (i) NZ$100.
   (ii) NZ$250.

(c) How many pounds are equal to NZ$600?

Exercise continues …

5   This conversion graph is for pounds to kilograms, for up to 100 pounds.

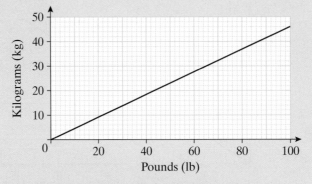

(a) Use the graph to find the number of kilograms equal to
    (i)   20 lb.
    (ii)   75 lb.

(b) Use the graph to find the number of pounds equal to
    (i)   40 kg.
    (ii)   25 kg.

(c) How many pounds are equal to 60 kg?

6   A newspaper states that €1 is equal to HK$6.3.

(a) On a piece of graph paper, mark axes horizontally for euros up to 50, using 2 cm to €10, and vertically for Hong Kong dollars up to 400, using 2 cm to HK$100.

(b) Plot the point (50, 315) and join to (0, 0).

(c) Find the number of Hong Kong dollars equal to
    (i)   €40.
    (ii)   €16.

(d) Find the number of euros equal to
    (i)   HK$150.
    (ii)   HK$220.

(e) How many euros are equal to HK$400?

7   Sue, James and their family live in England but have relatives in Malta.
    Before the Maltese lira was replaced by the euro in January 2008 they changed the Maltese lira they had back into pounds.
    One pound was equal to 0.64 Maltese lira.

(a) On a piece of graph paper, mark axes horizontally for pounds up to £100 (use a scale of 2 cm to £20), and vertically for Maltese lira up to 80 lira (use a scale of 2 cm to 20 lira).

(b) Plot the point (100, 64) and join it to (0, 0) with a straight line.

Exercise continues …

(c) (i) Sue received £45 when she exchanged her Maltese lira.
How many Maltese lira did she have?

(ii) James received £85 when he exchanged his Maltese lira.
How many Maltese lira did he have?

(d) (i) Dominic had 20 Maltese lira to exchange.
How many pounds did he receive?

(ii) Gabriella had 45 Maltese lira to exchange.
How many pounds did she receive?

(e) Georgina had 100 Maltese lira to exchange.
How many pounds did she receive?

8 In temperature 0°C (Celsius) is equal to 32°F (Fahrenheit) and 100°C is equal to 212°F.

(a) Draw a graph to show this conversion using a scale from 0 to 100 for temperatures
in degrees Celsius on the horizontal scale and from 0 to 225 for temperatures in
degrees Fahrenheit on the vertical axis.

(b) Use your graph to convert these temperatures to degrees Fahrenheit.

(i) 20°C

(ii) 64°C

(c) Use your graph to convert these temperatures to degrees Celsius.

(i) 50°F

(ii) 165°F

Challenge 10.1

Explain how you could use the graph you drew for question 8 to
convert 240°C to degrees Fahrenheit.

Challenge 10.2

Find a currency conversion or other conversion not used in this
chapter and draw a conversion graph.

Write a question asking for two conversions.

Exchange questions with a friend.

Travel graphs

Travel graphs show journeys by plotting distance covered against time.

Example 10.2

This graph shows Emily's journey to work.

Interpret the graph.
What could each stage represent?

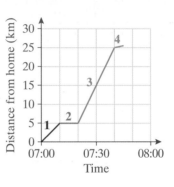

TIP

Always take time to work out what the scales represent.

As there are 60 minutes in an hour, each small square represents 10 minutes here.

Solution

Stage 1 is from 07:00 to 07:10 and Emily travels 5 km. She could be driving to the station.

Stage 2 is from 07:10 to 07:20 and there is no movement. Emily could be waiting for the train.

During stage 3 Emily travels 20 km. She could be on the train, arriving at 07:40.

Stage 4 of the journey lasts 5 minutes. Emily could be walking from the station to her office. The distance is about 500 m.

Example 10.3

Trudy left home at 8 a.m. and walked 300 m to the bus stop in 5 minutes.

She then had an 8-minute wait for the bus.

She caught the bus and it took 6 minutes to travel the 1.2 km to the school stop.

She got off and slowly walked the 100 m to school in a further 4 minutes.

(a) Draw a distance–time graph to illustrate this story.
Show time from 8 a.m. to 8.30 a.m. and use a scale of 10 small squares for 10 minutes.
Show distance from 0 to 3 km and use a scale of 10 small squares for 1 km.

(b) At what time did she arrive at school?

Solution

(a) 5 minutes is five small squares horizontally.
300 m is 0.3 km so it is three small squares vertically.
8 minutes' wait means move eight squares horizontally, i.e. to square 13 horizontally.
1.2 km in 6 minutes means move six squares horizontally and 12 vertically, i.e. to square 19 horizontally and 15 vertically.
100 m in 4 minutes means move four squares horizontally and one vertically, i.e. to square 23 horizontally and 16 vertically.

(b) She arrived at school at 8.23 a.m.

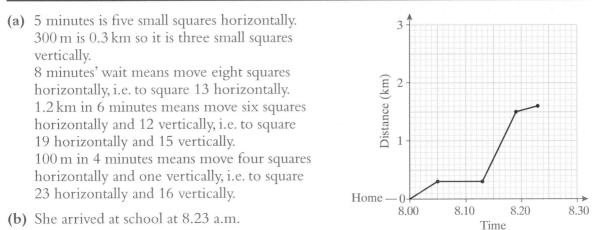

Exercise 10.2

1 This graph shows a car journey from London.
Describe and give an explanation for each stage.

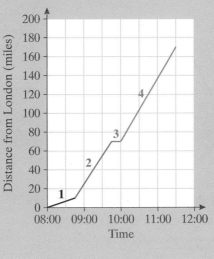

Exercise continues …

2 This graph shows the race between the
hare and the tortoise.
The blue line shows the hare's journey.
The red line shows the tortoise's.
Describe what happened. Knowing the
story could help!

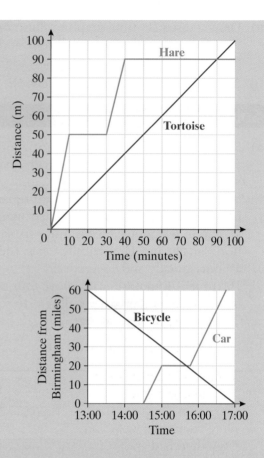

3 This graph shows two journeys between
Birmingham and Hereford.
The red line represents a bicycle.
The blue line represents a car.

(a) Describe the journeys.

(b) What happened where the lines cross?

4 This graph shows Emma's ride on her scooter.

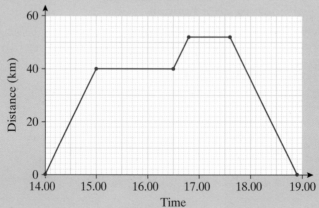

(a) How long did Emma stop for the first time?

(b) What was her furthest distance from her starting point?

(c) When did she arrive at the furthest distance?

(d) How long did she take to ride back?

Exercise continues …

5 This graph shows the journeys of two men, Brian and Malcolm, who both left Leeds. Brian left first.

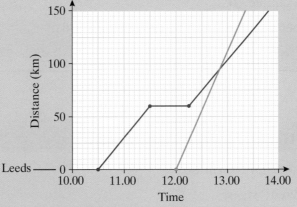

(a) How much later than Brian did Malcolm leave?

(b) How long did Brian stop for?

(c) How far had they travelled when Malcolm passed Brian?

(d) At what time did Malcolm pass Brian?

6 Jasmin and Patrick went on a walk.
They left home at 9.30 a.m. and walked 2 miles away from there in 45 minutes.
They then sat down for 15 minutes.
They continued away from home for another 1.5 miles in the next 30 minutes.
They then had another rest for 15 minutes.
They returned directly home and arrived there at 1:00 p.m.

(a) Draw a travel graph to show this journey.
Show time from 9.30 a.m. to 1.00 p.m. and use a scale of 2 cm for 30 minutes.
Show distance from 0 to 4 miles and use a scale of 2 cm for 2 miles.

(b) (i) How far did they walk altogether?
(ii) How long did it take them to walk back from the furthest point?

● **Challenge 10.3**

This graph shows the amount of fuel in a car's petrol tank.

(a) How many litres were used between 6 and 7 p.m?

(b) Describe what happened between 7.30 and 8 p.m.

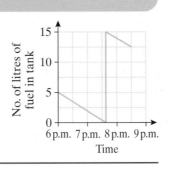

● Challenge 10.4

Look at the graph in Example 10.2.

The third stage is steeper than the first. The second stage is flat.

What does it mean?

● Challenge 10.5

Draw a graph of your journey to school.

How steep are the different stages?

● Challenge 10.6

Draw a graph with straight lines only, to show these journeys.

(a) A car travelling on a motorway to the nearest city along a clear road

(b) A car travelling along a busy motorway

(c) A shopping trip to a supermarket outside town

(d) A train travelling along a fast stretch between two cities

(e) A ferry crossing the English Channel

(f) A hiker walking in the Lake District, first along a flat piece of ground and then up a very steep slope

What you have learned

- When you plot or read graphs, you must make sure you know what the scales on the axes represent
- You can use conversion graphs to convert between different units and currencies
- A travel graph has time on the horizontal axis and distance on the vertical axis
- A horizontal line on a travel graph indicates no movement

Mixed exercise 10

1 This graph converts dollars ($) to euros (€).

Use the graph to find how many

(a) euros is $40.

(b) dollars is €35.

(c) dollars is €1.

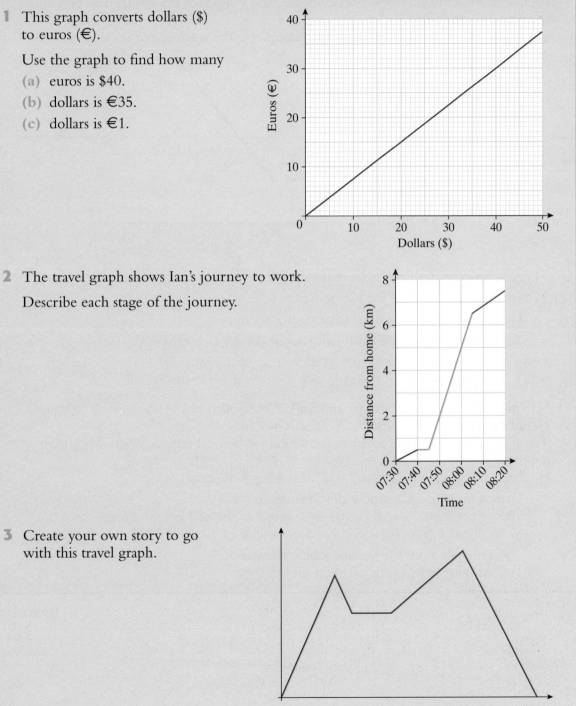

2 The travel graph shows Ian's journey to work.

Describe each stage of the journey.

3 Create your own story to go with this travel graph.

Mixed exercise 10 continues …

4 This graph shows the journeys of Pauline and Scott.
Pauline drove south from junction 29 of the M1, stopped at the service area at A and continued south.
Scott drove north from junction 21, also stopped at the services and continued north.

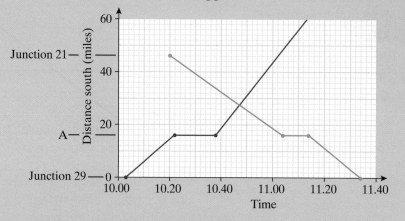

(a) At what time did Pauline leave junction 29?

(b) How far is it between the two junctions?

(c) How long did Pauline stop at the service area?

(d) What time was it when Scott and Pauline passed each other?

(e) (i) Who drove at the fastest speed for part of the journey?
    (ii) For which part of their journey did they drive at this speed?

5 Mr Jenkins left home at 7.15 a.m. and walked the 500 m to the station in 10 minutes.
He then waited for the train which left at 7.46 a.m.
The train took 10 minutes to cover the 12 km to the next station, where Mr Jenkins got off.
He then took a taxi for 2 km to his office, taking 8 minutes.

(a) Draw a travel graph to show this journey.
    Show time from 7 a.m. to 8.40 a.m. and use a scale of 1 cm for 20 minutes.
    Show distance from 0 to 15 km and use a scale of 1 cm for 5 km.

(b) (i) How long did Mr Jenkins wait at the station?
    (ii) At what time did he arrive at his office

Reflection

- Symmetry
- Reflections

- About special triangles and quadrilaterals
- The meaning of *horizontal* and *vertical*
- About three-dimensional shapes
- How to plot and read coordinates
- How to plot horizontal and vertical lines from their equations
- How to plot the lines $y = x$ and $y = -x$

Reflection symmetry

○ Discovery 11.1

Fold a piece of paper.
Cut a shape from the paper.
Open out the paper and look at the shape you have made.

Do this again with another piece of paper.

Look at the shapes that other people have made.
What is the same about them?
What is different about them?

Fold another piece of paper twice, with the folds at right angles.
Cut out a shape and open out the paper.
What are the shapes like this time?

Look at this shape.

It can be folded down the middle along the red line so that both sides match.

The red line is a **line of symmetry**. Some shapes have more than one line of symmetry.
This star has five lines of symmetry. One is shown and there are four more.

Reflection symmetry may also be called line symmetry.

Check up 11.1

Look around you.

Which objects or shapes in the room have reflection symmetry? List or sketch them.

How many lines of symmetry do they each have?

Example 11.1

Shade three more squares so that the red line is a line of symmetry.

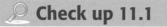

Solution

Count squares from the line of symmetry.
Make sure the shaded squares match on both sides of the line.
For example, in the first row you need to shade the second square on the right to match the second square on the left.

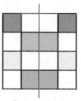

Example 11.2

Complete this shape so that it has two lines of symmetry.

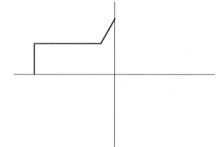

Solution

First use the vertical line of symmetry.

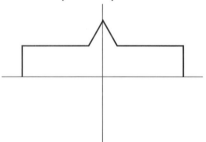

Use tracing paper to trace the shape and the red vertical line.
Turn the tracing paper over and line up the red lines.
Trace over the lines of the shape in its new position.
Remove the paper and complete the top part of the shape.

Now do the same with the horizontal line of symmetry.
This time, you need to copy the shapes on the left and right.

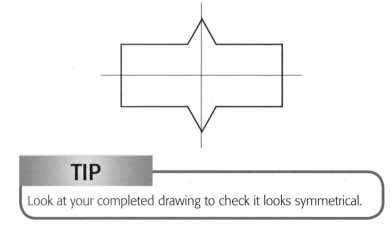

TIP

Look at your completed drawing to check it looks symmetrical.

Symmetry of three-dimensional shapes

● Discovery 11.2

Take a cube and place it against a mirror.
The image is an identical cube.

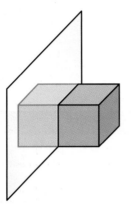

The object and image together form a cuboid twice as long as the cube.

The mirror is a **plane of symmetry** for the cuboid.

(a) Show that this cuboid has five planes of symmetry.

(b) The length, width and height of another cuboid are all different.
How many planes of symmetry does it have?

(c) How many planes of symmetry does a cube have?
Mark them on a diagram.

(d) Find the number of planes of symmetry for these shapes.

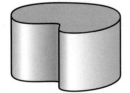

Prism

Regular pentagonal prism

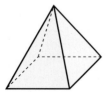

Square–based pyramid

Cylinder

Exercise 11.1

1 Copy these shapes.
On each shape, draw the lines of symmetry.

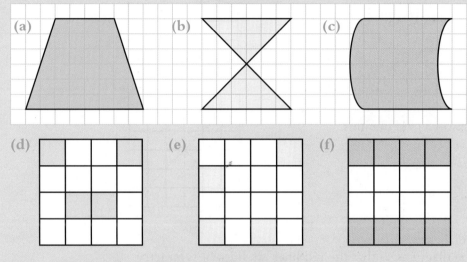

2 How many lines of symmetry do these shapes have?

(a) (b) (c)

3 Draw a square.
A square has four lines of symmetry.
Draw all the lines of symmetry on your square.

4 This triangle has one line of symmetry.
What type of triangle is it?

5 Copy this grid.
Shade more squares so that the diagonal
red line is the line of symmetry.

Exercise continues …

6 Copy this grid.
Shade more squares so that the grid has
two lines of symmetry.

7 Make a pattern with two diagonal lines
of symmetry.
Shade squares on a grid like this.

8 Copy these diagrams.
Complete the diagrams so that the red lines are lines of symmetry.

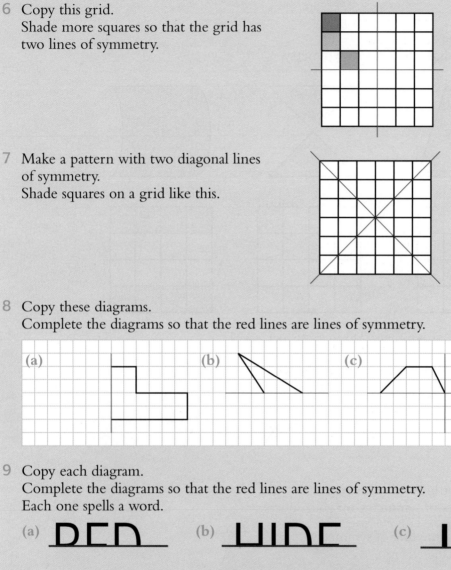

(a) (b) (c)

9 Copy each diagram.
Complete the diagrams so that the red lines are lines of symmetry.
Each one spells a word.

(a) RED (b) HIDE (c) ICE

(d) KICK (e) CODE (f) HOD

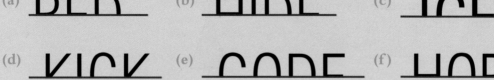

Challenge 11.1

Investigate the relationship between the number of planes of symmetry
of a prism and the number of lines of symmetry of its cross-section.

Reflections

You can **reflect** an object in a mirror line.

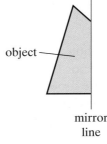

Reflecting an object is like completing a diagram so a shape has a line of symmetry.

Different colours have been used here to help you see what has changed. Usually the colour stays the same!

The **image** is the other side of the mirror line and is the opposite way round.

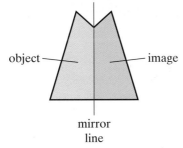

Each point on the image is the same distance away from the mirror line as the point matching it on the object.

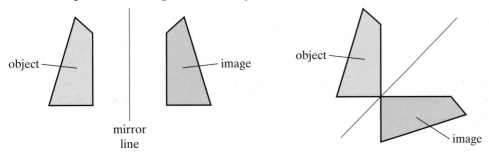

When an object is reflected, its image is **congruent** to it. That means it is the same shape and size.

However, the image is the opposite way round. To draw a reflection, trace the object and the mirror line, then turn the tracing paper over to draw the image.

◎ Discovery 11.3

Work in a group.
- Each draw the same object on a piece of squared paper.
- Each draw a different mirror line.
- Reflect the object in your mirror line.
- Compare your diagrams in the group.
- See how the position of the image changes when the mirror line changes.

> **TIP**
>
> You can count squares or use tracing paper to help you draw the image.

Example 11.3

Draw x- and y-axes from -5 to 5.

Plot the points $(2, 1)$, $(4, 1)$ and $(2, 4)$.

Join the points to form a triangle. Label it A.

Reflect triangle A in the x-axis. Label the image B.

Reflect triangle A in the y-axis. Label the image C.

Solution

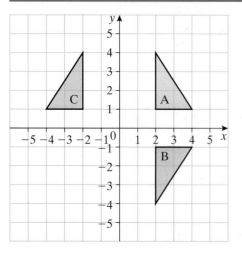

The x-axis is the horizontal axis.
The base of triangle A is one square above the x-axis so the base of triangle B will be one square below it.
The third vertex (or corner) is 4 squares above the x-axis so the third vertex of triangle B will be four squares below it.

The y-axis is the vertical axis.
The vertical side of triangle A is two squares to the right of the y-axis so the vertical side of triangle C will be two squares to the left of it.
The third vertex is four squares to the right of the y-axis so the third vertex of triangle C will be four squares to the left of it.

Example 11.4

Reflect this shape in the line $y = -x$.

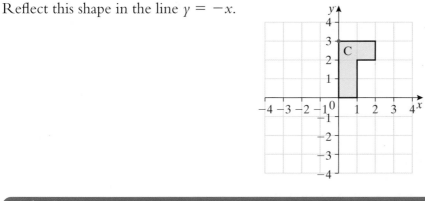

Solution

Draw the line $y = -x$ from $(-4, 4)$ to $(4, -4)$.

One point is on the line, so its image stays there. The images of the other points are each the same distance from the line as the original point, but on the opposite side. For example, point C and its image are both one and a half squares from the mirror line.

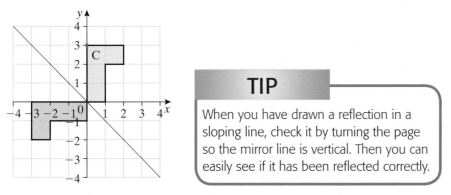

TIP

When you have drawn a reflection in a sloping line, check it by turning the page so the mirror line is vertical. Then you can easily see if it has been reflected correctly.

Recognising and describing reflections

A reflection is a **transformation**. It changes the shape. You will learn about other types of transformation later in this unit. You need to be able to recognise when a transformation is a reflection and describe it.

You can tell whether a shape has been reflected using tracing paper: if you trace the **object**, you will have to turn the tracing paper over to fit the tracing on to the **image**.

You also need to find the **mirror line**. You do this by measuring the distance between points on the object and image.

Example 11.5

Describe the transformation that maps shape ABC on to shape A′B′C′.

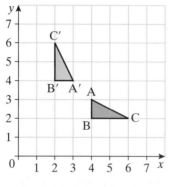

Solution

You can probably tell just by looking that the transformation is a reflection, but you could check using tracing paper.

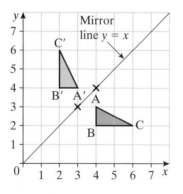

TIP

Check the line is correct by turning the page so that the mirror line is vertical.

To find the mirror line, put a ruler between two corresponding points (B and B′) and mark the midpoint of the line between them. The midpoint is $(3, 3)$.

Do the same for two other corresponding points (C and C′). The midpoint is $(4, 4)$.

Join the points to find the mirror line. The mirror line passes through $(1, 1)$, $(2, 2)$, $(3, 3)$, $(4, 4)$ … .
It is the line $y = x$.

The transformation is a reflection in the line $y = x$.

TIP

You must both state that the transformation is a reflection and give the mirror line.

The mirror line can be any straight line.

Exercise 11.2

1 Copy each shape on to squared paper and reflect it in the mirror line shown.

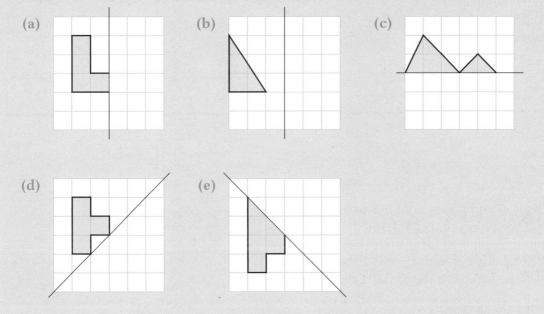

(a)

(b)

(c)

(d)

(e)

2 Copy each pair of shapes on to squared paper and draw the mirror line for the reflection.

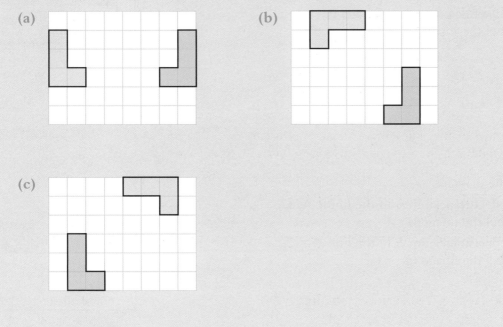

(a)

(b)

(c)

Exercise continues ...

3 Copy the diagram.
Reflect flag A in the mirror line.
Label the image B.
Reflect flag A in the *x*-axis.
Label the image C.

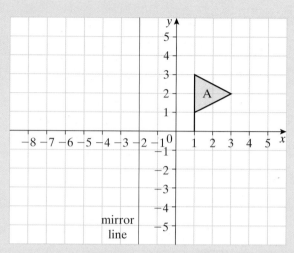

4 Draw *x*- and *y*-axes from −5 to 5.
Plot the points (2, 1), (4, 1), (4, 2) and (2, 5).
Join the points to form a trapezium. Label it A.
Reflect shape A in the *y*-axis. Label the image B.
Reflect shape A in the *x*-axis. Label the image C.

5 Copy the diagram.
  (a) Reflect flag A in the line *x* = 3.
      Label the image B.
  (b) Reflect flag A in the line *y* = −1.
      Label the image C.

6 Copy the diagram.
  (a) Reflect trapezium A in the line *y* = 3.
      Label the image B.
  (b) Reflect trapezium A in the line *x* = 2.
      Label the image C.

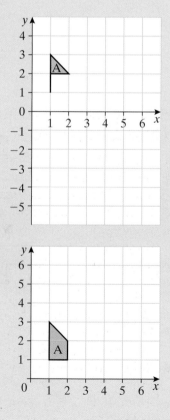

Exercise continues …

7 Draw x- and y-axes from -4 to 4.
Draw a triangle with vertices at $(1, 0)$, $(1, -2)$ and $(2, -2)$. Label it A.
Reflect triangle A in the line $y = 1$. Label it B.
Reflect triangle B in the line $y = x$. Label it C.

8 Draw x- and y-axes from -4 to 4.
Draw a triangle with vertices at $(1, 1)$, $(2, 3)$ and $(3, 3)$. Label it A.
Reflect triangle A in the line $y = 2$. Label it B.
Reflect triangle A in the line $y = -x$. Label it C.

9 Describe fully the transformation that maps
   (a) flag A on to flag B.
   (b) flag A on to flag C.
   (c) flag B on to flag D.

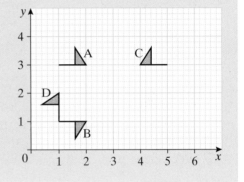

What you have learned

- A two-dimensional shape has reflection symmetry if it can be folded down the middle so that both sides match
- The fold line is a line of symmetry
- Some two-dimensional shapes have more than one line of symmetry
- A three-dimensional shape has reflection symmetry if it can be cut through the centre so that both halves match
- The cut line is a plane of symmetry
- Some three-dimensional shapes have more than one plane of symmetry
- When you reflect a shape, the image is on the other side of the mirror line
- When you reflect a shape, each point on the image is the same distance from the mirror line as the corresponding point on the object
- Two shapes are congruent if they are exactly the same shape and size
- When you reflect a shape, the object and image are congruent but the opposite way round
- To describe a reflection you need to say that it is a reflection and give the mirror line

Mixed exercise 11

1 How many lines of symmetry does each of these shapes have?

(a) 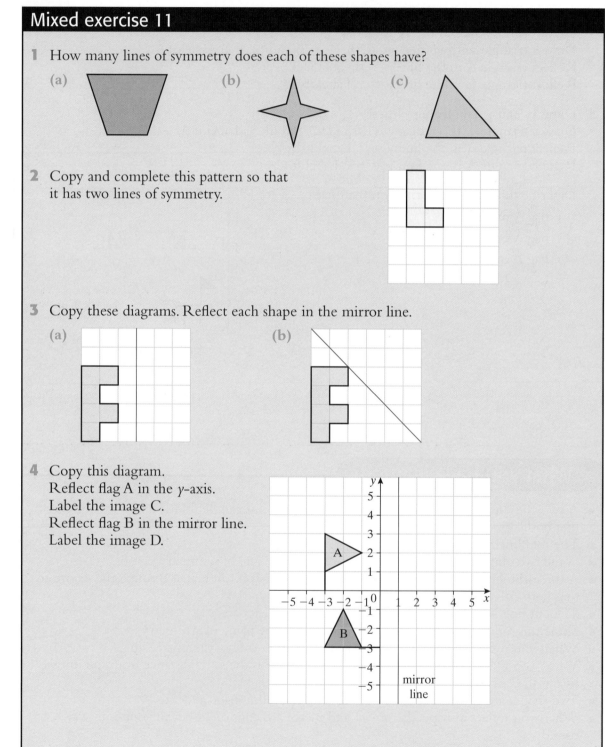 (b) (c)

2 Copy and complete this pattern so that it has two lines of symmetry.

3 Copy these diagrams. Reflect each shape in the mirror line.

(a) (b)

4 Copy this diagram.
Reflect flag A in the y-axis.
Label the image C.
Reflect flag B in the mirror line.
Label the image D.

Mixed exercise 11 continues …

5 Draw x- and y-axes from -5 to 5.
Plot the points $(2, 2)$, $(4, 2)$ and $(2, 5)$.
Join them to form a triangle. Label it A.
Reflect triangle A in the y-axis. Label the image B.
Reflect triangle A in the x-axis. Label the image C.
Reflect triangle A in the line $y = x$. Label the image D.
Reflect triangle A in the line $y = -x$. Label the image E.

6 Describe fully the transformation that maps

(a) triangle A on to triangle B.

(b) triangle A on to triangle C.

(c) triangle C on to triangle D.

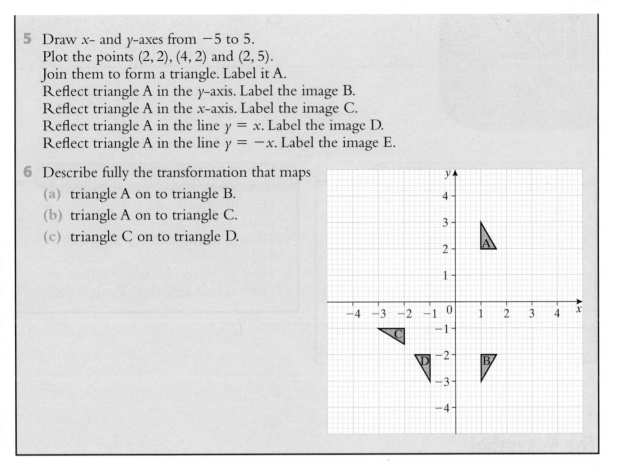

Chapter 12

Percentages

► This chapter is about

- Understanding what is meant by percentage
- Converting between fractions, decimals and percentages
- Calculating a percentage of a quantity
- Calculating percentage increase and percentage decrease
- Calculating one quantity as a percentage of another

► You should already know

- How to express a fraction in its lowest terms
- How to multiply and divide numbers by 100
- How to convert fractions to decimals
- How to find a fraction of a quantity

The % symbol

You have probably seen the % symbol many times in advertisements and newspapers.

● Discovery 12.1

Find as many examples of percentages as you can from newspapers, magazines and advertising material.

'Per cent' means out of 100, so 20% means 20 out of every 100.

Example 12.1

Find 20% of 300.

In 300 there are 3 hundreds.

20% means 20 out of every 100 so

$$20\% \text{ of } 300 = 3 \times 20$$
$$= 60.$$

Fractions, decimals and percentages

You probably already know that $50\% = \frac{1}{2} = 0.5$.

Statements like

> Half $(\frac{1}{2})$ the class are boys,
> 50% of the class are boys,
> 0.5 of the class are boys

are all saying the same thing.

Converting percentages to fractions

50% means 50 out of every 100.

You can write it as $\frac{50}{100}$.

In Chapter 3 you learned how to cancel fractions to their lowest terms.

$\frac{50}{100} = \frac{5}{10} = \frac{1}{2}$ Divide the numerator and denominator by 10 and then 5.

or $\frac{50}{100} = \frac{1}{2}$ Divide the numerator and denominator by 50.

You can turn other percentages into fractions in the same way.

$20\% = \frac{20}{100} = \frac{2}{10} = \frac{1}{5}$ Divide the numerator and denominator by 10 and then 2.

Check up 12.1

Change these percentages to fractions.

(a) 10%   (b) 30%   (c) 40%   (d) 60%   (e) 70%
(f) 80%   (g) 90%   (h) 25%   (i) 75%

TIP

You should learn these fraction equivalents.

Converting percentages to decimals

Look again at 50%.

50% means 50 out of every 100.

You can write it as $\frac{50}{100}$.

This is the same as $50.0 \div 100 = 0.500 = 0.5$ (moving the digits two places to the right).

In the same way,

$43\% = \frac{43}{100} = 43.0 \div 100 = 0.43$

and $3\% = \frac{3}{100} = 3.0 \div 100 = 0.03$.

> **TIP**
>
> A common error is to write 3% as 0.3, rather than 0.03.

Example 12.2

Change these percentages to fractions and decimals.

(a) 15%   **(b)** 5%   **(c)** 140%

Solution

(a) Fraction $15\% = \frac{15}{100} = \frac{3}{20}$ Divide numerator and denominator by 5.

    Decimal $15\% = 15 \div 100 = 0.15$

(b) Fraction $5\% = \frac{5}{100} = \frac{1}{20}$ Divide numerator and denominator by 5.

    Decimal $5\% = 5 \div 100 = 0.05$

(c) Fraction $\frac{140}{100} = \frac{14}{10} = \frac{7}{5}$ Divide numerator and denominator by 10 then by 2.

                 $= 1\frac{2}{5}$ Convert to a mixed number.

    Decimal $140\% = 140 \div 100$
                         $= 1.40 = 1.4$

⦿ Discovery 12.2

VAT (Value Added Tax) is a tax added on to the price of goods.

(a) Find out the current rate of VAT.

(b) Write the current rate of VAT as a fraction and as a decimal.

Converting fractions and decimals to percentages

To change fractions and decimals to percentages you reverse
the process and multiply by 100.
This is shown in the following example.

Example 12.3

Change these fractions and decimals to percentages.

(a) $\frac{2}{5}$ **(b)** $\frac{8}{25}$

(c) 0.37 **(d)** 0.06

Solution

(a) In Chapter 6 you learned that to multiply a fraction by a whole
number you multiply just the numerator by that number.

$\frac{2}{5} \times 100$

$2 \times 100 = 200$ so $\frac{2}{5} \times 100 = \frac{200}{5}$.

Then you find how many 5s there are in 200.

$\frac{200}{5} = 40$

So $\frac{2}{5}$ as a percentage is 40%.

(b) $\frac{8}{25} \times 100$

$8 \times 100 = 800$ so $\frac{8}{25} \times 100 = \frac{800}{25}$.

$\frac{800}{25} = 32$

So $\frac{8}{25}$ as a percentage is 32%.

(c) $0.37 \times 100 = 37$ You multiply the decimal by 100.

So 0.37 as a percentage is 37%.

(d) $0.06 \times 100 = 6$ You multiply the decimal by 100.

So 0.06 as a percentage is 6%.

You can compare fractions, decimals and percentages by converting
them all to the same sort of number.
Usually converting them to either decimals or percentages makes them
easiest to compare.
Choose whichever is less work!

Example 12.4

Put these numbers in order, smallest first.

$\frac{1}{4}$ 35%   0.41 $\frac{11}{40}$ 0.3

Solution

Change them all to decimals.

$\frac{1}{4} = 0.25$

$35\% = 0.35$

$0.41 = 0.41$

$\frac{11}{40} = 11 \div 40 = 0.275$

$0.3 = 0.3$

So the order is

    0.25   0.275   0.3   0.35   0.41

Now write the original numbers in order of size.

$\frac{1}{4}$ $\frac{11}{40}$ 0.3   35%   0.41

Example 12.5

Roper Foods increased the capacity of their standard packet of cereal by 15%.

Deacon Grocers increased the capacity of the same size packet by $\frac{1}{6}$.

Which company increased the standard packet size by more?

Solution

Convert $\frac{1}{6}$ to a percentage. $\frac{1}{6} = 16.666...\%$

16.666...% is bigger than 15% so Deacon Grocers increased the standard packet size by more.

> **TIP**
>
> Notice that 16.666... isn't exact.
> You could round it to 16.7 but to write the number exactly you have to use a fraction.
> You could write $16\frac{2}{3}\%$.

> **TIP**
>
> It is worth learning some basic equivalents.
>
> $\frac{1}{2} = 0.5 = 50\%$ $\frac{1}{5} = 0.2 = 20\%$ $\frac{1}{3} = 0.333... = 33.3...\%$
>
> $\frac{1}{4} = 0.25 = 25\%$ $\frac{2}{5} = 0.4 = 40\%$ (33% to the nearest 1%)
>
> $\frac{3}{4} = 0.75 = 75\%$ $\frac{3}{5} = 0.6 = 60\%$ $\frac{2}{3} = 0.666... = 66.6...\%$
>
> $\frac{4}{5} = 0.8 = 80\%$ (67% to the nearest 1%)

Exercise 12.1

1   Change these percentages to fractions.
    Write your answers in their lowest terms.
    (a) 35%   (b) 65%   (c) 8%   (d) 120%

2   Change these percentages to decimals.
    (a) 16%   (b) 27%   (c) 83%   (d) 7%
    (e) 31%   (f) 4%   (g) 17%   (h) 2%
    (i) 150%   (j) 250%   (k) 9%   (l) 12.5%

3   Change these decimals to percentages.
    (a) 0.62   (b) 0.56   (c) 0.04   (d) 0.165   (e) 1.32

4   Change these fractions to percentages.
    (a) $\frac{7}{10}$ (b) $\frac{3}{5}$ (c) $\frac{7}{20}$ (d) $\frac{10}{25}$ (e) $\frac{17}{50}$

5   Write three fractions that are the same as 40%.

6   Write three fractions that are the same as 60%.

7   At a local football match, $\frac{17}{20}$ of the crowd were home supporters.
    What percentage was this?

8   At Bradway school, $\frac{3}{5}$ of the teachers are female.
    What percentage is this?

9   In a survey about the colour of cars, 22% of the people said they preferred red cars
    and $\frac{3}{20}$ said they preferred silver cars.
    Which of the two colours was more popular?

10 David walks $\frac{7}{8}$ of a kilometre to school and Paula walks 0.87 km to school.
    Who walks further?

11 Put these numbers in order, smallest first.
    $\frac{4}{5}$ 88%   0.83 $\frac{17}{20}$ $\frac{7}{10}$

12 Put these numbers in order, smallest first.
    $\frac{3}{8}$ 35%   0.45 $\frac{2}{5}$ $\frac{5}{12}$

Exercise continues …

13 Put these numbers in order, smallest first.

$\frac{3}{5}$ 30%   0.7 $\frac{3}{4}$ $\frac{2}{3}$

14 Josh and David are in different classes.
In Josh's class, $\frac{4}{7}$ are boys.
In David's class, 45% are boys.
Whose class has the higher proportion of boys?

15 Intake United, Darnall Players and Handsworth Rovers have all played the same number of games.
Intake United have won $\frac{3}{8}$, Darnall Players have won 35% and Handsworth Rovers have won 0.4 of the games they have played.
Rank the teams in order of the number of matches they have won.

16 At Rayworth School, Class 11D boys were asked to choose their favourite sport.
$\frac{2}{7}$ chose football, 0.27 chose rugby and 28% chose gymnastics.
List the sports in the order, most popular first.

Finding a percentage of a quantity

In Chapter 3 you learned how to find a fraction of a quantity.
The next example reminds you how to do this.

Example 12.6

Find $\frac{3}{10}$ of £60.

Solution

You need to calculate $\frac{3}{10} \times 60$.

$60 \div 10 = 6$ First you divide 60 by 10 to find $\frac{1}{10}$ of 60.

$6 \times 3 = 18$ Then you multiply your answer by 3 to find $\frac{3}{10}$.

Answer: £18

You can find 30% of £60 using the fraction or decimal equivalents.

Method 1: Using fractions

$30\% = \frac{30}{100} = \frac{3}{10}$.

Then you calculate $\frac{3}{10} \times 60 = £18$ as in Example 12.6.

This is the same as saying
$$10\% \text{ of } £60 = £6$$
so $$30\% \text{ of } £60 = £6 \times 3 = £18.$$

Method 2: Using decimals
$$30\% = 0.30 = 0.3$$
$$0.3 \times 60 = £18$$

Example 12.7

Calculate 15% of £68.

Solution

Method 1: Using fractions

You can break the percentage down into parts that are easier to calculate.

$$15\% = 10\% + 5\%$$

$10\% \text{ of } £68 = £68 \div 10$ $10\% = \frac{1}{10}$ so divide £68 by 10.
$$\qquad\qquad\quad = £6.80$$

$5\% \text{ of } £68 = £6.80 \div 2$ $5\% = \frac{1}{2}$ of 10% so divide your answer by 2.
$$\qquad\qquad\quad = £3.40$$

$£6.80 + £3.40 = £10.20$ $10\% + 5\% = 15\%$

Method 2: Using decimals

Using your calculator: $£68 \times 0.15 = £10.20$

Exercise 12.2

Do not use your calculator for questions 1 to 4.

1 (a) Find 20% of £80.   (b) Find 40% of £25.   (c) Find 35% of £60.

2 Shamir invests £120 and earns 5% interest in the first year.
   Calculate the interest.

3 60% of the students in a school are girls. There are 400 students in the school.
   How many are girls?

Exercise continues ...

4 15% of the takings at a concert were given to a charity. The takings were £8400. How much did the charity receive?

You may use your calculator for questions 5 to 9.

5 (a) Find 17% of £48.
   (b) Find 48% of £180.

6 Phoebe invests £450 and earns 4% interest in the first year. Calculate the interest.

7 Find 120% of 32 metres.

8 76% of the crowd at a football match were adults.
   There were 28 000 people in the crowd. How many were adults?

9 Jane pays tax at 22% on earnings of £380. How much tax does she pay?

Challenge 12.1

Dan has £450 to invest.

Bank A offers 6% interest per year.

Bank B offers 3% interest every six months.

With which bank would Dan get most interest in one year?

How much is the difference?

Percentage increase and decrease

You can calculate percentage increases by first calculating the increase and then adding this to the original amount.

Similarly, you can calculate percentage decreases by first calculating the decrease and then subtracting this from the original amount.
This is shown in the following examples.

Example 12.8

A shop increased its annual sales of computers by 20% in 2009.
In 2008 it sold 1200 computers.

Without using your calculator, work out how many computers were sold in 2009.

Solution

10% of 1200 = 120

20% of 1200 = 120 × 2 = 240

Sales in 2009 = 1200 + 240 = 1440.

Example 12.9

A company selling insurance offers a 15% reduction for policies arranged on the internet.

The normal cost of a policy is £360.

What is the cost of a policy arranged on the internet?

Solution

Without using a calculator

> 10% of 360 = 36
>
> 5% of 360 = 18

so the reduction is

> 36 + 18 = 54.

Using a calculator

> 360 × 0.15 = 54

Cost of policy arranged on the internet

> = £360 − £54 = £306.

> **TIP**
>
> When companies offer a reduction like this, it is often called a **discount**.

Exercise 12.3

Do not use your calculator for questions 1 to 6.

1 Increase £400 by these percentages.
  - (a) 20%
  - (b) 45%
  - (c) 6%
  - (d) 80%

2 Decrease £240 by these percentages.
  - (a) 30%
  - (b) 15%
  - (c) 3%
  - (d) 60%

Exercise continues …

3 Simon earns £12 000 per year. He receives a salary increase of 4%.
   Find his new salary.

4 Bills can be paid by direct debit (monthly payments direct from a bank account).
   An electricity company offers a discount of 5% for payments by direct debit.
   The normal bill is £36 per month. How much is the bill if it is paid by direct debit?

5 VAT on fuel bills is charged at 8%. Lee's gas bill is £120 before VAT.
   What is the bill after VAT has been added?

6 An electrical goods shop is having a sale.

Complete the table to find the sale price of these articles.

Item	Original price (£)	Reduction (£)	Sale price (£)
Television	150		
Washing machine	360		
DVD player	40		
Computer system	550		

You may use your calculator for questions 7 to 12.

7 Increase £68 by these percentages.
   (a) 12%   (b) 26%   (c) 7%   (d) 64%

8 Decrease £312 by these percentages.
   (a) 18%   (b) 32%   (c) 9%   (d) 78%

9 The value of a car fell by 13% in the first year. It cost £8500 when new.
   What was its value after 1 year?

Exercise continues ...

10 76% more students studied ICT in 2009 than in 1999.
46 000 studied ICT in 1999.
How many studied ICT in 2009?

11 A sofa costs £340 before VAT.
What is the price after VAT is added?
(You will need the VAT rate and its decimal equivalent that you found in Discovery 12.2.)

12 An energy company offers a 6% discount if both gas and electricity are purchased from them.
Jane's gas bill is £42 and her electricity bill is £34.
How much will her total bill be if she purchases both from the same company and receives the 6% discount?

Calculating one quantity as a percentage of another

Dan got sixteen out of twenty in his last maths test.
What is this as a percentage?

Dan's teacher marked the paper $\frac{16}{20}$.

This is the fraction of the total marks that Dan scored.

As you found out earlier, to convert a fraction into a percentage you multiply by 100.

So Dan's percentage mark is $\frac{16}{20} \times 100$.

The easiest way to do this is first cancel $\frac{16}{20}$ to $\frac{8}{10}$.

Then do $\frac{8}{10} \times 100$.

$8 \times 100 = 800$ and $800 \div 10 = 80$.

On a calculator you can simply do $16 \div 20 \times 100 = 80$.

So, Dan's percentage mark is 80%.

In general, to find A as a percentage of B,

first write as a fraction, $\frac{A}{B}$,

then do $\frac{A}{B} \times 100$.

Example 12.10

Find 18 as a percentage of 40.

Solution

Write 18 as a fraction of 40 $\frac{18}{40}$.

Then multiply by 100 $\frac{18}{40} \times 100 = \frac{9}{20} \times 100$ By cancelling

$9 \times 100 = 900$

$900 \div 20 = 45$

Answer: 45%

Example 12.11

Find 40 centimetres as a percentage of 2 metres.

Solution

First change to the same units $\quad 2\,m = 200\,cm$

Then write as a fraction and multiply by 100

$\frac{40}{200} \times 100 = \frac{20}{100} \times 100 = 20\%$

Example 12.12

Sophie bought a house for £90 000 and sold it for £110 000.

Find her profit as a percentage of what she paid for the house.

Solution

Profit $= 110\,000 - 90\,000 = 20\,000$

You need 20 000 as a percentage of 90 000.

Write as a fraction $\quad\frac{20\,000}{90\,000}$

Multiply by 100 $\quad\frac{20\,000}{90\,000} \times 100$

Using a calculator

$20\,000 \div 90\,000 \times 100 = 22.2\%$ (to 1 decimal place)

Exercise 12.4

Do not use your calculator for questions 1 to 5.

1 Calculate £2 as a percentage of £20.

2 Calculate 5 metres as a percentage of 20 metres.

3 Calculate 80p as a percentage of £2.

4 Jane earns £5 per hour. She receives a pay increase of 25p per hour.
   Calculate her pay increase as a percentage of £5.

5 A television originally costing £150 is reduced by £30.
   Calculate the reduction as a percentage of the original price.

You may use your calculator for questions 6 to 10.

6 Calculate £26 as a percentage of £200.

7 Calculate 3 metres as a percentage of 40 metres.

8 In a school there are 800 students. 425 of them are girls.
   What percentage of the students are girls?
   Give your answer to the nearest whole number.

9 Graham scored 66 out of 80 in an English test. What is this as a percentage?

10 A car was bought for £9000. A year later it was sold for £7500.
    Calculate the loss in value as a percentage of £9000.
    Give your answer to the nearest whole number.

Challenge 12.1

A shop increased all its prices by 20%.
Later, in a sale, the shop reduced its prices by 20%.

A washing machine was originally priced at £350.
Dean said that meant that the washing machine was the same price in
the sale as it was originally.
Kim said no, it will be cheaper in the sale than £350.

Who was right? Explain your answer.
If Kim is right, calculate the reduction as a percentage of the original
£350.

What you have learned

- 'Percentage' means the number of parts per 100
- To change a percentage to a fraction you write the percentage as the numerator of the fraction and the denominator as 100 and then simplify if possible
- To change a percentage to a decimal you divide the percentage by 100
- To change a fraction or a decimal to a percentage you multiply by 100
- The fraction and decimal equivalents of some percentages

Fraction	Decimal	Percentage
$\frac{1}{10}$	0.1	10%
$\frac{1}{5}$	0.2	20%
$\frac{3}{10}$	0.3	30%
$\frac{2}{5}$	0.4	40%
$\frac{1}{2}$	0.5	50%
$\frac{3}{5}$	0.6	60%
$\frac{7}{10}$	0.7	70%
$\frac{4}{5}$	0.8	80%
$\frac{9}{10}$	0.9	90%
$\frac{1}{4}$	0.25	25%
$\frac{3}{4}$	0.75	75%
$\frac{1}{3}$	0.333...	33.3...%
$\frac{2}{3}$	0.666...	66.6...%

- Usually the best method to find a percentage of a quantity without a calculator is to convert the percentage to a fraction and find the fraction of the quantity
- Usually the best method to find a percentage of a quantity with a calculator is to convert the percentage to a decimal and multiply the quantity by the decimal
- To calculate a percentage increase first calculate the increase and then add it to the original amount
- To calculate a percentage decrease first calculate the decrease and then subtract it from the original amount

Mixed exercise 12

Do not use your calculator for questions **1** to **7**.

1 Change these percentages to decimals.

   (a) 27%   (b) 96%   (c) 2%   (d) 16.5%   (e) 350%

2 Convert these into percentages.

   (a) 0.08   (b) $\frac{7}{10}$ (c) 0.72   (d) $\frac{16}{40}$ (e) 1.23

3 Put these numbers in order, smallest first.

   $\frac{182}{1000}$ 18.1%   0.1801 $\frac{9}{50}$

4 Which of the following is nearest to 65%?
Show how you decide.

   $\frac{25}{40}$ 0.655 $\frac{14}{20}$ 0.643

5 Leo's water bottle contains $\frac{3}{4}$ of a litre of water.
James' contains 0.72 of a litre.
Which bottle contains more water?

6 Calculate 30% of £28.

7 Juliet puts £150 into a bank account that earns 5% interest in the first year.
Calculate the interest.

You may use your calculator for questions **8** to **13**.

8 A CD costs £12 before VAT.
If the rate of VAT is 17.5%, how much is the VAT?

9 Carl earns £16 000 per year. He receives a salary increase of 3%.
What is his new salary?

10 In a sale all prices are reduced by 15%.
Calculate the sale price of a toaster originally priced at £35.

11 An antique was bought for £120. It was sold at a profit of 80%.
How much was it sold for?

12 Kate earns £240 per week. She receives a pay increase of £10 per week.
Calculate the increase as a percentage of £240. Give your answer to 1 decimal place.

13 In Karim's survey, 17 out of 60 people had watched more than 3 hours of
television yesterday.
What is this as a percentage? Give your answer to the nearest whole number.

Chapter 13 Rotation

<table>
<tr><td>

▶ **This chapter is about**

- Rotation symmetry
- Rotating shapes
- Describing rotations

</td><td>

▶ **You should already know**

- The meaning of *horizontal* and *vertical*
- How to plot and read coordinates
- About 90° and 180° angles
- About special triangles and quadrilaterals

</td></tr>
</table>

Rotation symmetry

⊙ Discovery 13.1

Look at this shape.

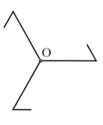

Trace the shape.
Place a pencil point at O and turn the paper round until it fits on to itself again.

This shape fits on to itself three times in a complete turn.
It has rotation symmetry of order 3.

This shape has no rotation symmetry.

There is only one position it fits on to itself in a complete turn.

It has rotation symmetry of order 1.

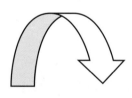

🔍 Check up 13.1

Look around you.

Which objects or shapes in the room have rotation symmetry? List or sketch them.

What order of rotation symmetry do they each have?

Example 13.1

Write underneath each shape its order of rotation symmetry.

For each shape that has rotation symmetry, mark with a dot its centre of rotation.

Write also how many lines of symmetry it has.

Solution

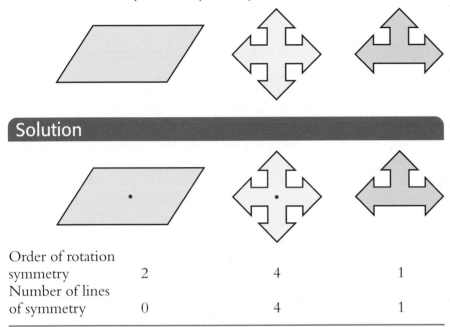

Order of rotation symmetry	2	4	1
Number of lines of symmetry	0	4	1

Exercise 13.1

1 Describe the rotation symmetry of these letters.

(a) H   (b) A   (c) N

(d) S   (e) B   (f) X

2 Describe the rotation symmetry of these.

(a) θ (b) Φ (c) ℕ

(d) Ø   (e) ✳   (f) ♣

3 Describe the rotation symmetry of these shapes.

(a) A square   (b) An equilateral triangle

(c) A rectangle   (d) An isosceles triangle

Exercise continues …

4 Make a copy of each of these shapes.
  If there is rotation symmetry, show the centre and state the order of rotation symmetry

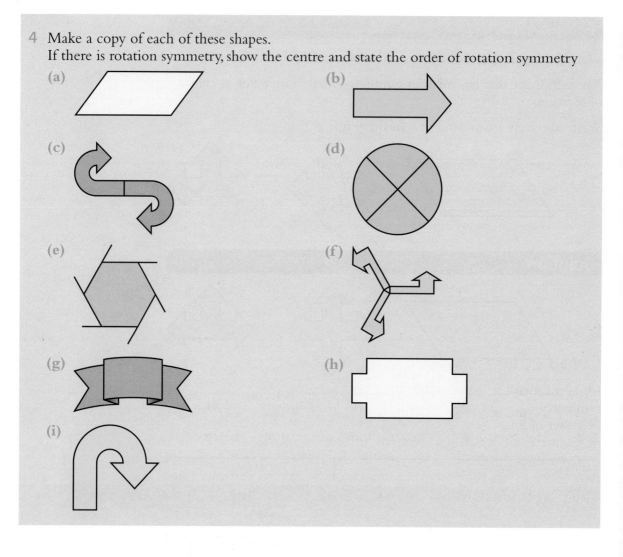

(a)

(b)

(c)

(d)

(e)

(f)

(g)

(h)

(i)

● **Challenge 13.1**

Write your name in capital letters.

Which letters in your name have rotation symmetry?

Write down the order of rotation symmetry for each one.

Which letters in your name have reflection symmetry?

Draw the lines of symmetry for each one.

Challenge 13.2

(a) I am thinking of a special quadrilateral.
It has four lines of symmetry.
It has rotation symmetry of order 4.
What shape am I thinking of?

(b) I am thinking of a special quadrilateral.
It has one line of symmetry.
It has no rotation symmetry.
What shape am I thinking of?

(c) I am thinking of a special quadrilateral.
It has two lines of symmetry.
It has rotation symmetry of order 2.
What shape am I thinking of?

(d) What other quadrilaterals do you know?
Working in pairs, describe them to each other in a similar way.
You may also like to describe regular polygons.

Completing shapes

In Chapter 11 you learned how to complete shapes so that they had reflection symmetry.

You can also complete shapes so that they have rotation symmetry.

Example 13.2

Shade more squares so that the pattern has rotation symmetry of order 2.

Solution

You can check that your diagram is correct by rotating it through a half turn.

Exercise 13.2

1 Copy this grid.
  Shade more squares so that the pattern has
  rotation symmetry of order 2.

2 Copy this grid.
  Shade more squares so that the pattern has
  rotation symmetry of order 4.

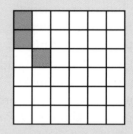

3 Shade squares on a 4 × 4 grid so that your pattern has rotation symmetry of order 2
  but no reflection symmetry.

4 Copy this diagram.

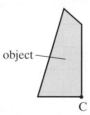

  Complete it so that it has rotation symmetry of order 2.

Drawing rotations

In Chapter 11 you learned that a reflection is a **transformation**. The
second transformation you are going to learn about is **rotation**.

Rotations

You can **rotate** an object about a point, C. The point is called the
centre of rotation.

object —

C

Rotating an object is like completing one step towards making a
drawing with rotation symmetry.

Trace the object then use a pencil to keep C still. Turn the tracing paper through the angle of rotation required.

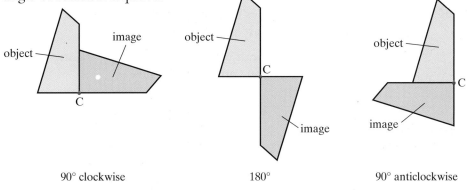

90° clockwise   180°   90° anticlockwise

Since 180° is a half-turn, you can rotate clockwise or anticlockwise.
90° anticlockwise is the same as 270° clockwise.

Each point on the image is the same distance away from the centre of rotation as the point matching it on the object. This is true wherever the centre of rotation.

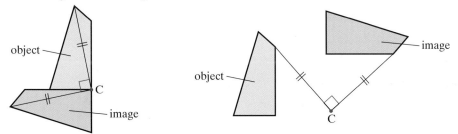

When an object is rotated, its image is **congruent** to it. Unlike reflections, the image is the same way round. You turn the tracing paper round for a rotation; for a reflection you turn it over.

Rotations are **anticlockwise**, unless you are told otherwise.

◉ Discovery 13.2

Work in a group.
- Each draw the same object on a piece of squared paper.
- Trace over your object.
- Each mark a different centre of rotation.
- Use a pencil to keep the centre of rotation still.
  Turn your tracing paper though 90° clockwise.
- Compare your diagrams in the group.
- See how the position of the image changes when the centre of rotation changes.

You could repeat the activity using a rotation of 180° or 90° anticlockwise.

TIP

You can count squares or use tracing paper to help you draw the image.

Example 13.3

Rotate the shape through 90° about the
point C(1, 2).

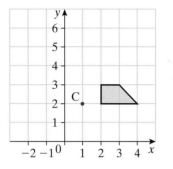

Solution

Angles of rotation are anticlockwise unless
you are told otherwise.

You can rotate the shape using tracing paper
or you can count squares.

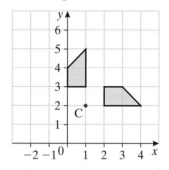

Exercise 13.3

1 Copy the diagram.

   (a) Rotate shape T through a half-turn
        about the origin. Label it A.

   (b) Rotate shape T through 90° clockwise
        about the origin. Label it B.

   (c) Rotate shape T through 90° anticlockwise
        about the origin. Label it C.

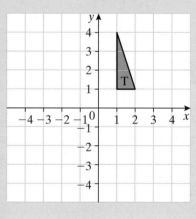

2 Copy the diagram.

   (a) Rotate shape T through a half-turn
        about the origin. Label it A.

   (b) Rotate shape T through 90° clockwise
        about the origin. Label it B.

   (c) Rotate shape T through 90° anticlockwise
        about the origin. Label it C.

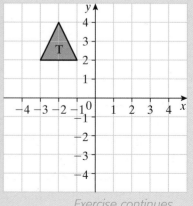

Exercise continues …

3 Copy the diagram.

(a) Rotate shape T through 90° anticlockwise about the origin. Label it A.

(b) Rotate shape T through 90° clockwise about the origin. Label it B.

(c) Rotate shape T through 180° about the origin. Label it C.

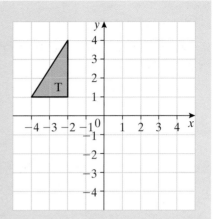

4 Copy the diagram.

(a) Rotate shape T through 90° anticlockwise about the origin. Label it A.

(b) Rotate shape T through 90° clockwise about the origin. Label it B.

(c) Rotate shape T through 180° about the origin. Label it C.

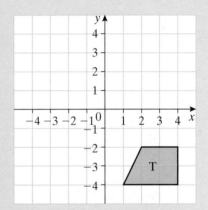

5 Copy the diagram.
Rotate the shape through a half-turn about the point (4, 3).

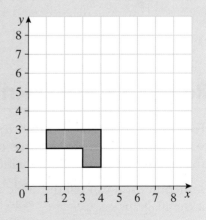

TIP

Tracing paper is always stated as optional extra material in examinations. When doing transformation questions, always ask for it.

Exercise continues …

6 Copy the diagram.
Rotate the shape through a half-turn about
the point (5, 3).

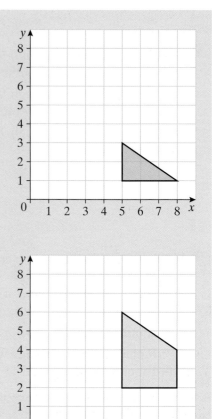

7 Copy the diagram.
Rotate the shape through 90° anticlockwise about
the point (5, 2).

Recognising and describing rotations

To describe a rotation you have to say the transformation
is a rotation and give the **centre of rotation** and the
angle and **direction** of the rotation.

Sometimes you can tell the angle of rotation just by looking
at the diagram.

If you can't, you need to identify a pair of sides that
correspond in the object and image and measure the
angle between them. You may need to extend the lines.

You can usually find the centre of rotation by counting
squares or using tracing paper.

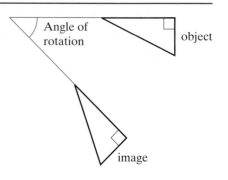

Example 13.4

Describe fully the single transformation
that maps flag A on to flag B.

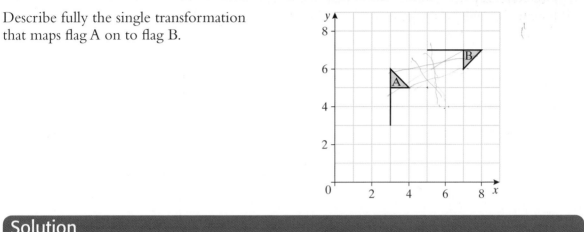

Solution

It is clear that the transformation is a rotation and that the angle is 90° clockwise.
Clockwise rotations are described as negative.
This is a rotation of $-90°$.

Use tracing paper and a pencil to find the centre of rotation.
Trace flag A and use the pencil to hold the tracing to the diagram at a point.

Rotate the tracing paper and see if the tracing fits over flag B.
Keep trying different points until you find the centre of rotation.
Here, the centre of rotation is $(6, 4)$.

The transformation is a rotation of 90° clockwise about $(6, 4)$.

Exercise 13.4

1 Describe fully the transformation
that maps
(a) triangle A on to triangle B.
(b) triangle A on to triangle C.
(c) triangle A on to triangle D.

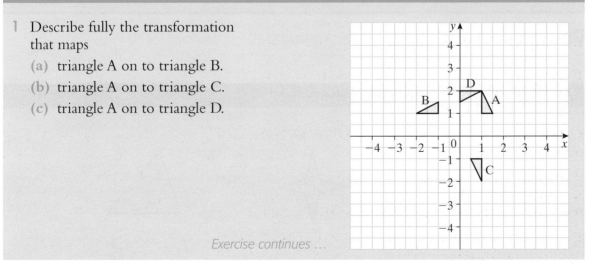

Exercise continues …

2   Describe fully the transformation that maps
    (a)   triangle A on to triangle B.
    (b)   triangle A on to triangle C.
    (c)   triangle A on to triangle D.
    (d)   triangle A on to triangle E.
    (e)   triangle B on to triangle E.

    Hint: Some of these transformations are reflections.

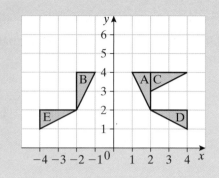

What you have learned

- A shape has rotation symmetry if it fits on to itself more than once in a complete turn
- The order of rotation symmetry is the number of times it fits on to itself in a complete turn
- If a shape fits on to itself only once in a complete turn, it has no rotation symmetry; you can say it has rotation symmetry of order 1
- You can rotate an object about a point; the point is the centre of rotation
- When you rotate a shape, each point on the image is the same distance from the centre of rotation as the corresponding point on the object
- Two shapes are congruent if they are exactly the same shape and size
- When you rotate a shape, the object and image are congruent and the same way round
- To describe a rotation you must state that the transformation is a rotation and give the centre of rotation and the angle and direction of rotation

Mixed exercise 13

1   Describe the rotation symmetry of these shapes.

(a)   (b)   (c)

Mixed exercise 13 continues ...

2 Copy and complete this pattern so that it has rotation symmetry of order 4.

3 Copy the diagram.

(a) Rotate shape A through 90° anticlockwise about the origin.
Label it B.

(b) Rotate shape A through 180° about the point (2, −1).
Label it C.

4 Which of these shapes are rotations of shape A?

5 Describe fully the transformation that maps

(a) flag A on to flag B.

(b) flag A on to flag C.

(c) flag A on to flag D.

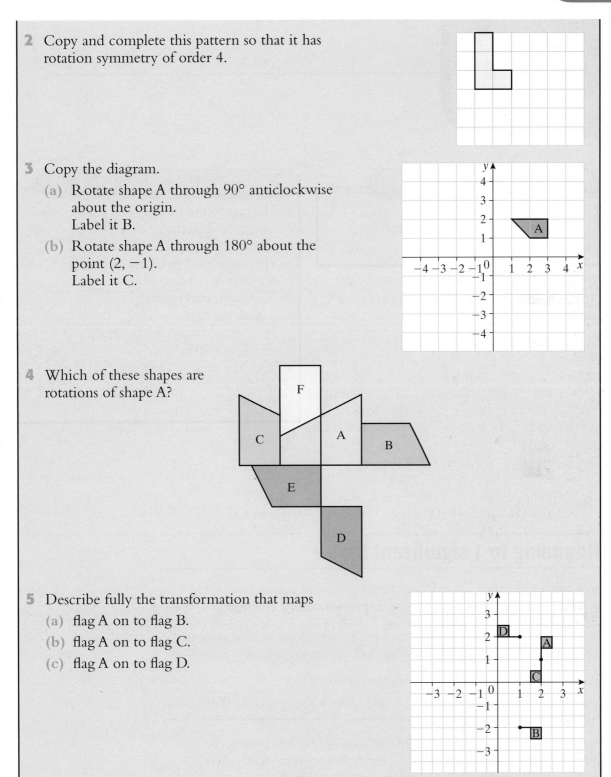

Estimation

- Rounding to 1 significant figure
- Estimating the answer to a problem
- Checking answers

- How to round numbers to the nearest integer and to the nearest 10, 100, 1000,...
- How to round numbers to a given number of decimal places
- How to round numbers to 1 significant figure
- How to add, subtract, multiply and divide numbers
- That the area of a rectangle can be calculated by multiplying the length by the width

Rounding to 1 significant figure

In Unit A Chapter 12 you learned how to round numbers to 1 significant figure and that is very important to this chapter.

Check up 14.1

Round these numbers to 1 significant figure.
(a) 5432   (b) 104   (c) 306   (d) 15 090

You also need to be able to apply this to numbers less than 1. As before, the first significant figure is the first non-zero digit when starting from the left of the number and moving right.

Example 14.1

Round these numbers to 1 significant figure.

(a) 0.372   (b) 0.002 043

(c) 0.000 6904   (d) 0.999

Solution

(a) 0.4

The second significant figure is 7 so you add 1 to the first significant figure, 3, and do not write any more zeros.

(b) 0.002

The second significant figure is the 0 after the 2 so you leave the 2 as it is.

(c) 0.0007

The second significant figure is 9 so you add 1 to the 6 to get 7.

(d) 1

The second significant figure is the second 9 so you add 1 to the first 9 which becomes 10 tenths which equal 1.

Estimating the answer to a problem

There are two reasons for estimating the answer to a problem. The first is to check that the answer on the calculator or computer is correct and the second is when you only need an approximate answer.

One way to estimate the answer to a problem is to round each number to 1 significant figure, work out the calculation without using a calculator and then give your answer correct to 1 or 2 significant figures only.

Example 14.2

Jim buys 24 golf balls at £3.75 each.
What is the approximate cost?

Solution

20 × 4 = £80   Round to 1 significant figure.

Example 14.3

Estimate the result of each of these calculations by rounding to 1 significant figure.

(a) 3.7×2.1 **(b)** 47×63

(c) 4.32×8.76 **(d)** 2.95×0.321

(e) $87 \div 38$ **(f)** 4.21×0.016

(g) $0.815 \div 3.41$

Solution

(a) $4 \times 2 = 8$

(b) $50 \times 60 = 3000$

(c) $4 \times 9 = 36$

(d) $3 \times 0.3 = 0.9$

(e) $90 \div 40$ is the same as $9 \div 4 = 2.25$

$90 \div 40 = 2.25$ or 2.3 or 2 Estimates should really have only one or two digits, so here round to 1 decimal place or 1 significant figure.

(f) $4 \times 0.02 = 0.08$ $4 \times 2 = 8$ and there are 2 decimal places.

(g) $0.8 \div 3 = 0.2666$
             $= 0.27$ or 0.3

> **TIP**
>
> Do not try to be too accurate with your answer. Give 1 or 2 significant figures only.

It is sometimes useful to know if the actual answer is higher or lower than the estimate. This depends on whether you rounded the original numbers up or down.

Example 14.4

Estimate the result of each of these calculations.
In each case, say whether the estimated result is bigger or smaller than the actual result and explain your answer.

(a) 4.7×3.9 **(b)** 516×2.34 **(c)** $79.6 \div 23.7$

Solution

(a) $5 \times 4 = 20$
The estimated result is bigger than the actual result because both the rounded values used in the calculation are bigger than the actual values.

(b) $500 \times 2 = 1000$
The estimated result is smaller than the actual result because both the rounded values used in the calculation are smaller than the actual values.

(c) $80 \div 20 = 4$
The estimated result is bigger than the actual result because the rounded value of the number being divided is bigger than the actual value and the rounded value of the number it is divided by is smaller than the actual value.

Exercise 14.1

1 Round these numbers to 1 significant figure.

   (a) 0.539   (b) 0.666   (c) 0.000 518

   (d) 0.008 609   (e) 0.98   (f) 3.963

For each of questions 2 to 23, show the approximations you used to get your estimate.

2 At the school fete, Tony sold 245 ice-creams at 85p each.
Estimate his takings.

3 A rectangle measures 5.8 cm by 9.4 cm.
Estimate its area.

4 Sally bought 18 packets of crisps at 32p each.
Estimate how much she spent.

5 A cuboid has edges of 3.65 cm, 2.44 cm and 2.2 cm.
To find the volume of a cuboid you multiply the length by the width by the height.
Estimate the volume of this cuboid.

6 At the Arts Theatre, tickets cost £7.95.
412 tickets were sold for the production of 'Anything Goes'.
Estimate the amount the theatre took for the tickets.

7 At a theatre they sold coffee in the interval at 68p a cup.
They sold 73 cups.
Estimate the amount they took for cups of coffee.

8 A rectangle has an area of 63.53 cm² and its length is 9.61 cm.
Estimate the width of the rectangle.

Exercise continues ...

9   On a long-haul flight an aeroplane flew 3123 miles in 8.4 hours.
    Estimate how far it flew on average in each hour.

10 Peter went to see his local football team 23 times during the season.
    Tickets cost £18.50 for each game.
    Estimate how much he spent.

11 Ian played 23 innings at cricket during the season.
    He scored a total of 734 runs.
    Estimate how many runs he scored on average per innings.

12 At Dave's ice-cream stall he sold 223 ice-creams at 67p each.
    Estimate how much he took.

13 Martyn rode 327 miles in 12 days.
    Estimate how far he rode on average each day.

14 To find the area of a circle of radius 2.68 metres, you need to work out
    3.14 × 2.68 × 2.68
    Estimate the result of this calculation.

15 Kate has £30.
    How many CDs, at £7.99 each, can she buy?

16 Estimate the result of each of these calculations.
    (a) 5.89 × 1.86   (b) 19.25 ÷ 3.8   (c) 36.87 × 22.87   (d) 9.7 ÷ 3.5
    (e) 2.14 × 0.82   (f) 3.14 × 7.92   (g) 289 ÷ 86   (h) 4.93 × 0.025

17 Estimate the result of each of these calculations.
    (a) 14 × 2.65   (b) 912 × 26   (c) 54.6 × 2.91   (d) 45 ÷ 6.8
    (e) 287 ÷ 9.81   (f) 3.462 × 0.34   (g) 0.27 ÷ 4.8   (h) 2.1 × 9.4 × 3.7

18 28 members of the Wednesday club went for a meal together.
    The total cost was £389.52.
    Estimate how much each paid on average.

19 The flat roof of Darrell's garage is in the shape of rectangle measuring 5.41 metres by
    3.64 metres.
    A company charges £8.75 per square metre to repair the roof.
    Estimate how much it will cost to repair the roof.

Exercise continues ...

20 Estimate the result of each of these calculations.
In each case, say whether the estimated result is bigger or smaller than the actual result and explain your answer.

    (a) 4.21×81.6 (b) $189 \div 11.4$ (c) 16.3×897

21 The sides of a rectangle are 4.7 cm and 6.8 cm long.
    (a) Estimate the area of the rectangle.
    (b) Is the estimated area greater or smaller than the actual area?
        Explain your answer.

22 The area of a rectangle is 17.3 cm² and the length is 6.3 cm.
    (a) Estimate the width.
    (b) Is the estimated width greater or less than the actual width?
        Explain your answer.

23 Kirsty is going to paint the floor of the community hall.
The hall measures 9.7 m by 5.1 m.
A one-litre tin of floor paint covers 6.5 m².
Estimate how many one-litre tins of paint she needs to buy.

⬤ Challenge 14.1

The attendance at a Premiership football match was 60 000.
What could the exact attendance have been?

Challenge 14.2

Jane bought a car. She said, 'I bought this car for around fifteen thousand pounds and it has a 2.5 litre engine'.

What could the exact price of the car and size of the engine have been?

Checking answers

There are many methods for checking the answers to calculations. Rounding each number to 1 significant figure to find an estimate of the answer is one of them. Here are some more methods.

Using common sense

Does your answer sound sensible in the context of the problem? In practical problems, your own experience should tell you if an answer is wrong by a significant factor. For example, you know that the height of a man is not 18.94 metres. It is probably ten times too big. This is called **different order of magnitude**.

Using number facts

Discovery 14.1

You can use a calculator for this task.
Multiply 400 by the numbers in each of these sets.

Set A:	5	1.1	3.2	1.003	1.4	1.2
Set B:	0.5	0.999	0.6	0.9	0.7	0.95

What happens to the number 400?
Does it get bigger or smaller?

What conclusions can you come to? In particular, what happens when any number is multiplied by a number which is

(a) greater than 1?   (b) less than 1?   (c) 1 itself?

Discovery 14.2

You can use a calculator for this task.
Divide 400 by the numbers in each of the sets in Discovery 14.1.

What conclusions can you come to? In particular, what happens when any number is divided by a number which is

(a) greater than 1?   (b) less than 1?   (c) 1 itself?

Look at the answer to this calculation.

$752 \div 24 = 18\,048$

When 752 is divided by a number that is greater than 1, the result should be less than 752.

Instead, it is more. It looks as if the ⊗ key was pressed by mistake, instead of the ⊘ key. Using number facts can help to spot errors like this.

Here are some number facts you should have found in Discoveries 14.1 and 14.2.

Starting with any positive number,
- multiplying by a positive number greater than 1 gives a result that is larger than the number.
- multiplying by a positive number smaller than 1 gives a smaller result.
- dividing by a positive number greater than 1 gives a smaller result.
- dividing by a positive number smaller than 1 gives a larger result.

Here are some other number facts that you can use to check a calculation.

- odd × odd = odd   even × odd = even   even × even = even
- + × + = +   + × − = −   − × − = +
  and similarly for division
- Multiplying any number by 5 will give a result that ends in 0 or 5.
- The last digit in a multiplication comes from multiplying the last digits of the numbers.

● Challenge 14.3

Make up some multiplication and division calculations to test whether this statement is true.

'In multiplication and division calculations, rounding each number to 1 significant figure will always give an answer which is correct to 1 significant figure.'

Using inverse operations

Without a calculator, it is difficult to work out the square root of most numbers. However, if you know the square numbers, you can tell whether an answer is sensible.

Example 14.5

Show how you can tell that the answer to this calculation is wrong.

$\sqrt{35}$ = 9.52 to 2 decimal places

Solution

6^2 = 36 so $\sqrt{35}$ must be less than 6.

Using estimates

Example 14.6

Kate has £25 birthday money to spend.
She sees CDs at £7.99.
How many of them can she buy?

Solution

You do not need to know the exact answer to 25 ÷ 7.99 so do a quick estimate.

Use £8 instead of £7.99.

3 × 8 = 24 so 25 ÷ 8 = 3 'and a bit'.

So Kate can buy three CDs.

Exercise 14.2

1   Which of these answers might be correct and which are definitely wrong?
    Show how you decided.
    (a) 39.6 × 18.1 = 716.76
    (b) 175 ÷ 1.013 = 177.275
    (c) 8400 × 9 = 756 000
    (d) A lift takes 9 people, so a party of 110 people will need 12 trips.
    (e) Henry has £100 birthday money to spend and reckons that he can afford 5 DVDs
         costing £17.99 each.

Exercise continues ...

2 Look at these calculations. The answers are all wrong.
For each calculation, show how you can tell this quickly, without using a calculator to work it out.

(a) $-6.2 \div -2 = -3.1$

(b) $12.4 \times 0.7 = 86.8$

(c) $31.2 \times 40 = 124.8$

(d) $\sqrt{72} = 9.49$ to 2 d.p.

(e) $0.3^2 = 0.9$

(f) $16.2 \div 8.1 = 20$

(g) $125 \div 0.5 = 25$

(h) $6.4 \times -4 = 25.6$

(i) $24.7 + 6.2 = 30.8$

(j) $76 \div 0.5 = 38$

(k) $(-0.9)^2 = -0.81$

(l) $\sqrt{1000} = 10$

(m) $1.56 \times 2.5 = 0.39$

(n) $360 \div 15 = 2400$

3 Use estimates to calculate a rough total cost for each of these.

(a) Seven packs of crisps at 22p each

(b) Nine CDs at £13.25 each

(c) 39 theatre tickets at £7.20 each

(d) Five T-shirts at £5.99 and two pairs of socks at £1.99

(e) Three meals at £5.70 and two drinks at 99p

Challenge 14.4

Which of the numbers in these statements are likely to be exact, and which have been rounded.

(a) Yesterday, I spent £14.62.

(b) My height is 180 cm.

(c) Her new dress cost £40.

(d) The attendance at the Arsenal match was 32 000.

(e) The cost of building the new school is £27 million.

(f) The value of π is 3.142.

(g) The Olympic Games were held in Beijing in 2008.

(h) There were 87 people at the meeting.

Challenge 14.5

Look for some examples of rounded and unrounded numbers in newspapers, magazines, advertising material and on TV.

Which, if any, of the rounded numbers do you think have been rounded up, and which have been rounded down? Why?

Can you think of any cases when someone might want to round figures up rather than down?

What you have learned

- The first significant figure of a number is the first non-zero digit when starting from the left of the number and moving right
- To round numbers to 1 significant figure you look at the second significant figure; if it is less than 5 you leave the first significant figure as it is; if it is 5 or more you add 1 to the first significant figure
- When rounding a number greater than 10 to 1 significant figure, you use zeros to replace all but the first significant figure in the number so that the size of the number is correct
- When rounding a number less than 1 to 1 significant figure, you do not include any digits after the first corrected significant figure
- You can estimate the answer to a problem by rounding the numbers in the calculation to 1 significant figure
- You can use your own experience to check if the answer to a practical problem is about the right size or order of magnitude
- You can use number facts to check the answer to some calculations
- You can use inverse operations to check the answer to some calculations
- You can use an estimate of one or more of the numbers in a calculation to make an estimate of the answer

Mixed exercise 14

1 Round these numbers to 1 significant figure.
(a) 41   (b) 29 184   (c) 8162   (d) 756 324   (e) 9871
(f) 0.0015   (g) 4.05   (h) 0.145   (i) 0.00 993   (j) 0.749

2 There are 16 478 people at a football match.
How many people are there to 1 significant figure?

3 Work out an estimate of the answer to each of these calculations.
Show the approximations you use.

(a) $\sqrt{84}$ (b) $\dfrac{1083}{8.2}$ (c) 7.05^2

(d) $43.7 \times 18.9 \times 29.3$ (e) $\dfrac{2.46}{18.5}$ (f) $\dfrac{29}{41.6}$

(g) 917×38 (h) $\dfrac{283 \times 97}{724}$ (i) $\dfrac{614 \times 0.83}{3.7 \times 2.18}$

(j) $\dfrac{6.72}{0.051 \times 39.7}$ (k) $\sqrt{39 \times 80}$ (l) $65.4 \div 3.9$

(m) $\dfrac{194 \div 3.9}{27.3}$

4 Which of these calculations are correct and which are wrong?
Show how you decide.
(a) $17.35 \div 0.75 = 13.0125$ (b) $78\,000 \times 9.5 = 7\,410\,000$
(c) $6713 \times 45 = 302\,058$ (d) $\sqrt{50} - \sqrt{20} = 5.477$

5 For each of these problems, show the approximations you used to get your estimate.

(a) A piece of land measures 127 m by 67 m and it is valued at £1.95 per square metre.
Estimate the value of the land.

(b) Natasha knows that when her car's fuel tank is half empty, it will take 6 gallons to fill it.
The cost of fuel is £1.10 per litre.
A gallon is approximately 4.5 litres.
She only has £30 in cash.
Is this enough?

Chapter

15

Enlargement

▶ This chapter is about

- Enlarging shapes using a scale factor
- Enlarging shapes using a scale factor and a centre of enlargement
- Recognising and describing enlargements
- Fractional scale factors

▶ You should already know

- How to measure lengths accurately
- How to measure angles accurately
- How to plot and read coordinates
- How to recognise and describe reflections and rotations

Scale factor

Enlargements are another type of transformation.

◉ Discovery 15.1

Measure the lengths and angles of the two shapes.
Copy and complete the tables.

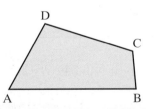

Side	Length	Side	Length
AB		A′B′	
BC		B′C′	
CD		C′D′	
DA		D′A′	

Angle	Size	Angle	Size
Angle A		Angle A′	
Angle B		Angle B′	
Angle C		Angle C′	
Angle D		Angle D′	

(a) What can you say about the lengths of the sides of the two shapes?

(b) What can you say about the angles of the two shapes?

- In an **enlargement** each length of a shape is the same number of times bigger than the lengths of the original shape.
- The original shape is called the **object**. The new shape is called the **image**.
- The lengths of the object and the lengths of the image are **proportional**.
- The number of times the lengths of the image are bigger than the object is called the **scale factor** of the enlargement.
- The angles of the object and the image are the same size. Only the lengths change.
- The object and image of an enlargement are **similar**.

The scale factor of an enlargement tells you how many times bigger the image is than the object. If the shapes are drawn on squared paper you can count units. Otherwise you will have to measure the lengths using a ruler.

In an enlargement the object and the image are similar. All the sides are enlarged by the same scale factor. The angles in the object and the image stay the same.

Example 15.1

Draw an enlargement of this shape using a scale factor of 2.

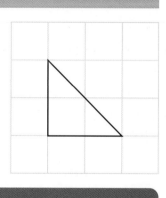

Solution

The vertical and horizontal sides are 2 squares long in the object.

You need to draw vertical and horizontal sides of length
2 × 2 = 4 in the image.

Join the two sides you have drawn to make a triangle.

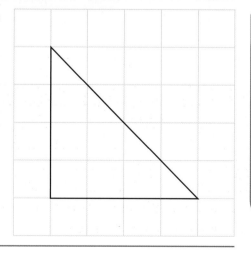

TIP

You can measure the length of the sloping side in the image and check that it is twice as long as the sloping side in the object. If it is not, you know your diagram must be wrong.

Example 15.2

Find the scale factor of this enlargement.

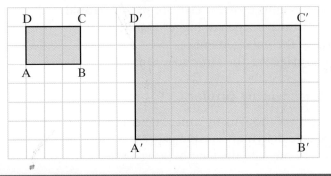

Solution

Side	Length	Side	Length
AB	3	A'B'	9
BC	2	B'C'	6

The sides in the image are three times as long as the sides in the object. The scale factor is 3.

Example 15.3

For each of these pairs of shapes, is the larger shape an enlargement of the smaller shape? Give a reason for your answer.

(a)

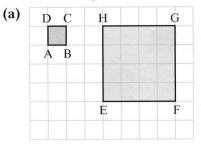

(b)

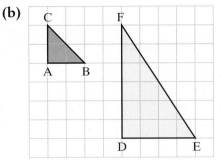

(c)

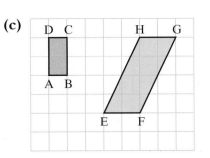

Solution

(a) The sides of the smaller square are 1 square long.
The sides of the larger square are 4 squares long.
The angles are the same.
This is an enlargement because the shapes are similar.

(b)

Side	Length	Side	Length
AB	2	DE	4
AC	2	DF	6

This is not an enlargement because the scale factors used to enlarge the sides are different.

> **TIP**
>
> You only need to measure two sides to start with. Make sure they are corresponding sides though. In this case, the sides making the right angle have been compared.

(c)

Angle	Size	Angle	Size
Angle A	90°	Angle E	Not 90°
Angle B	90°	Angle F	Not 90°

This is not an enlargement because the angles in the two shapes are different.

> **TIP**
>
> You did not need to measure the sides of these shapes because you can see that the angles in the two shapes are different. This means the shapes are not similar so it cannot be an enlargement. This is true even though the sides of the parallelogram are twice as long as the sides of the rectangle.

Check up 15.1

Write your first name or initials on squared paper.

Enlarge each letter by a scale factor of 3.

Exercise 15.1

1 For each of these shapes:
  • copy the shape on to squared paper.
  • draw an enlargement of the shape using the scale factor given.

(a) Scale factor 2   (b) Scale factor 3   (c) Scale factor 3

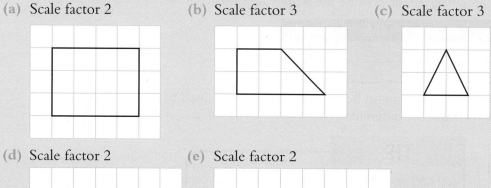

(d) Scale factor 2   (e) Scale factor 2

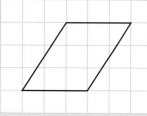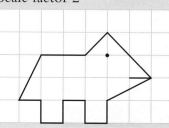

2 Work out the scale factor of enlargement of each of these pairs of shapes.

(a)   (b)

(c)   (d)

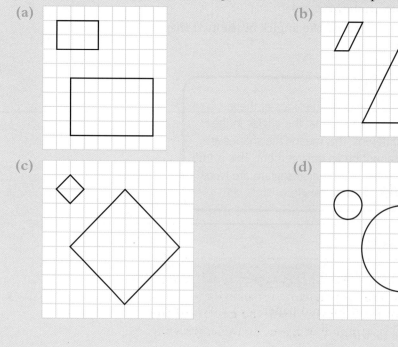

Exercise continues ...

3 For each of these pairs of shapes, is the larger shape an enlargement of the smaller shape? Give a reason for your answer.

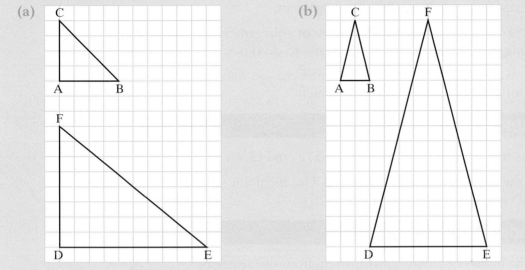

(a)

(b)

○ **Discovery 15.2**

Work in groups of four or five.

(a) Each person draws a circle accurately on a piece of paper.
Compare your diagram with the others drawn by your group.
Are the circles congruent?
Are the circles similar?
What is the relationship between all circles?

(b) Repeat the activity with these shapes.
    (i)   Rectangles
    (ii) Squares
    (iii) Equilateral triangles
    (iv) Isosceles triangles
Make a display of your findings.

Centre of enlargement

The enlargements you have seen so far in this chapter could have been drawn in any position on the paper. More usually you are asked to draw an enlargement from a given point, the **centre of the enlargement**. In this case the enlargement has to be the correct size *and* in the correct position.

For an enlargement with scale factor 2, each corner of the image must be twice as far from the centre of enlargement as the corresponding corner of the original shape. Example 15.4 shows you how to draw an enlargement from a given centre of enlargement.

You might also be asked to find the centre of enlargement of a shape and its image. Example 15.5 shows you how to do this.

When you are asked to describe an enlargement you must give the scale factor and the centre of the enlargement.

Example 15.4

Draw the triangle with vertices at (3, 2), (5, 2) and (2, 4).

Enlarge the triangle by a scale factor of 3. Use the point (1, 1) as the centre of enlargement.

Solution

Count the number of units across then up from the centre of enlargement to one of the corners of the triangle.

Then multiply the distances by 3 to find the position of that corner in the image.

Mark its position.

Repeat for the other two corners.

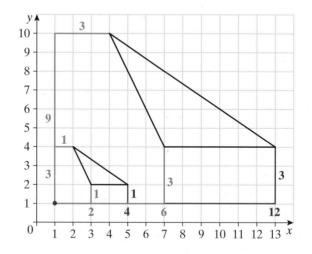

TIP

Measure each length of the image to make sure that it is three times (or whatever the scale factor is) bigger than the length in the original shape.

Example 15.5

Find the scale factor and the centre of enlargement of these shapes.

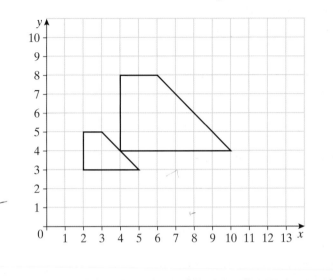

Solution

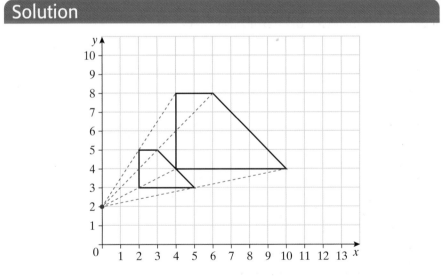

You find the scale factor as you have done before, by measuring corresponding sides in the object and image.

The scale factor is 2.

To find the centre of enlargement, you join the corresponding corners of the two shapes and extend the lines until they cross.
The point where they cross is the centre of the enlargement.

The centre of enlargement is the point $(0, 2)$.

Exercise 15.2

1 Copy each of the shapes on to squared paper.
 Enlarge each of them by the scale factor given.
 Use the dot as the centre of the enlargement.

(a) Scale factor 3

(b) Scale factor 2

(c) Scale factor 4

(d) Scale factor 3

(e) Scale factor 2

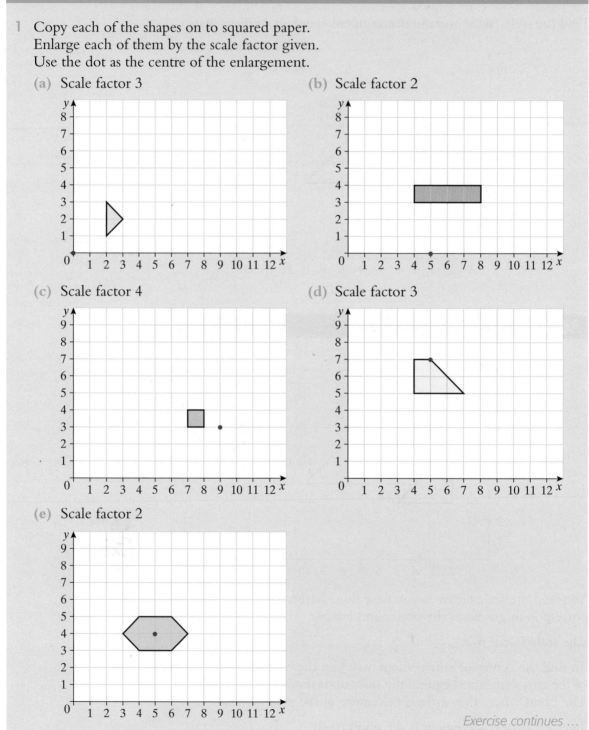

Exercise continues ...

2 Copy each of these diagrams on to squared paper. For each of these diagrams find
(i)   the scale factor of the enlargement.
(ii) the coordinates of the centre of the enlargement.

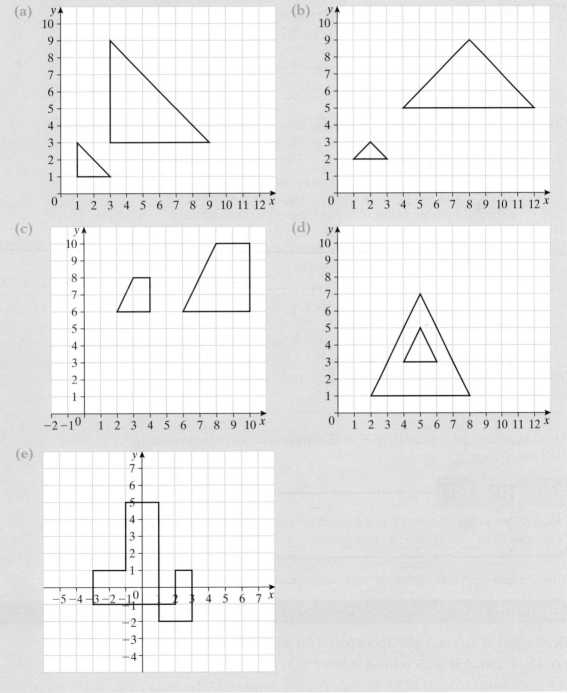

Fractional scale factors

◉ Discovery 15.3

(a) Draw a pair of axes and label them 0 to 6 for x and y.
   (i) Draw a triangle with vertices at $(1, 2)$, $(3, 2)$ and $(3, 3)$.
      Label it A.
   (ii) Enlarge the triangle by scale factor 2, with the origin as the centre of enlargement.
      Label it B.

(b) (i) Think about what happens to the lengths of the sides of an object when it is enlarged by scale factor 2.
      What do you think will happen to the lengths of the sides of an object if it is enlarged by scale factor $\frac{1}{2}$?
   (ii) Think about the position of the image when an object is enlarged by scale factor 2. What happens to the distance between the centre of enlargement and the object?
      What do you think will be the position of the image if an object is enlarged by scale factor $\frac{1}{2}$?

(c) Draw a pair of axes and label them 0 to 6 for x and y.
   (i) Draw a triangle with vertices at $(2, 4)$, $(6, 4)$ and $(6, 6)$.
      Label it A.
   (ii) Enlarge the triangle by scale factor $\frac{1}{2}$, with the origin as the centre of enlargement. Label it B.

(d) Compare your diagram with the diagram you drew for part (a). What do you notice?

An enlargement with scale factor $\frac{1}{2}$ is the **inverse** of an enlargement with scale factor 2.

> ## TIP
> Although the image is smaller than the object, an enlargement with scale factor $\frac{1}{2}$ is still called an enlargement.

You can also draw enlargements with other fractional scale factors.

Example 15.6

Draw a pair of axes and label them 0 to 8 for both x and y.

(a) Draw a triangle with vertices at $P(5, 1)$, $Q(5, 7)$ and $R(8, 7)$.

(b) Enlarge the triangle PQR by scale factor $\frac{1}{3}$, centre $C(2, 1)$.

Solution

The sides of the enlargement are $\frac{1}{3}$ the lengths of the original.

The distance from the centre of enlargement, C, to P is 3 across. So the distance from C to P′ is $3 \times \frac{1}{3} = 1$ across.

The distance from C to Q is 3 across and 6 up. So the distance from C to Q′ is $3 \times \frac{1}{3} = 1$ across and $6 \times \frac{1}{3} = 2$ up.

The distance from C to R is 6 across and 6 up. So the distance from C to R′ is $6 \times \frac{1}{3} = 2$ across and $6 \times \frac{1}{3} = 2$ up.

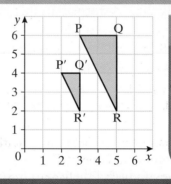

Example 15.7

Describe fully the single transformation that maps triangle PQR on to triangle P′Q′R′.

TIP

You must state that the transformation is an enlargement and give the scale factor and centre of enlargement.

Solution

It is obvious that the shape has been enlarged. The length of each side of triangle P′Q′R′ is half the length of the corresponding side of triangle PQR, so the scale factor is $\frac{1}{2}$.

To find the centre of enlargement, join the corresponding corners of the two triangles and extend the lines until they cross.

The point where they cross is the centre of enlargement, C. Here, C is at $(1, 2)$.

The transformation is an enlargement with scale factor $\frac{1}{2}$, centre $(1, 2)$.

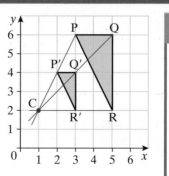

TIP

To check the centre of enlargement, find the distance from the centre of enlargement to a corresponding pair of points. For example, the distance from C to P is 2 across and 4 up, so the distance from C to P′ should be 1 across and 2 up.

Exercise 15.3

1 Draw a pair of axes and label them 0 to 6 for both x and y.

  (a) Draw a triangle with vertices at $(4, 2)$, $(6, 2)$ and $(6, 6)$.
Label it A.

  (b) Enlarge triangle A by scale factor $\frac{1}{2}$, with the origin as the centre of enlargement.
Label it B.

  (c) Describe fully the transformation that maps triangle B on to triangle A.

2 Draw a pair of axes and label them 0 to 8 for both x and y.

  (a) Draw a triangle with vertices at $(4, 5)$, $(4, 8)$ and $(7, 8)$.
Label it A.

  (b) Enlarge triangle A by scale factor $\frac{1}{3}$, with centre of enlargement $(1, 2)$.
Label it B.

  (c) Describe fully the transformation that maps triangle B on to triangle A.

3 Draw a pair of axes and label them 0 to 8 for both x and y.

  (a) Draw a triangle with vertices at $(0, 2)$, $(1, 2)$ and $(2, 1)$.
Label it A.

  (b) Enlarge triangle A by scale factor 4, with the origin as the centre of enlargement.
Label it B.

  (c) Describe fully the transformation that maps triangle B on to triangle A.

4 Draw a pair of axes and label them 0 to 8 for both x and y.

  (a) Draw a triangle with vertices at $(4, 3)$, $(4, 5)$ and $(6, 2)$.
Label it A.

  (b) Enlarge triangle A by scale factor $1\frac{1}{2}$, with centre of enlargement $(2, 1)$.
Label it B.

  (c) Describe fully the transformation that maps triangle B on to triangle A.

5 Describe fully the transformation that maps

  (a) triangle A on to triangle B.

  (b) triangle B on to triangle A.

  (c) triangle A on to triangle C.

  (d) triangle C on to triangle A.

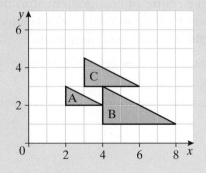

Exercise continues …

Hint: In questions 6 to 8, not all the transformations are enlargements.

6 Describe fully the transformation that maps

(a) triangle A on to triangle B.

(b) triangle A on to triangle C.

(c) triangle A on to triangle D.

(d) triangle E on to triangle A.

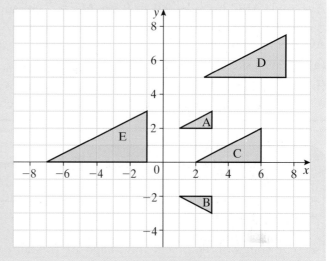

7 Describe fully the transformation that maps

(a) flag A on to flag B.

(b) flag A on to flag C.

(c) flag A on to flag D.

(d) flag A on to flag E.

(e) flag F on to flag E.

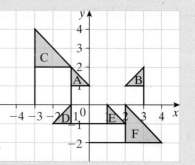

8 Describe fully the transformation that maps

(a) triangle A on to triangle B.

(b) triangle A on to triangle C.

(c) triangle C on to triangle D.

(d) triangle E on to triangle F.

(e) triangle G on to triangle F.

(f) triangle F on to triangle G.

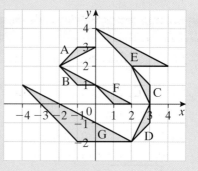

Challenge 15.1

A triangle ABC has sides AB = 9 cm, AC = 7 cm and BC = 6 cm.
A line XY is drawn parallel to BC through a point X on AB and Y on
AC. AX = 5 cm.

(a) Draw a sketch of the triangles.

(b) (i) Describe fully the transformation that maps ABC on to AXY.
  (ii) Work out the length of XY correct to 2 decimal places.

What you have learned

- In an enlargement, the lengths of the sides in the object and image are proportional
- The number of times the lengths of the image are bigger than the lengths of the object is
  the scale factor
- The angles of the object and image are the same
- Two shapes are similar if their angles are the same and the lengths of their sides are
  proportional
- The object and image of an enlargement are similar
- When a centre of enlargement is used, the distance from each point on the image to
  the centre is the distance from the corresponding point on the object to the centre
  multiplied by the scale factor
- To describe an enlargement you must state that the transformation is an enlargement and
  give the scale factor and centre of enlargement
- If the scale factor of an enlargement is less than 1, the image will be smaller than the
  object

Mixed exercise 15

1 For each of these shapes:
  - copy the shape on to squared paper.
  - draw an enlargement of the shape using the scale factor given.
  (a) Scale factor 3   (b) Scale factor 2

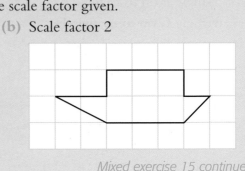

Mixed exercise 15 continues …

2 Copy each of the shapes on to squared paper.
Enlarge each of them by the scale factor given.
Use the dot as the centre of the enlargement.

(a) Scale factor 2

(b) Scale factor 4

3 Copy each of these diagrams on to squared paper. For each of these diagrams find
(i) the scale factor of the enlargement.
(ii) the coordinates of the centre of the enlargement.

(a)

(b)

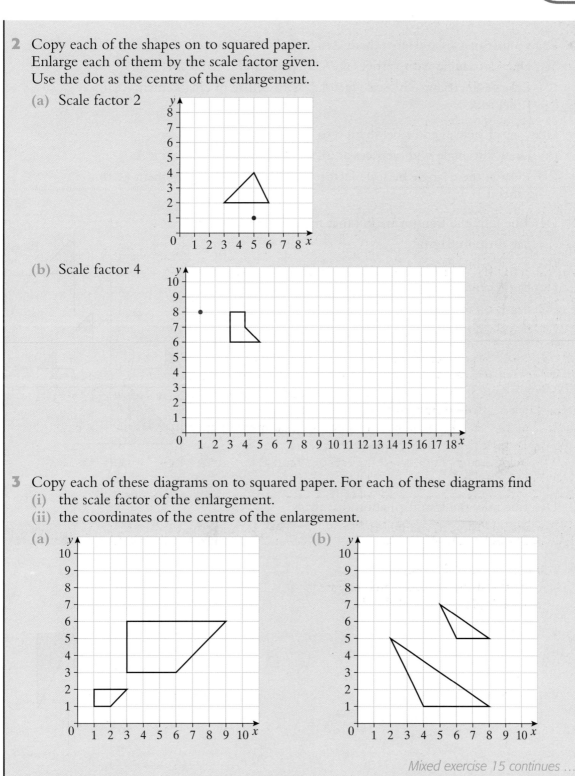

Mixed exercise 15 continues …

4 Draw a pair of axes and label them 0 to 9 for both x and y.

(a) Draw a triangle with vertices at $(6, 3)$, $(6, 6)$ and $(9, 3)$. Label it A.

(b) Enlarge the triangle by scale factor $\frac{1}{3}$, with centre of enlargement $(0, 0)$. Label it B.

5 Draw a pair of axes and label them 0 to 8 for both x and y.

(a) Draw a triangle with vertices at $(6, 4)$, $(6, 6)$ and $(8, 6)$. Label it A.

(b) Enlarge the triangle by scale factor $\frac{1}{2}$, with centre of enlargement $(2, 0)$. Label it B.

6 Describe fully the transformation that maps

(a) flag A on to flag B.

(b) flag A on to flag C.

(c) flag A on to flag D.

(d) flag E on to flag C.

(e) flag B on to flag C.

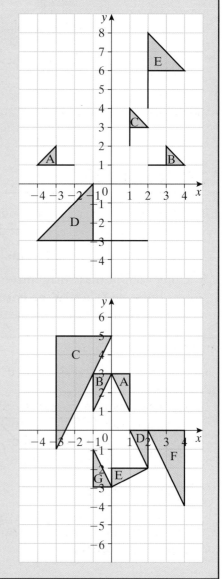

7 Describe fully the transformation that maps

(a) triangle A on to triangle B.

(b) triangle B on to triangle C.

(c) triangle D on to triangle E.

(d) triangle F on to triangle D.

(e) triangle A on to triangle G.

Chapter 16

Scatter diagrams and time series

Scatter diagrams

A scatter diagram is used to find out whether there is **correlation**, or a relationship, between two sets of data.

Data are presented as pairs of values each of which is plotted as a coordinate point on a graph.

Here are some examples of what a scatter diagram could look like and how you might interpret them.

Strong positive correlation

Here, one quantity increases as the other increases.
This is called **positive correlation**.
The trend is bottom left to top right.
When the points are closely in line, you say that the correlation is strong.

Weak positive correlation

Here the points again display positive correlation.
The points are more scattered so you say that the correlation is **weak**.

Strong negative correlation

Here, one quantity decreases as the other increases.
This is called **negative correlation**.
The trend is from top left to bottom right.
Again, the points are closely in line, so you say
that the correlation is **strong**.

Weak negative correlation

Here the points again display negative
correlation.
The points are more scattered so the
correlation is **weak**.

No correlation

When the points are totally scattered and
there is no clear pattern you say that there
is **no correlation** between the two quantities.

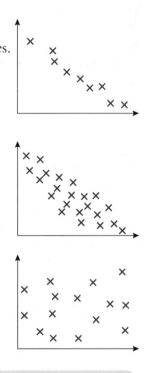

Discovery 16.1

Do tall people have large feet?

Survey the people in your class or in your year and record their height
and foot size. (Foot length might be better than shoe size but you
could use either).

Plot height on the horizontal axis and foot size on the vertical axis.
What do you notice?

Try other year groups.

If a scatter diagram shows correlation, you can draw a **line of best fit** on it.

Try putting your ruler in various positions on the scatter diagram until
you have a slope which matches the general slope of the points. There
should be roughly the same number of points on each side of the line.

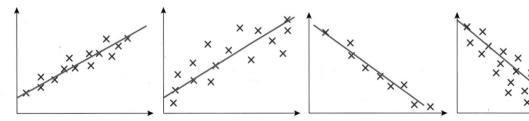

You cannot draw a line of best fit on a scatter diagram with no correlation.

You can use the line of best fit to predict a value when only one of the pair of quantities is known.

Example 16.1

The table shows the weights and heights of 12 people.

Height (cm)	150	152	155	158	158	160	163	165	170	175	178	180
Weight (kg)	56	62	63	64	57	62	65	66	65	70	66	67

(a) Draw a scatter diagram to show these data.

(b) Comment on the strength and type of correlation between these heights and weights.

(c) Draw a line of best fit on your scatter diagram.

(d) Tom is 162 cm tall. Use your line of best fit to estimate his weight.

Solution

(a), (c)

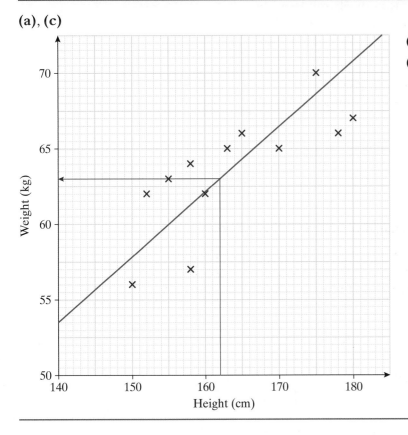

(b) Weak positive correlation.

(d) Draw a line from 162 cm on the Height axis, to meet your line of best fit.

Now draw a horizontal line and read off the value where it meets the Weight axis.

Tom's probable weight is about 63 kg.

Exercise 16.1

1 The scatter diagram shows the numbers of sun beds hired out and the hours of sunshine at Brightsea.

   Comment on the results shown by the scatter diagram.

2 A firm noted the numbers of days of 'sick-leave' taken by its employees in a year, and their ages. The results are shown in the graph.

   Comment on the results shown by the scatter diagram.

3 A teacher thinks that there is a correlation between how far back in class a student sits and how well they do at maths. To test this, she plotted their last maths grade against the row they sit in.

   Here is the graph she drew.

   Was the teacher right?
   Give your reasons.

4 The scatter diagram shows the positions of football teams in the league and their mean crowd numbers, in thousands.

   Comment on the scatter diagram.

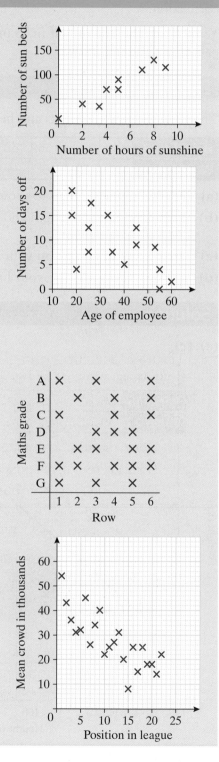

Exercise continues …

5   The scatter diagram shows the ages of people
and the numbers of lessons they took before
they passed their driving tests.

Comment on the scatter diagram.

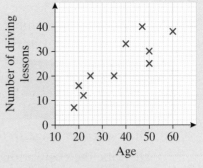

6   The table shows the number of bad peaches per box after different delivery times.

Delivery time (hours)	10	4	14	18	6
Number of bad peaches	2	0	4	5	2

(a) Draw a scatter diagram to show this information.

(b) Describe the correlation shown in the scatter diagram.

(c) Draw a line of best fit on your scatter diagram.

(d) Use your line of best fit to estimate the number of bad peaches expected after a
12 hour delivery time.

7   The table shows the marks of 15 students taking Paper 1 and Paper 2 of a maths exam.
Both papers were marked out of 40.

Paper 1	36	34	23	24	30	40	25	35	20	15	35	34	23	35	27
Paper 2	39	36	27	20	33	35	27	32	28	20	37	35	25	33	30

(a) Draw a scatter diagram to show this information.

(b) Describe the correlation shown in the scatter diagram.

(c) Draw a line of best fit on your scatter diagram.

(d) Joe scored 32 on Paper 1 but was absent for Paper 2.
Use your line of best fit to estimate his score on Paper 2.

8   The table shows the engine size and petrol consumption of nine cars.

Engine size (litres)	1.9	1.1	4.0	3.2	5.0	1.4	3.9	1.1	2.4
Petrol consumption (mpg)	34	42	23	28	18	42	27	48	34

(a) Draw a scatter diagram to show this information.

(b) Describe the correlation shown in the scatter diagram.

(c) Draw a line of best fit on your scatter diagram.

(d) Another car has an engine size of 2.8 litres.
Use your line of best fit to estimate the petrol consumption of this car.

Exercise continues ...

9 Tracy thinks that the larger your head, the cleverer you are.
   The table shows the number of marks scored in a test by ten students, and the
   circumference of their heads.

Circumference of head (mm)	600	500	480	570	450	550	600	460	540	430
Mark	43	33	45	31	25	42	23	36	24	39

   (a) Draw a scatter diagram to show this information.

   (b) Describe the correlation shown in the scatter diagram.

   (c) Is Tracy correct?

   (d) Can you think of any reasons why the comparison may not be valid?

10 In Kim's game, 20 objects are placed on a table and you are given a certain time to
    look at them. They are then removed or covered up and you have to recall as many as
    possible. The table shows the lengths of time given to nine people and the numbers of
    items they remembered.

Time in seconds	Number of items
20	9
25	8
30	12
35	10
40	12
45	15
50	13
55	16
60	18

   (a) Draw a scatter diagram to show this information, with time in seconds on the
        horizontal axis.

   (b) Comment on the diagram.

   (c) Draw a line of best fit.

   (d) Use your line of best fit to estimate the number of items remembered if 32
        seconds are allowed.

   (e) Why should the diagram not be used to estimate the number of items
        remembered in 3 seconds?

Exercise continues …

11 In Jane's class, a number of students have part-time jobs. Jane thinks that the more time they spend on their jobs, the worse they will do at school. She asked ten of them how many hours a week they spent on their jobs, and found their mean marks in the last examinations. Her results are shown in the table.

Student	Time on part-time job (hours)	Mean mark in examination
A	9	50
B	19	92
C	13	52
D	3	70
E	15	26
F	20	10
G	5	80
H	17	36
I	6	74
J	22	24

(a) Plot a scatter diagram to show Jane's results, with time in hours on the horizontal axis.

(b) Do the results confirm Jane's views? Are there any exceptions?

(c) Draw a line of best fit for the relevant points.

(d) Estimate the mean score of a student who spent 12 hours on their part-time job.

Challenge 16.1

(a) Dan thinks that the more time he spends on his school work, the less money he will spend. Sketch a scatter diagram that shows this.

(b) Fiona thinks that the more she practises, the more goals she will score at hockey. Sketch a scatter diagram to show this.

Challenge 16.2

Investigate one of these.
- Petrol consumption and the size of a car's engine
- A person's height and their head circumference
- House price and the distance from the town centre
- Age of car and second-hand price

Time series

The table shows the takings at Annabella's café over a four-week period.

	M	Tu	W	Th	F	Sa
Week 1	£270	£302	£268	£342	£406	£432
Week 2	£286	£298	£237	£352	£425	£486
Week 3	£254	£283	£251	£364	£397	£459
Week 4	£273	£304	£258	£372	£410	£501

The graph below represents these data. The points are joined with broken lines, since there is no meaning to 'in-between' values.

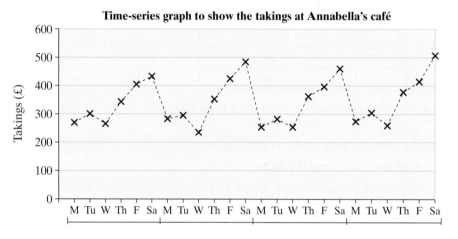

Time-series graph to show the takings at Annabella's café

The graph shows the variation in the takings during the week and that they have a cyclic pattern.

For instance, the takings are highest on Saturdays and lowest on Wednesdays.

Example 16.2

This graph shows the number of units of electricity used each quarter by the Peters' family over 5 years.

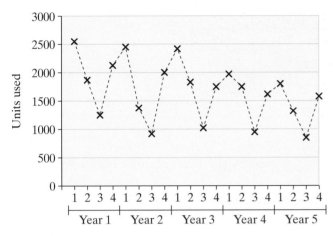

(a) Describe the seasonal variation.

(b) Describe the long-term trend.

Solution

(a) Each year, the greatest use of electricity by the Peters' family is in the first quarter (winter) with the least use being in the third quarter (summer).

(b) The general trend is for slightly reduced use of electricity.

Exercise 16.2

1 The table shows the number of people visiting a leisure centre each day for a 4-week period.

	Su	M	Tu	W	Th	F	Sa
Week 1	1037	542	731	1084	832	905	1617
Week 2	1405	741	750	905	794	927	1392
Week 3	1605	763	801	928	937	1017	1854
Week 4	2047	694	728	861	904	935	1532

(a) Plot a time-series graph of these figures.

(b) Describe the trend over this period.

Exercise continues …

2 The table shows the number of people visiting a doctors' surgery each day for a 4-week period.

	M	Tu	W	Th	F
Week 1	105	71	63	84	92
Week 2	126	86	71	94	115
Week 3	142	91	84	88	104
Week 4	115	82	73	84	91

(a) Plot a time-series graph of these figures.

(b) Describe the trend over this period.

3 These figures show the number of units of gas used by the Peters' family over a period of 3 years.

	1st quarter	2nd quarter	3rd quarter	4th quarter
Year 1	201	157	54	135
Year 2	191	161	63	173
Year 3	207	141	71	158

Describe the seasonal variation.

4 The graph shows a shop's quarterly sales of rainwear.

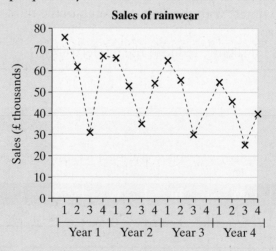

Sales of rainwear

(a) Describe the seasonal variation.

(b) Describe the long-term trend.

Exercise continues ...

5 The graph shows the variation in currency conversion rates during part of 2009 and 2010.

Exchange rate: British pounds to 1 US dollars

(a) On approximately what date did the US dollar buy fewest British pounds?

(b) Approximately, what was the most British pounds that 1 dollar would buy during February 2010?

(c) On 14 April 2010, the exchange rate was 0.531 745 British pounds to 1 US dollar. How many US dollars did 1 British pound buy then? Give your answer correct to the nearest cent.

What you have learned

- Scatter diagrams show the correlation between two variables

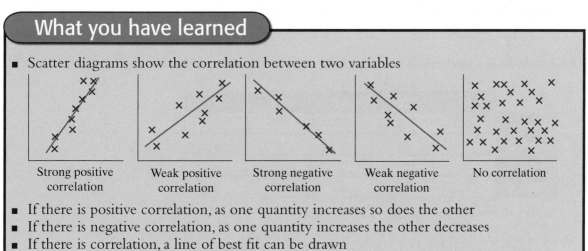

| Strong positive correlation | Weak positive correlation | Strong negative correlation | Weak negative correlation | No correlation |

- If there is positive correlation, as one quantity increases so does the other
- If there is negative correlation, as one quantity increases the other decreases
- If there is correlation, a line of best fit can be drawn
- There should be approximately the same number of points on each side of the line of best fit
- The line of best fit can be used to estimate the value of one quantity if the value of the other is known
- The line of best fit should only be used to estimate values within the range of the data given

- A time-series graph has time on the horizontal axis
- A time-series graph represents values over a period of time
- A time-series graph may have a cyclical pattern
- A time-series graph may show a long-term trend

Mixed exercise 16

1 A pet shop owner carried out a survey to investigate the average weight of a breed of rabbit at various ages. The table shows his results.

Age of rabbit (months)	1	2	3	4	5	6	7	8
Average weight (g)	90	230	490	610	1050	1090	1280	1560

(a) Draw a scatter diagram to show this information.

(b) Describe the correlation shown in the scatter diagram.

(c) Draw a line of best fit on your scatter diagram.

(d) Use your line of best fit to estimate
   (i) the weight of a rabbit of this breed which is $4\frac{1}{2}$ months old.
   (ii) the weight of a rabbit of this breed which is 9 months old.

(e) If the line of best fit was extended you could estimate the weight of a rabbit of this breed which is 20 months old.
   Would this be sensible? Give a reason for your answer.

2 The table shows the number of people visiting a dentists' surgery each day for a 4-week period.

	M	Tu	W	Th	F
Week 1	35	42	63	24	51
Week 2	33	24	60	26	48
Week 3	38	44	66	29	46
Week 4	35	47	65	27	49

(a) Plot a time-series graph of these figures.

(b) One of the dentists works part-time. On which day does he not work?

(c) On one day in this period, a dentist was ill and her patients' appointments were postponed. When was this?

Straight lines and inequalities

▸ **You should already know**

- How to substitute values into algebraic expressions
- How to recognise and plot the equations of lines parallel to the axes
- How to draw a pair of axes and label them with a suitable scale
- How to draw conversion graphs
- How to write an equation to represent a real-life situation
- That parallel lines are always the same distance apart and never meet
- How to solve equations

Drawing straight-line graphs

You can draw a graph of a straight line from its equation. You need to be able to substitute values into algebraic expressions. You learned how to do this in Unit A Chapter 5.

🔍 Check up 17.1

Substitute the value $x = -2$ into each of these expressions.

(a) $6x + 3$ (b) $5 - 2x$ (c) $x - 4$

You met the equations of lines parallel to the axes in Unit A Chapter 7.

🔍 Check up 17.2

Draw and label x- and y-axes from -6 to 6.

Draw and label the lines $x = -3$, $x = 3$, $y = 4$ and $y = -5$.

TIP

To check you have drawn $y = 4$ and not $x = 4$, find the coordinates of some of the points on the line: the y-coordinate should be 4.

The most common straight-line graphs have equations of the form $y = 3x + 2$, $y = 2x - 3$, etc.

This can be written in a general form as

$y = mx + c$ where m and c are numbers.

To draw a straight-line graph, work out three pairs of coordinates by substituting values of x into the formula to find y.

You can draw a straight line with only two points, but you should always work out a third point as a check.

Example 17.1

Draw the graph of $y = -2x + 1$ for values of x from -4 to 2.

Solution

Find the values of y when $x = -4, 0$ and 2.

$y = -2x + 1$

When $x = -4$
$y = -2 \times -4 + 1$
$y = 9$

When $x = 0$
$y = -2 \times 0 + 1$
$y = 1$

When $x = 2$
$y = -2 \times 2 + 1$
$y = -3$

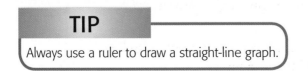

The y values needed are -3 to 9.
Draw the axes and plot the points $(-4, 9)$, $(0, 1)$ and $(2, -3)$.
Join them with a straight line. Label it $y = -2x + 1$.

TIP

Always use a ruler to draw a straight-line graph.

Exercise 17.1

1 Draw the graph of $y = 4x$ for values of x from -3 to 3.

2 Draw the graph of $y = x + 3$ for values of x from -3 to 3.

3 Draw the graph of $y = 3x - 4$ for values of x from -2 to 4.

4 Draw the graph of $y = 4x - 2$ for values of x from -2 to 3.

5 Draw the graph of $y = -3x - 4$ for values of x from -4 to 2.

6 Draw the graph of $y = -2x + 5$, for values of x from -2 to 4.

7 Draw the graph of $y = -x + 1$, for values of x from -3 to 3.

8 (a) Draw the graph of the equation $C = 5n + 15$ for values of n from 0 to 20.
   (b) From the graph find the value of n when $C = 80$.
        Give your answer correct to the nearest whole number.

9 The cost £C of printing a leaflet is given by the formula $C = 5 + 0.2n$, where n is the number of copies.
   (a) Draw the graph of $C = 5 + 0.2n$ for values of n from 0 to 200.
   (b) From the graph find the value of n when $C = 30$.
        Give your answer correct to the nearest whole number.

10 A plasterer works out his daily charge (£C) using the equation $C = 12n + 40$, where n is the number of hours he works on a job.
   (a) Draw the graph of C against n for values of n from 0 to 10.
   (b) From your graph find how many hours he works when the charge is £130.

Harder straight-line graphs

Sometimes you will be asked to draw graphs with equations of a different form.

For equations such as $2y = 3x + 1$, work out three points as before, remembering to divide by 2 to find the y value.

Example 17.2

Draw the graph of $2y = 3x + 1$ for values of x from -3 to 3.

Solution

Find the values of y when $x = -3, 0$ and 3.

The equation of the line is $2y = 3x + 1$.

When $x = -3$
$2y = 3 \times -3 + 1$
$2y = -8$
$y = -4$

When $x = 0$
$2y = 3 \times 0 + 1$
$2y = 1$
$y = \frac{1}{2}$

When $x = 3$
$2y = 3 \times 3 + 1$
$2y = 10$
$y = 5$

The values of y needed are -4 to 5.

Draw the axes and plot the points $(-3, -4), (0, \frac{1}{2})$ and $(3, 5)$.
Join them with a straight line. Label it $2y = 3x + 1$.

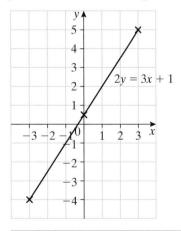

For equations such as $4x + 3y = 12$, work out y when $x = 0$, and x when $y = 0$. These are easy to work out: you can find a third point as a check after you have drawn the line.

Example 17.3

Draw the graph of $4x + 3y = 12$.

Solution

Find the value of y when $x = 0$ and the value of x when $y = 0$.

$4x + 3y = 12$

When $x = 0$
$3y = 12$ $4 \times 0 = 0$ so the x term 'disappears'.
$y = 4$

When $y = 0$
$4x = 12$ $3 \times 0 = 0$ so the y term 'disappears'.
$x = 3$

The values of x needed are 0 to 3. The values of y needed are 0 to 4.

Draw axes and plot the points $(0, 4)$ and $(3, 0)$.
Join them with a straight line. Label it $4x + 3y = 12$.

TIP

Take care when plotting the points. Do not put $(0, 4)$ at $(4, 0)$ by mistake.

Choose a point on the line you have drawn and check it by substituting the x and y values into the equation.

For example, the line passes through $(1\frac{1}{2}, 2)$.

$4x + 3y = 12$
$4 \times 1\frac{1}{2} + 3 \times 2 = 6 + 6 = 12$ ✓

Exercise 17.2

1 Draw the graph of $3x + 5y = 15$

2 Draw the graph of $2x + 5y = 10$

3 Draw the graph of $7x + 2y = 14$

4 Draw the graph of $3x + 2y = 15$

5 Draw the graph of $2y = 5x + 3$, for $x = -3$ to 3

6 Draw the graph of $3y = 2x + 6$, for $x = -3$ to 3

7 Draw the graph of $2y = 3x - 5$, for $x = -2$ to 4

8 Draw the graph of $2y = 5x - 8$, for $x = -2$ to 4

The graphical solution of simultaneous equations

○ Discovery 17.1

To hire a mini-bus, Delaney's Cabs charge £20 plus £2 a mile and Tracey's Cars charge £50 plus £1.50 a mile.

C is the charge and n is the number of miles.

For Delaney's Cabs the equation is $C = 2n + 20$.

For Tracey's Cars the equation is $C = 1.5n + 50$.

(a) On the same graph, draw a line for each equation for values of n up to 100.

(b) When is it cheaper to use Delaney's Cabs and when is it cheaper to use Tracey's Cars?

(c) When do they charge the same amount?

To answer part **(c)** of Discovery 17.1 you were in fact solving two simultaneous equations, $C = 2n + 20$ and $C = 1.5n + 50$. The solution is where the two lines meet and tells you when the two cab companies charge the same amount.

Exercise 17.3

1   (a) On the same grid, draw the graphs of $y = 8$ and $y = 4x + 2$, for $x = -3$ to 3.
    (b) Write down the coordinates of the point where the two lines cross.

2   (a) On the same grid, draw the graphs of $y = x + 3$ and $y = 4x - 3$, for $x = -2$ to 3.
    (b) Write down the coordinates of the point where the two lines cross.
    (c) Write down the simultaneous equations that you solved in part (b).

3   (a) On the same grid, draw the graphs of $y = 2x + 3$ and $2x + y = 7$, for $x = 0$ to 5.
    (b) Write down the coordinates of the point where the two lines cross.
    (c) Write down the simultaneous equations that you solved in part (b).

4   (a) On the same grid, draw the graphs of $y = 3x - 2$ and $4x + y = 12$, for $x = 0$ to 5.
    (b) Write down the coordinates of the point where the two lines cross.
    (c) Write down the simultaneous equations that you solved in part (b).

Exercise continues …

5 Mr and Mrs Smith took their three grandchildren to the cinema.
It cost £33.
Mr Jones took his two children to the same show.
It cost £19.
The cost of an adult's ticket is £x and a child's ticket is £y.

(a) Explain why $2x + 3y = 33$ and $x + 2y = 19$.

(b) Draw the graphs of the equations in part (a) on the same axes.

(c) Use your graph to find the price of an adult's ticket and the price of a child's ticket.

6 The sum of two numbers is 9 and the difference 4.
Call the numbers x and y.

(a) Write down two equations for x and y.

(b) Draw the graphs of the equations from part (a) for values from 0 to 10.

(c) Use your graph to find the two numbers.

The equation of a line

The **gradient** of a line is the mathematical way of measuring its steepness or rate of change.

$$\text{gradient} = \frac{\text{increase in } y}{\text{increase in } x}$$

To find the gradient of a line, mark two points on the line, then draw in the horizontal and the vertical to form a triangle as shown.

$$\text{gradient} = \frac{6}{2} = 3$$

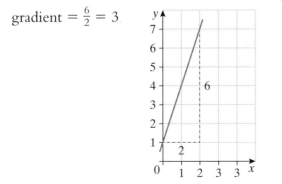

Here the gradient $= \frac{-8}{2}$ or $\frac{8}{-2}$.

Both give the answer -4.

TIP

Check you have the correct sign, positive or negative, for the slope of the line.

TIP

Choose two points far apart on the graph, so that the x-distance between them is an integer. If possible, choose points where the graph crosses gridlines. This makes reading values and dividing easier.

Lines with a positive gradient slope forwards.

Lines with a negative gradient slope backwards.

Horizontal lines have a gradient of zero.

● Discovery 17.2

Using graph-drawing software, draw the following lines on the same axes. Use a different colour for each line if possible.

A $y = 2x$ $y = 2x + 1$ $y = 2x + 2$ $y = 2x + 3$ $y = 2x + 4$
   $y = 2x - 1$ $y = 2x - 2$

What do you notice?
Check your ideas with these lines: $y = x$ $y = x + 1$
                                       $y = x + 2$ $y = x + 3$

Do they work for this set?

B $y = -x$ $y = -x + 1$ $y = -x + 2$ $y = -x + 3$

Clear the screen of lines already drawn, then draw these two sets.

C $y = x$ $y = 2x$ $y = 3x$ $y = 4x$
D $y = -x$ $y = -2x$ $y = -3x$ $y = -4x$

What do you notice about the lines from **C** and **D**?
Now plot some lines of your own and see if your conclusions are correct.

In Discovery 17.2 you should have found that when the equation of a line is written in the form $y = mx + c$, where m and c are numbers, then m is the gradient of the line and c is the value of y where the line crosses the y-axis. In other words, the line passes through $(0, c)$. The point $(0, c)$ is called the **y-intercept** because it is where the line intercepts, or crosses, the y-axis.

Using these facts you can find the gradient and the y-intercept directly from the equation of the line without having to plot the graph.

Example 17.4

The equation of a straight line is $5x + 2y = 10$.
Find its gradient and y-intercept.

Solution

Rearranging the equation:
$$2y = -5x + 10$$
$$y = -2.5x + 5$$

So the gradient is -2.5 and the y-intercept is 5.

Parallel lines

You may have noticed from the work earlier in the chapter that

Lines with the same gradient are parallel.

For example, these lines are all parallel.
$y = 2x$
$y = 2x + 3$
$y = 2x + 4$

Example 17.5

State an equation for a line which is parallel to each of these.
(a) $y = 4x + 1$ **(b)** $y = 5 - \frac{1}{2}x$

Solution

(a) The gradient of the line $= 4$.
    Any parallel line will also have gradient 4.
    So one possible line is $y = 4x - 3$.

(b) The gradient of the line $= -\frac{1}{2}$.
    Any parallel line will also have gradient $-\frac{1}{2}$.
    So one possible line is $y = -\frac{1}{2}x + 2$ or $y = 2 - \frac{1}{2}x$.

Exercise 17.4

1   Find the gradient of each of these lines and their y-intercept.
    (a) $y = 3x - 2$ **(b)** $y = 5x - 3$ **(c)** $y = 2 + 5x$
    (d) $y = 7 + 2x$ **(e)** $y = 7 - 2x$ **(f)** $y = 9 - 3x$

Exercise continues ...

2 Find the gradient of each of these lines and their y-intercept.

(a) $y + 2x = 5$ (b) $y - 5x = 1$ (c) $4x + 2y = 7$

(d) $3x + 2y = 8$ (e) $6x + 5y = 10$ (f) $2x + 5y = 15$

3 On the same diagram, sketch and label the graphs of these three equations. You do not need to do an accurate plot.

$$y = 2x + 1 \qquad y = 2x - 3 \qquad y = -4x + 1$$

4 On the same diagram, sketch and label the graphs of these three equations. You do not need to do an accurate plot.

$$y = 3x + 2 \qquad y = 3x - 2 \qquad y = -x + 2$$

5 Write down equations of lines parallel to each of these lines.

(a) $y = 5x$ (b) $y = -x$ (c) $y = -3x + 7$

(d) $y = 6 - 4x$ (e) $y = \frac{1}{3}x + 4$ (f) $2y = x + 4$

(g) $2x + y = 8$ (h) $3x + 2y = 12$ (i) $5x - 2y = 10$

● Challenge 17.1

When a road goes up a steep slope, a road sign gives the gradient of the slope.

Investigate the different ways that gradients are given on road signs.

Is this the same on road signs in the rest of Europe?

How do these gradients relate to the mathematical gradient used in this chapter?

Inequalities

If you want to buy a packet of sweets costing 79p, you need at least 79p.

You may have more than that in your pocket. The amount in your pocket must be greater than or equal to 79p.

If the amount in your pocket is x, then this can be written as $x \geqslant 79$. This is an inequality.

The symbol $\geqslant$ means 'greater than or equal to'.
The symbol $>$ means 'greater than'.
The symbol $\leqslant$ means 'less than or equal to'.
The symbol $<$ means 'less than'.

On a number line you use an open circle to represent $>$ and $<$ and a solid circle to represent $\geqslant$ and $\leqslant$.

Inequalities are solved in a similar way to equations.

Example 17.6

Solve the inequality $2x - 1 > 8$.
Show the solution on a number line.

Solution

$2x - 1 > 8$
$\quad 2x > 9 \qquad$ Add 1 to each side.
$\quad x > 4.5 \quad$ Divide each side by 2.

$$-5 \quad -4 \quad -3 \quad -2 \quad -1 \quad 0 \quad 1 \quad 2 \quad 3 \quad 4 \quad 5$$

> **TIP**
>
> The inequality sign is $>$ so you use an open circle on the number line.

Negative inequalities work a bit differently. It is best to make sure that you do not end up with a negative x term.

Example 17.7

Solve the inequality $7 - 3x \leqslant 1$.
Show the solution on a number line.

Solution

You could start by subtracting 7 from each side, but this would give a negative x term.
$7 - 3x \leqslant 1$
$\quad 7 \leqslant 1 + 3x \qquad$ Add $3x$ to each side.
$\quad 6 \leqslant 3x \qquad\qquad$ Subtract 1 from each side.
$\quad 2 \leqslant x \qquad\qquad\;$ Divide each side by 3.
$\quad x \geqslant 2 \qquad\qquad\;$ Rewrite the inequality so that x is on the left.
$\qquad\qquad\qquad\qquad\;$ Remember to turn the inequality sign round as well.

$$-5 \quad -4 \quad -3 \quad -2 \quad -1 \quad 0 \quad 1 \quad 2 \quad 3 \quad 4 \quad 5$$

> **TIP**
>
> The inequality sign is $\geqslant$ so you use a solid circle on the number line.

Exercise 17.5

For each of questions 1 to 6, solve the inequality and show the solution on a number line.

1 $x - 3 > 10$

2 $x + 1 < 5$

3 $5 > x - 8$

4 $2x + 1 \leqslant 9$

5 $3x - 4 \geqslant 5$

6 $10 \leqslant 2x - 6$

For each of questions 7 to 20, solve the inequality.

7 $5x < x + 8$

8 $2x \geqslant x - 5$

9 $4 + x < -5$

10 $2(x + 1) > x + 3$

11 $6x > 2x + 20$

12 $3x + 5 \leqslant 2x + 14$

13 $5x + 3 \leqslant 2x + 9$

14 $8x + 3 > 21 + 5x$

15 $5x - 3 > 7 + 3x$

16 $6x - 1 < 2x$

17 $5x < 7x - 4$

18 $9x + 2 \geqslant 3x + 20$

19 $5x - 4 \leqslant 2x + 8$

20 $5x < 2x + 12$

What you have learned

- You find a point on a line by substituting a value of x or y into the equation of the line
- You can draw a straight-line graph by finding two points on the line and drawing a line through them; it is sensible to use a third point to check that you have not made a mistake
- The point at which two lines cross is the solution to the simultaneous equations represented by those lines
- The gradient of a line is a measure of its steepness and equals $\dfrac{\text{increase in } y}{\text{increase in } x}$
- The y-intercept of a line is the point at which it crosses the y-axis
- When an equation of a line is written in the form $y = mx + c$, m is the gradient of the line and c is the value of y where the line crosses the y-axis
- Lines with the same gradient are parallel
- The symbol $\geqslant$ means 'greater than or equal to', $>$ means 'greater than', $\leqslant$ means 'less than or equal to' and $<$ means 'less than'
- $x \geqslant 4$, $x > 3$, $y \leqslant 6$ and $y < 7$ are inequalities
- The solution to an inequality can be shown on a number line; you use an open circle to represent $>$ and $<$ and a solid circle to represent $\geqslant$ and $\leqslant$
- You solve inequalities by performing the same operation on both sides of the inequality, in a similar way to solving equations

Mixed exercise 17

1 Draw the graphs of these lines.
(a) $y = 3x + 1$
(b) $y = 4 - 2x$

2 Draw the graphs of these lines.
(a) $2x + y = 8$
(b) $3x + 2y = 12$

3 The difference between two numbers is 5 and their sum is 1.
Call the numbers x and y.
(a) Write down two equations for x and y.
(b) On a set of axes labelled from -5 to 5 for both x and y, draw the graphs of the equations from part (a).
(c) Use your graph to find the two numbers.

4 Find the gradient and y-intercept of each of these lines.
(a) $y = 3x + 5$
(b) $y = x - 6$
(c) $y = 2x - 12$
(d) $y = 4 - 2x$

5 Write down the equations of two lines parallel to each of these lines.
(a) $y = x - 7$
(b) $y = 5x + 3$
(c) $y = 2 - 3x$

6 Solve each of these inequalities and show the solution on a number line.
(a) $5x + 1 \leqslant 11$
(b) $10 + 3x \leqslant 5x + 4$
(c) $7x + 3 < 5x + 9$
(d) $6x - 8 > 4 + 3x$
(e) $5x - 7 > 7 - 2x$

Congruence and transformations

▶ This chapter is about

- Congruence
- Translating shapes
- Combining transformations

▶ You should already know

- The terms *object* and *image* as they apply to transformations
- How to reflect a shape in a mirror line
- How to recognise and describe reflections
- How to rotate a shape about a point
- How to recognise and describe rotations
- How to enlarge a shape using a scale factor and a centre of enlargement
- How to recognise and describe enlargements

Congruence

Shapes that are exactly the same size and shape are said to be **congruent**.
If you cut out two congruent shapes one shape would fit exactly on top of the other.

When you reflect a shape the object and image are congruent.
Similarly, when you rotate a shape the object and image are congruent.

In this tangram triangles A and B are congruent.

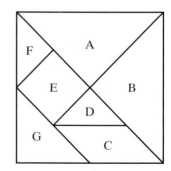

🔍 Check up 18.1

(a) Which other shapes in the tangram above are congruent?
(b) What can you say about triangles A and G in the tangram?

Exercise 18.1

1   Which of the shapes below are congruent to shape A?

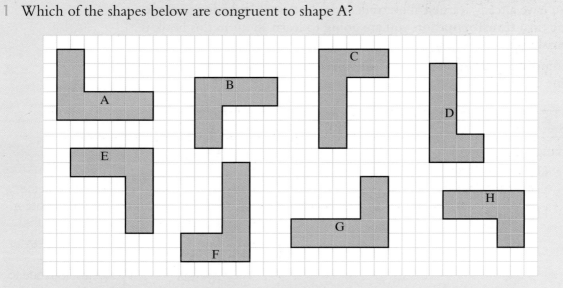

2   In this question, sides that are equal are marked with corresponding lines and angles that are equal are marked with corresponding arcs.

Which of these pairs of triangles are congruent?

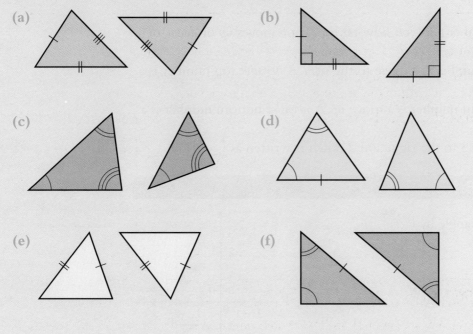

Translations

You have already learned about reflections, rotations and enlargements. The final **transformation** you are going to learn about in this book is **translation**.

A translation moves all points of an object the same distance in the same direction. The object and image are congruent.

◉ Discovery 18.1

Triangle B is a translation of triangle A.

(a) How can you tell it is a translation?

(b) How far across has it moved?

(c) How far down has it moved?

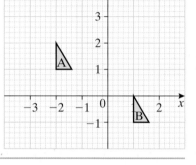

TIP

Take care with the counting. Choose a point on both the object and image and count the squares from one to the other.

How far a shape moves in a translation is written as a **column vector**.

The *top* number tells you how far the shape moves *across*, or in the *x*-direction.

The *bottom* number tells you how far the shape moves *up* or *down*, or in the *y*-direction.

A *positive* top number is a move to the *right*. A *negative* top number is a move to the *left*.

A *positive* bottom number is a move *up*. A *negative* bottom number is a move *down*.

A translation of 3 to the right and 2 down is written as $\begin{pmatrix} 3 \\ -2 \end{pmatrix}$.

Example 18.1

Translate the triangle by $\begin{pmatrix} -3 \\ 4 \end{pmatrix}$.

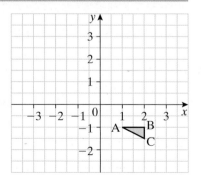

Solution

$\begin{pmatrix} -3 \\ 4 \end{pmatrix}$ means move 3 units left, and 4 units up.

Point A moves from $(1, -1)$ to $(-2, 3)$.

Point B moves from $(2, -1)$ to $(-1, 3)$.

Point C moves from $(2, -1.5)$ to $(-1, 2.5)$.

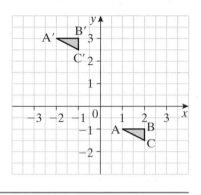

To describe a translation you need to state that the transformation is a translation and give the column vector.

Example 18.2

Describe fully the transformation that maps shape A on to shape B.

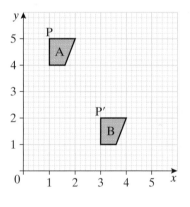

> ### TIP
>
> Try not to confuse the words *transformation* and *translation*.
>
> *Transformation* is the general name for all changes made to shapes.
>
> *Translation* is the particular transformation where all points of an object move the same distance in the same direction.

Solution

It is clearly a translation as the shape stays the same way up.

To find the movement choose one point on the object and image and count the squares moved.

For example, P moves from $(1, 5)$ to $(3, 2)$. This is a movement of 2 to the right and 3 down.

The transformation is a translation of $\begin{pmatrix} 2 \\ -3 \end{pmatrix}$.

Exercise 18.2

1 Draw a pair of axes and label them -2 to 6 for x and y.

   (a) Draw a triangle with vertices at $(1, 2)$, $(1, 4)$, and $(2, 4)$. Label it A.

   (b) Translate triangle A by vector $\begin{pmatrix} 2 \\ 1 \end{pmatrix}$. Label it B.

   (c) Translate triangle A by vector $\begin{pmatrix} 4 \\ -2 \end{pmatrix}$. Label it C.

   (d) Translate triangle A by vector $\begin{pmatrix} -2 \\ -3 \end{pmatrix}$. Label it D.

2 Draw a pair of axes and label them -2 to 6 for x and y.

   (a) Draw the trapezium with vertices at $(2, 1)$, $(4, 1)$, $(3, 2)$ and $(2, 2)$. Label it A.

   (b) Translate trapezium A by vector $\begin{pmatrix} 2 \\ 3 \end{pmatrix}$. Label it B.

   (c) Translate trapezium A by vector $\begin{pmatrix} -4 \\ 0 \end{pmatrix}$. Label it C.

   (d) Translate trapezium A by vector $\begin{pmatrix} -3 \\ 2 \end{pmatrix}$. Label it D.

3 Describe the transformation that maps

   (a) triangle A on to triangle B.

   (b) triangle A on to triangle C.

   (c) triangle A on to triangle D.

   (d) triangle B on to triangle D.

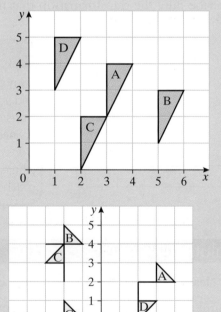

4 Describe the transformation that maps

   (a) flag A on to flag B.

   (b) flag A on to flag C.

   (c) flag A on to flag D.

   (d) flag A on to flag E.

   (e) flag A on to flag F.

   (f) flag E on to flag G.

   (g) flag B on to flag E.

   (h) flag C on to flag D.

   Hint: Not all the transformations are translations.

Challenge 18.1

Draw a pair of axes and label them −6 to 6 for x and y.

(a) Draw a shape in the positive region near the origin. Label it A.

(b) Translate shape A by vector $\begin{pmatrix} 2 \\ 1 \end{pmatrix}$. Label it B.

(c) Translate shape B by vector $\begin{pmatrix} 3 \\ -2 \end{pmatrix}$. Label it C.

(d) Translate shape C by vector $\begin{pmatrix} -6 \\ -1 \end{pmatrix}$. Label it D.

(e) Translate shape D by vector $\begin{pmatrix} 1 \\ 2 \end{pmatrix}$. Label it E.

(f) What do you notice about shapes A and E?
Can you suggest why this happens?
Try to find other combinations of translations for which this happens.

Combining transformations

Sometimes when one transformation is followed by another, the result is equivalent to a single transformation.

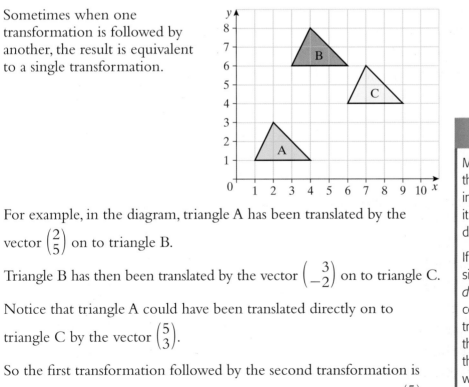

For example, in the diagram, triangle A has been translated by the vector $\begin{pmatrix} 2 \\ 5 \end{pmatrix}$ on to triangle B.

Triangle B has then been translated by the vector $\begin{pmatrix} 3 \\ -2 \end{pmatrix}$ on to triangle C.

Notice that triangle A could have been translated directly on to triangle C by the vector $\begin{pmatrix} 5 \\ 3 \end{pmatrix}$.

So the first transformation followed by the second transformation is equivalent to the single transformation: translation by the vector $\begin{pmatrix} 5 \\ 3 \end{pmatrix}$.

TIP

Make sure you do the transformations in the right order, as it usually makes a difference.

If you are asked for a single transformation, *do not* give a combination of two transformations as this does not answer the question and will usually score no marks.

Example 18.3

Find the single transformation that is equivalent to a reflection in the line $x = 1$, followed by a reflection in the line $y = -2$.

Solution

Choose a simple shape and draw a diagram.
An asymmetrical shape such as a right-angled triangle or a flag is usually best.

In the diagram, reflecting the object flag A in the line $x = 1$ gives flag B.

Reflecting flag B in the line $y = -2$ gives flag C.

The transformation that maps A directly on to C is a rotation through 180°.

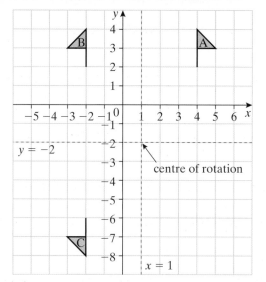

The centre of rotation is $(1, -2)$, which is where the mirror lines cross.

The transformation is a rotation through 180° about the centre of rotation $(1, -2)$.

Use tracing paper to check this.

A rotation of 180° is the only rotation for which you do not need to state the direction, as 180° clockwise is the same as 180° anticlockwise.

Exercise 18.3

1 (a) Plot the points $(1, 2)$, $(1, 4)$ and $(2, 4)$ and join them to form a triangle. Label it A.

Translate triangle A by the vector $\begin{pmatrix} 5 \\ 2 \end{pmatrix}$. Label the image B.

(b) On the same grid, translate triangle B by the vector $\begin{pmatrix} 2 \\ -4 \end{pmatrix}$. Label the image C.

(c) Describe fully the single transformation that is equivalent to a translation by the vector $\begin{pmatrix} 5 \\ 2 \end{pmatrix}$ (A on to B) followed by a translation by the vector $\begin{pmatrix} 2 \\ -4 \end{pmatrix}$ (B on to C).

Exercise continues …

2 (a) On a new grid, plot the points $(0, 2)$, $(1, 4)$ and $(3, 2)$ and join them to form a triangle.
Label it D.

Translate triangle D by the vector $\begin{pmatrix} -4 \\ 2 \end{pmatrix}$.

Label it E.

(b) On the same grid, translate triangle E by the vector $\begin{pmatrix} 8 \\ 0 \end{pmatrix}$.

Label the image F.

(c) Describe fully the single transformation that is equivalent to a translation by the vector $\begin{pmatrix} -4 \\ 2 \end{pmatrix}$ (D on to E) followed by a translation by the vector $\begin{pmatrix} 8 \\ 0 \end{pmatrix}$ (E on to F).

3 (a) Draw a set of axes. Label the x-axis from 0 to 13 and the y-axis from 0 to 15.
Plot the points $(1, 2)$, $(2, 4)$ and $(1, 3)$ and join them to form a triangle.
Label it G.
Enlarge triangle G with scale factor 2 and centre the origin.
Label the image H.

(b) On the same grid, enlarge triangle H with scale factor 3 and centre of enlargement $(0, 5)$.
Label the image I.

(c) Describe fully the single transformation that is equivalent to an enlargement, scale factor 2, centre the origin (G on H) followed by an enlargement scale factor 3, centre $(0, 5)$ (H on to I).

4 (a) Copy the diagram.

Enlarge the flag J with scale factor $1\frac{1}{2}$ and centre of enlargement $(1, 2)$.
Label the image K.

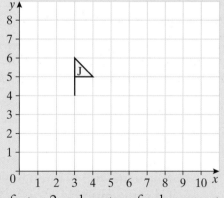

(b) On the same grid, enlarge the flag K with scale factor 2 and centre of enlargement $(2, 8)$.
Label the image L.

(c) Describe fully the single transformation that is equivalent to an enlargement, scale factor $1\frac{1}{2}$, centre the point $(1, 2)$ (J on to K) followed by an enlargement, scale factor 2, centre $(2, 8)$ (K on to L).

Exercise continues …

5 (a) Plot the points (1, 0), (1, −2) and (2, −2) and join them to form a triangle.
Label it A.
Reflect triangle A in the line $y = 1$.
Label the image B.

(b) Reflect triangle B in the line $y = x$.
Label the image C.

(c) Describe fully the single transformation that is equivalent to a reflection in the line $y = 1$ (A on to B) followed by a reflection in the line $y = x$ (B on to C)

6 (a) On a new grid, plot the points (2, 5), (3, 5) and (1, 3) and join them to form a triangle.
Label it D.
Reflect triangle D in the line $x = \frac{1}{2}$.
Label the image E.

(b) On the same grid, reflect triangle E in the line $y = -x$.
Label the image F.

(c) Describe fully the single transformation that is equivalent to a reflection in the line $x = \frac{1}{2}$ (D on to E) followed by a reflection in the line $y = -x$ (E on to F).

7 (a) Copy the diagram.

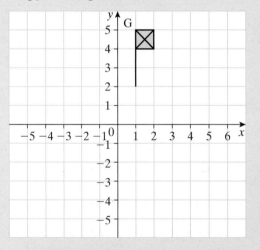

Rotate the flag G through 90° clockwise about the point (1, 2).
Label the image H.

(b) On the same grid, rotate the flag H through 180° about the point (2, −1).
Label the image I.

(c) Describe fully the single transformation that is equivalent to a rotation through 90° clockwise about the point (1, 2))G on to H) followed by a rotation through 180° about the point (2, −1) (H on to I).

Exercise continues ...

8 (a) Plot the points (0, 1), (0, 4) and (2, 3) and join them to form a triangle.
Label it J.
Rotate triangle J through 90° anticlockwise about the point (2, 3).
Label the image K.

(b) On the same grid, rotate triangle K through 90° clockwise about the point (2, −1).
Label the image L.

(c) Describe fully the single transformation that is equivalent to a rotation through
90° anticlockwise about the point (2, 3) (J on to K) followed by a rotation
through 90° clockwise about the point (2, −1) (K on to L).

9 Copy the diagram.

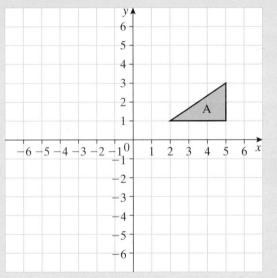

(a) Reflect triangle A in the x-axis. Label the image B.
(b) Reflect triangle B in the y-axis. Label the image C.
(c) Describe fully the single transformation that will map triangle A on to triangle C.

In questions 10 to 16, carry out the transformations on a simple shape of your choice.

10 Describe fully the single transformation that is equivalent to a reflection in the line $x = 1$
followed by a reflection in the line $x = 5$.

11 Describe fully the single transformation that is equivalent to a reflection in the line $y = 2$
followed by a reflection in the line $y = 6$.

12 Describe fully the single transformation that is equivalent to an enlargement, scale
factor 2 and centre the origin, followed by a translation by the vector $\binom{3}{2}$.

Exercise continues ...

13   Describe fully the single transformation that is equivalent to a rotation through 90° clockwise about the origin, followed by a translation by the vector $\begin{pmatrix} 4 \\ 0 \end{pmatrix}$.

14   Describe fully the single transformation that is equivalent to a reflection in the x-axis followed by a rotation through 90° anticlockwise about the origin.

15   Describe fully the single transformation that is equivalent to a reflection in the y-axis followed by a rotation through 90° anticlockwise about the origin.

16   Describe fully the single transformation that is equivalent to a reflection in the line $y = x$ followed by a reflection in the line $y = -x$.

● Challenge 18.2

Look at your answers to Exercise 18.3 questions **10** and **11**.

Try to make a general statement about the result of reflection in a mirror line followed by reflection in a parallel mirror line.

● Challenge 18.3

Look at your answers to Exercise 18.3 questions **5** and **6**.

Try to make a general statement about the result of reflection in a mirror line followed by reflection in an intersecting mirror line.

What you have learned

- Two shapes are congruent if they are exactly the same shape and size
- When a shape is translated the object and image are the same way round
- When a shape is translated all the points of the object move the same distance in the same direction
- When a shape is translated the object and the image are congruent
- To describe a translation you state that the transformation is a translation and give the column vector
- The top number of a column vector tells you how far the shape moves in the x-direction; the bottom number tells you how far it moves in the y-direction
- You can combine transformations; make sure you do the transformations in the order given
- If you are asked to describe a single transformation, do not give a combination of transformations

Mixed exercise 18

1 Divide this L-shape into four congruent smaller L-shapes.

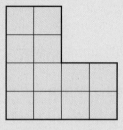

2 Copy the diagram.

(a) Translate shape A by vector $\begin{pmatrix} 1 \\ -6 \end{pmatrix}$. Label it B.

(b) Translate shape A by vector $\begin{pmatrix} -3 \\ 0 \end{pmatrix}$. Label it C.

(c) Translate shape A by vector $\begin{pmatrix} -5 \\ -4 \end{pmatrix}$. Label it D.

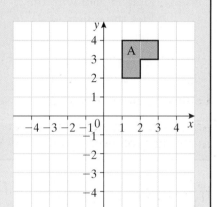

3 Describe fully these single transformations.

(a) Flag A on to flag B

(b) Flag B on to flag C

(c) Flag A on to flag D

(d) Flag D on to flag B

(e) Flag E on to flag A

(f) Flag A on to flag F

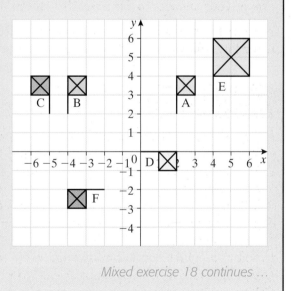

Mixed exercise 18 continues ...

4 Draw a grid with the x-axis from 0 to 10 and the y-axis from 0 to 7.

   (a) Plot the points $(1, 4), (1, 6)$ and $(2, 6)$ and join them to form a triangle.
       Label it A.
       Reflect triangle A in the line $y = x$.
       Label the image B.

   (b) Rotate triangle B through 90° anticlockwise about the point $(5, 5)$.
       Label the image C.

   (c) Describe fully the single transformation that maps triangle A on to triangle C.

5 (a) Draw a grid with the x-axis from 0 to 12 and the y-axis from 0 to 6.
       Plot the points $(4, 1), (6, 1)$ and $(4, 2)$ and join them to form a triangle.
       Label it D.
       Translate triangle D by $\binom{2}{3}$.
       Label the image E.

   (b) Enlarge triangle E with scale factor 2 and centre of enlargement $(5, 7)$.
       Label the image F.

   (c) Describe fully the single transformation that maps triangle D on to triangle F.

Unit C Contents

Two-dimensional representation of solids

Recognising nets

A **net** is a flat shape that can be folded to make a three-dimensional shape.

Here are two possible nets for a cube.

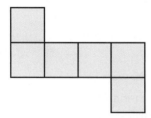 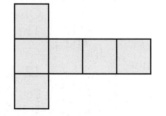

Challenge 1.1

Can you draw three other nets which will fold to make a cube?

Exercise 1.1

In this exercise, look at each net and say what shape it makes.
Then make up the net to check your answer.

> **TIP**
>
> Here are some tips for making three-dimensional shapes.
> - Put tabs on every other edge.
> - Use card rather than paper, if possible, to construct a shape that will last!
> - When using card, score the edges before folding.
> - Use glue that is suitable for the material you are using. If possible, use quick-drying glue.

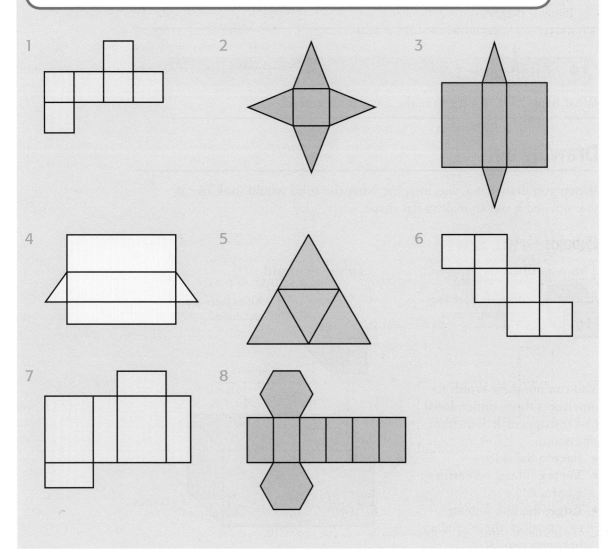

● **Challenge 1.2**

Can you see what shapes these nets will make? You can make them if you wish.

(a)

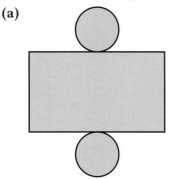

(b)

● **Challenge 1.3**

What happens if you try to make a net for a sphere?

Drawing nets

When you draw a net, you imagine what the solid would look like if you opened it out to make a flat shape.

Cuboids

This is a **cube**.

It has six faces: they are all squares.

This is a **cuboid**.

All its faces are rectangles…

…or squares

You can use these words to describe a three-dimensional (3-D) shape, such as a cube or cuboid.

- **Face**: a flat side
- **Vertex** (plural – **vertices**): a corner
- **Edge**: the line joining two vertices, the boundary between two faces

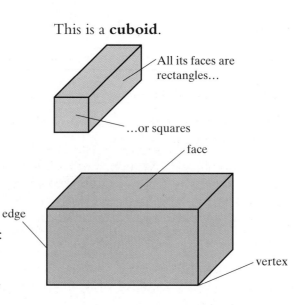

face

edge

vertex

Example 1.1

Draw a net of this cuboid.

All the lengths are given in centimetres.

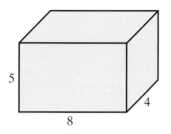

Solution

These two possible nets are not drawn to size but the lengths are marked.

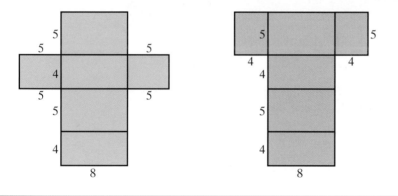

The edges that meet together are the same length.

Try drawing one of the nets in Example 1.1 cutting it out and folding it together to make the cuboid.

The faces around the main rectangle can be placed in other positions. Try this and see which arrangements will fold up to make the cuboid. Use squared paper to help you.

As a cuboid has six faces, its net needs six rectangles. However, not every arrangement of the six rectangles will fold together to make the cuboid.

If you are asked to draw the net of a box with no lid, then the net will be made up of five rectangles.

Example 1.2

This net can be folded to make a cuboid.
When it is folded, which point or points will meet with

(a) point A? **(b)** point D?

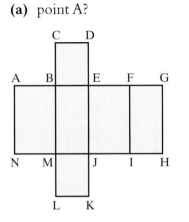

Solution

(a) Points C and G **(b)** Point F

Example 1.3

For each of these arrangements, say whether it is the net of cuboid.
They are drawn on squared paper to size.

(a) **(b)** **(c)**

(d) **(e)** **(f)**

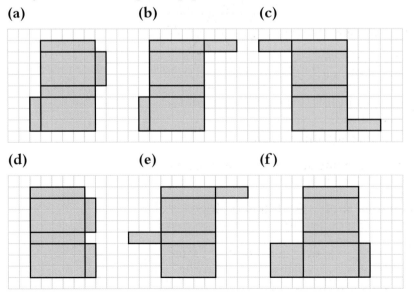

Solution

(a) Yes	**(b)** Yes	**(c)** No
(d) No	**(e)** Yes	**(f)** No

Pyramids

This is a **tetrahedron**.

It has four faces: they are all triangles

This is a **square-based pyramid**.

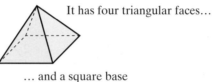

It has four triangular faces...

... and a square base

Example 1.4

This is a regular tetrahedron.

Each edge is 5 cm long.

Draw its net.

5 cm

Solution

Here is a possible net.

It is not drawn to size but the lengths are marked.

To draw it you need to use compasses.

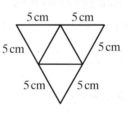

5 cm   5 cm

5 cm   5 cm

5 cm   5 cm

Draw a line 5 cm long.

Open your compasses to 5 cm.

Place the point on one end of the line and draw an arc.

Place the point on the other end of the line and draw an arc crossing the first.

Join the point where the arcs cross to both ends of the base line to make a triangle.

Using the three sides of the first triangle as the bases, construct three more triangles.

Exercise 1.2

1　Draw a net for each of these cuboids on squared paper. All lengths are given in centimetres.

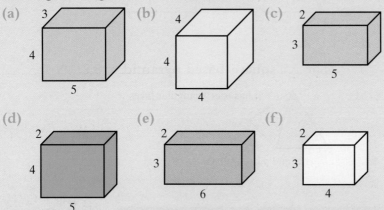

(a) 3　4　5

(b) 4　4　4

(c) 2　3　5

(d) 2　4　5

(e) 2　3　6

(f) 2　3　4

> **TIP**
> If you are not sure whether a net is correct, cut it out and try folding it up.

2　Draw a net for each of these pyramids. All lengths are given in centimetres. You will need to use compasses.

> **TIP**
> Draw the base for part **(b)** using squared paper. Then draw the triangular faces using compasses.

(a) 5　5　5　4　3　3

(b) 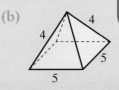 4　4　5　5

3　For a cuboid, state

(a) the number of edges.

(b) the number of vertices.

(c) the number of faces.

> **TIP**
> Give yourself practice in selecting and describing shapes to help you remember the words needed.

4　When this net is folded to make a cuboid, which point or points will meet with

(a) point A?

(b) point D?

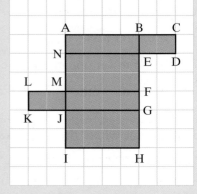

Exercise continues …

5 A cuboid packing case has a base 40 cm by 40 cm and a height of 80 cm. It has a top. Draw a net for the case. Use a scale of 1 cm to 20 cm.

6 The box containing my radio is 20 cm by 10 cm by 5 cm. It has a top. Draw a net of the box on squared paper. Use a scale of 1 cm to 5 cm.

7 A dice is a cube with 2 cm edges.
The numbers on the faces are 1, 2, 3, 4, 5 and 6.
The two numbers on each pair of opposite faces add up to 7.
Draw a net of the dice and write the numbers on the faces.

8 A box containing paper has a base 15 cm by 20 cm, and it is 5 cm high.
It has no top.
Draw a net for the box. Use a scale of 1 cm to 5 cm.

9 This is the box for my computer monitor.
It says MONITOR on each of the four sides.
It has a lid.
Draw a net for the box on squared paper.
Use a scale of 1 cm to 10 cm.
Write the word MONITOR the correct way
up on the correct faces.

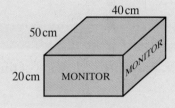

10 The box containing my pens is a cuboid with no lid.
The base is a square of side 6 cm, and it is 15 cm high.
Draw a net for the box on squared paper. Use a scale of 1 cm to 3 cm.

TIP

You will sometimes be asked to draw a net accurately and sometimes just to make a sketch. Even with a sketch, it is best to make the edges that are the same length look as though they are equal.

● Challenge 1.4

The sketch shows the edges of a cube.
Each vertex is labelled.

(a) How many edges do you think you will use to follow a route that visits all the vertices but does not go over any of the edges more than once?

(b) Now try to find a route as described in part (a). How many edges did you actually use?

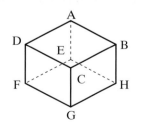

Isometric drawings

An **isometric drawing** is a way of representing three-dimensional shapes.

For these drawings, you will need a ruler, a pencil and **isometric** paper. This has a grid made of dots arranged in triangles.

You have to make sure you get the paper the right way round.

This is the correct way round.   This is the wrong way round.

Example 1.5

Make an isometric drawing of this shape.

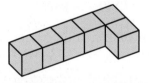

Solution

This is an isometric drawing of the shape.

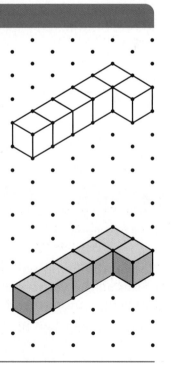

You can make the drawing look more realistic and easier to understand by using different shadings for the sides facing in the three different directions.

Exercise 1.3

1 Make an isometric drawing of a centimetre cube.

2 Make an isometric drawing of each of these shapes.

(a) 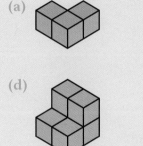   (b)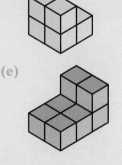

> **TIP**
>
> Make sure that the isometric paper is the right way round.

3 Make an isometric drawing of each of these shapes.

(a) A 2 by 2 by 2 cube   (a) A 4 by 3 by 2 cuboid

4 Make an isometric drawing of each of these shapes.

(a) 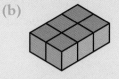   (b)   (c)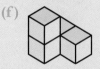

(d)   (e)   (f)

5 How many cubes make up this shape?

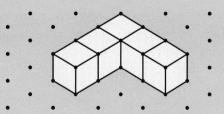

6 This shape is made from five cubes.
Make isometric drawings of all the other
different shapes that can be made using
just five cubes in one layer.
(There are 11 more to find.)

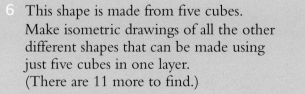

Exercise continues …

7   Make an isometric drawing of each of these solid shapes.
    Use a scale of 1 centimetre to represent 2 units.

8   Make an isometric drawing of each of these shapes.
    (a) Use a scale of 1 cm to 1 m.   (b) Use a scale of 1 cm to 5 m.

    (c) Use a scale of 1 cm to 6 cm.

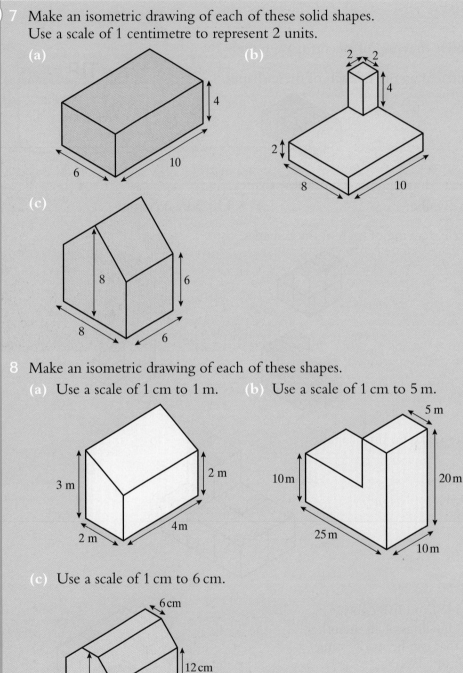

Plans and elevations

This diagram is part of a builder's plan for a housing estate.

It shows the shapes of the houses as seen from above. You may have heard the term 'bird's eye view' for this sort of picture. The mathematical term is **plan view**.

From the plan you can tell only what shape the buildings are from above. You cannot tell whether they are bungalows, two-storey houses or even blocks of flats.

The view of the front of an object is called a **front elevation**, the view from the side is called a **side elevation** and the view from the back is called a **back elevation**.

An elevation shows you the height of an object.

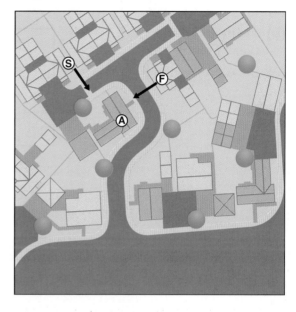

Example 1.6

For house A, sketch

(a) a possible view from F.

(b) a possible view from S.

Solution

(a) View from F   **(b) View from S**

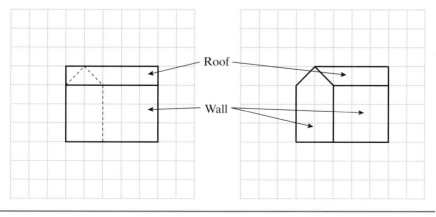

Example 1.7

For this shape, draw

(a) the plan.

(b) the front elevation (view from F).

(c) the side elevation from S.

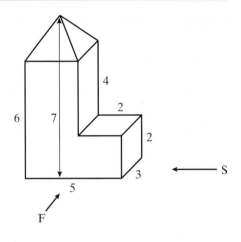

Solution

(a) Plan

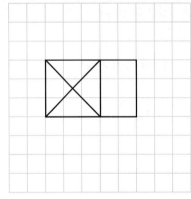

The cross shows the edges of the pyramid on top of the tower. The rectangle on the right is the flat top of the lower part of the shape.

(b) Front elevation

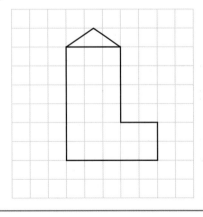

(c) Side elevation

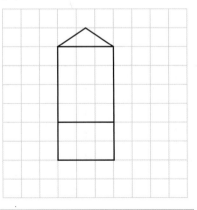

Example 1.8

Draw the plan view, front elevation and side elevation of this child's building block.

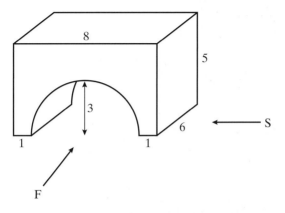

Solution

Plan

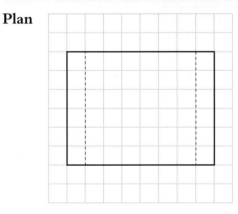

You can use broken lines to show hidden detail.

In the plan, the broken lines show the sides of the tunnel at floor level.

In the side elevation, the broken lines show the top of the tunnel.

Front elevation

Side elevation

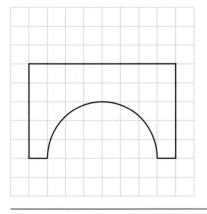
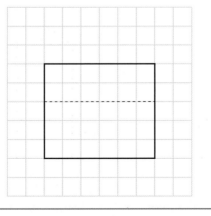

Exercise 1.4

1 Draw the plan view, front elevation and side elevation of each of these objects.

(a)

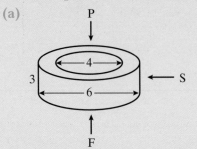

(b)

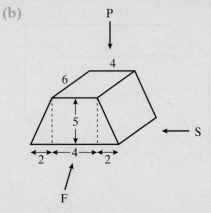

(c)

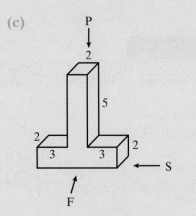

(d)

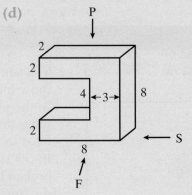

(e)

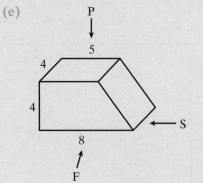

(f)

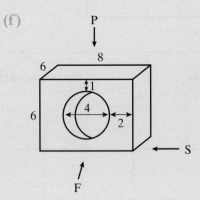

Exercise continues …

2   A shape is made from seven centimetre cubes.
    Here are the plan view and the front and
    side elevations.

    Make an isometric drawing of the shape.

Plan	Front	Side

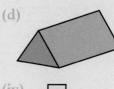

What you have learned

- A net is a flat shape that can be folded to make a three-dimensional shape
- Isometric drawings can be used to represent three-dimensional shapes; horizontal faces appear flattened, vertical edges remain vertical
- A plan view of an object is the shape of the object viewed from above
- An elevation of an object is the shape of the object viewed from the front, back or side

Mixed exercise 1

1   Here are four three-dimensional shapes and their nets.
    Match each shape to its net and name the three-dimensional shape.

(a) (b) (c) (d)

(i) (ii) (iii) (iv)

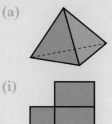

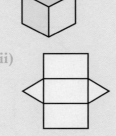

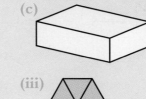

2   This net is folded to make a cube.

    (a) Which vertex will join to N?
    (b) Which line will join to CD?
    (c) Which line will join to IH?

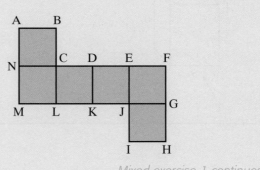

Mixed exercise 1 continues …

3 Draw a full-size net for this cuboid.

4 Make isometric drawings of these shapes.
(Make sure your paper is the right way round.)

(a) (b) (c) 3 cm 4 cm 1 cm

5 Draw the plan view, front elevation and side elevation of these objects.

(a) (b)

(c) (d)

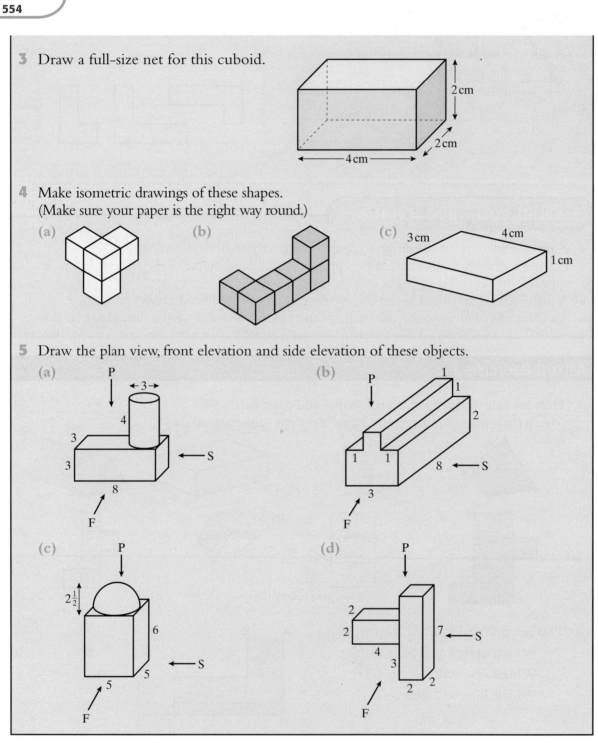

Probability 1

The language of probability and the probability scale

You can use words such as these to indicate how likely it is for an event to happen.

    certain   *very likely*   *an even chance*   *unlikely*   *impossible*

Here are some examples.

- In the UK, it is very likely to rain this month.
- It is certain that some people will celebrate Christmas on 25th December.

You can place these words on a probability scale.

| Impossible | Very unlikely | Unlikely | Evens | Likely | Very likely | Certain |

Example 2.1

Draw a probability scale.

Put arrows to show the chance of each of the following events happening.

(a) Getting a head when you toss a coin.

(b) Someone in the class will pass a bus stop on their way home.

(c) You will go to sleep in the next week.

Solution

You draw a probability scale like the one on the previous page. Decide which word to use for each of the situations and draw an arrow to it.

The position of the arrow for **(b)** will depend on your school, how near a bus route it is, and so on.

Impossible	Very unlikely	Unlikely	Evens	Likely	Very likely	Certain

(a)   (b)   (c)

Challenge 2.1

Work in groups.

Think of events whose chances of happening are impossible.

Repeat for the other words on the probability scale. Try to find five events for each word.

Often, you need to be more precise about how likely an event is to happen. For instance, a weather forecaster may say that there is a 0.2 chance of rain tomorrow in Birmingham. People may use this information to help them plan what they do the next day.

Now you use a probability scale numbered from 0 to 1.

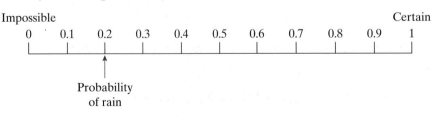

Impossible   Certain

0   0.1   0.2   0.3   0.4   0.5   0.6   0.7   0.8   0.9   1

Probability
of rain

On this scale,

- 0 is the probability of an event which is impossible.
- 1 is the probability of an event which is certain to happen.

Example 2.2

Draw a probability scale.

Put arrows to show the chance of each of the following events happening.

(a) Picking a dark chocolate out of a box of white chocolates.

(b) The next person you meet has a birthday in August.

(c) The next vehicle to pass the school is a car.

Solution

You draw a probability scale like the one on the previous page.

Decide how likely you think each situation is and draw an arrow in that position.

The position of the arrow for **(c)** will depend on what type of road your school is on.

Impossible   Certain

0   0.5   1

(a) (b)   **(c)**

● Challenge 2.2

On a piece of paper, write an event which has not yet happened. For example, 'It will snow here tomorrow.'

Imagine a probability line drawn across the classroom, with 0 at one end and 1 at the other.

As a class, decide where each person should stand on the line with the piece of paper for their event.

Exercise 2.1

1 Choose the best probability word from those below to complete these sentences.

   Impossible   Unlikely   Evens   Likely   Certain

   (a) It is .... that it will snow in the UK on Midsummer's Day.

   (b) It is .... that I will get an odd number when I roll an ordinary dice.

   (c) It is .... that in a family with three children, at least one is a boy.

Exercise continues …

2 Copy this scale.

Impossible   Very   Unlikely   Evens   Likely   Very likely   Certain
              unlikely

Put arrows to show the chance of each of the following events happening.

(a) Getting wet when you swim.

(b) You watch 12 hours of TV on a school day.

(c) You eat in the next 5 hours.

3 There are 20 pens in a box.
There are 6 red pens, 4 black pens and 10 blue pens.
A pen is taken out without looking.
Choose the correct probability word to complete these sentences.

(a) It is .... that the pen is red.

(b) It is .... that the pen is blue.

(c) It is .... that the pen is green.

4 For each part of this question, write numbers for all the cards to make the following
statements true when a card is turned over.

(a) It is certain to be a 1.   (b) It is impossible to be a 1.

(c) It is unlikely to be a 1.   (d) It is very likely to be a 1.

Calculating probabilities

◉ Discovery 2.1

Work in pairs and throw a dice 60 times. Record how many 6s you get.

Did you get as many as you expected?

Pool the results for the whole class and discuss them.

Many board games need an ordinary dice. Sometimes you need to
throw a 6 to start the game. You are unlikely to do this first time. You
can work out the probability of throwing a 6.

There are six numbers on a dice and only one is 6.

The probability of getting a six is 1 out of 6, which is $\frac{1}{6}$.

Similarly, there are three odd numbers on a dice.

The probability of getting an odd number is 3 out of 6, which is $\frac{3}{6} = \frac{1}{2}$ or 0.5.

When the **outcomes** of an **event** are equally likely, such as a dice landing on any of its faces, a coin showing either heads or tails or a playing card being one of the 52 different cards, you can calculate probabilities.

> **TIP**
>
> When you are writing probabilities, always cancel fractions if possible.

$$\text{The probability of an event happening} = \frac{\text{The number of ways the event can happen}}{\text{The total number of possible outcomes}}.$$

Example 2.3

A bag contains ten sweets.
Five are green, two are yellow and three are orange.
A sweet is taken out without looking.

Calculate the probability that it is

(a) green.   (b) orange.   (c) red.

Draw arrows on a probability scale to show these probabilities.

Solution

(a) There are five green sweets out of a possible ten.
   The probability of a green sweet is $\frac{5}{10} = 0.5$.

(b) There are three orange sweets, so the probability of an orange sweet is $\frac{3}{10} = 0.3$.

(c) There are no red sweets, so the probability of a red sweet is $\frac{0}{10} = 0$.

When you draw a probability scale, choose a scale that is suitable for the situation. Here you have ten possible outcomes so a suitable way to mark your scale is in tenths.

> **TIP**
>
> Never give your answer as '3 in 10' or '3 out of 10'. Always give probabilities as fractions or decimals, such as $\frac{3}{10}$ or 0.3.

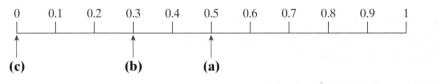

Example 2.4

A bag contains ten sweets.
Five are green, two are yellow and three are orange.
A sweet is taken out without looking.

Find the probability that the sweet is not orange.

Solution

There are three orange sweets in the bag and ten sweets altogether.

So there are seven sweets which are not orange.

The probability of the sweet not being orange is $\frac{7}{10} = 0.7$.

Notice that the probability of the sweet not being orange is $1 - 0.3$, and 0.3 is the probability that the sweet is orange.

This is an example of an important result.

The probability of an event not happening $= 1 -$ The probability of the event happening.

Check up 2.1

(a) A weather forecaster says that there is a 0.2 chance that it will rain tomorrow in Guildford.

What is the probability that it will not rain tomorrow in Guildford?

(b) Think of some other pairs of probabilities like the one above and the one in Example 2.4.

Work in pairs with one person writing the probability of something happening and the other writing the probability of it not happening.

The probability of two events

When you are finding the probability of two, or more, events, it helps to make a list or table of the possible outcomes.

Example 2.5

Gareth throws two dice. Show the outcomes on a grid (possibility space) and use the grid to find the probability that Gareth scores

(a) a double.   **(b)** a total of 11.   **(c)** less than 5.

Solution

There are 36 possible outcomes.

(a) The diagonal gives the six doubles.

$$P(\text{double}) = \tfrac{6}{36} = \tfrac{1}{6}$$

(b) The grid gives the total scores.

$$P(\text{total of 11}) = \tfrac{2}{36} = \tfrac{1}{18}$$

(c) $P(\text{less than 5}) = \tfrac{6}{36} = \tfrac{1}{6}$

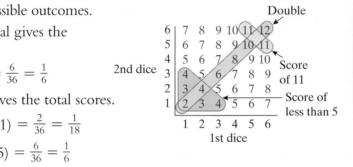

> **TIP**
>
> P(event) is a short way to write the probability of the event happening.

Example 2.6

Rachel is selecting a main course and a sweet from this menu. List the possible outcomes in a table.

> **MENU**
>
> **Main Course**   **Sweet**
> Sausage & Chips   Apple Pie
> Ham Salad   Fruit Salad
> Vegetable Lasagne

If she is equally likely to select any of the choices for each course, what is the probability that she selects Vegetable Lasagne and Apple Pie?

Solution

There are six possible outcomes. Vegetable Lasagne and Apple Pie occurs only once.

Main course	Sweet
Sausage & Chips	Apple Pie
Sausage & Chips	Fruit Salad
Ham Salad	Apple Pie
Ham Salad	Fruit Salad
Vegetable Lasagne	Apple Pie
Vegetable Lasagne	Fruit Salad

$P(\text{Vegetable Lasagne and Apple Pie}) = \tfrac{1}{6}$

● Challenge 2.3

Work in pairs.

One person chooses four main course meals that they like. The other chooses three desserts that they like.

List in a table like this, all the two-course meals you could make from your choices.

First Course	Second Course

Exercise 2.2

1 These six cards are laid face down and mixed up. Then a card is picked.

| 1 | 2 | 1 | 1 | 4 | 3 |

Copy this probability scale.

Impossible   Certain
0   0.5   1

Put arrows to show the probability of each of the following events happening. The number on the card is

(a) 1. (b) less than 5. (c) an even number.

2 There are ten sweets in a bag. Five are red, three are green and the others are yellow. A sweet is taken out without looking.

(a) Use a probability word to complete this sentence.

It is .... that the sweet is yellow.

(b) Draw a probability scale numbered 0 to 1.
Mark the probability of each of these statements on your scale. Use arrows and label them R, G and B.
R: The sweet is red.
G: The sweet is green.
B: The sweet is blue.

Exercise continues ...

3 This spinner is fair.
Calculate the probability of it landing on

(a) 4.

(b) an even number.

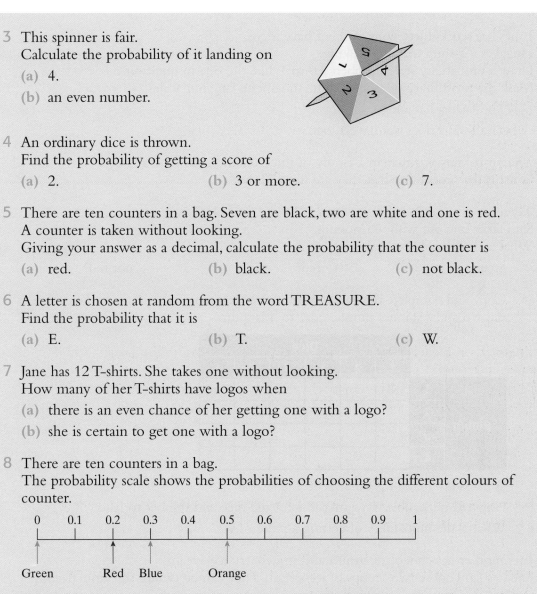

4 An ordinary dice is thrown.
Find the probability of getting a score of

(a) 2.   (b) 3 or more.   (c) 7.

5 There are ten counters in a bag. Seven are black, two are white and one is red.
A counter is taken without looking.
Giving your answer as a decimal, calculate the probability that the counter is

(a) red.   (b) black.   (c) not black.

6 A letter is chosen at random from the word TREASURE.
Find the probability that it is

(a) E.   (b) T.   (c) W.

7 Jane has 12 T-shirts. She takes one without looking.
How many of her T-shirts have logos when

(a) there is an even chance of her getting one with a logo?

(b) she is certain to get one with a logo?

8 There are ten counters in a bag.
The probability scale shows the probabilities of choosing the different colours of
counter.

How many counters of each colour are there in the bag?

9 Alison has 12 sweets in a bag. 7 of them are red and the other 5 are purple.

(a) She takes a sweet without looking.
Which colour is she more likely to get?

(b) Alison takes four sweets out of the bag.
There is now an even chance of getting a red sweet.
How many sweets of each colour has she taken out?

Exercise continues ...

10 James has ten T-shirts. Four of them have logos.
He takes a T-shirt without looking.
Draw a probability scale numbered 0 to 1, like the one in question 8.
Mark the probability of each of these statements on your scale.

(a) The T-shirt has a logo.

(b) The T-shirt does not have a logo.

11 The probability is $\frac{1}{8}$ that, in a family of three children, they are all boys.
What is the probability that they are not all boys?

12 There are five green balls, seven red balls and four black balls in a bag.
Sam takes one out without looking.
What is the probability that it is

(a) green?   (b) red?   (c) not red?

13 (a) Copy and complete this table to show the total score when one ordinary dice is
thrown and a spinner numbered 1 to 4 is spun.

		Dice					
		1	**2**	**3**	**4**	**5**	**6**
Spinner	**1**	2	3				
	2						
	3						
	4						

(b) What is the highest possible score?

(c) How many possible ways are there for the dice and spinner to land?

(d) What is the probability of getting a score of 9?

14 Jan's mother has chocolate, vanilla and strawberry ice-cream.
She lets Jan choose two scoops of ice-cream. They can be two of the same flavour.

(a) Copy and complete this table to show all the possible combinations of flavours Jan
can choose.
Two have been done for you. They count as different.

First scoop	C	V							
Second scoop	V	C							

(b) Jan likes all the flavours equally and picks a combination at random.
Find the probability that Jan has one scoop of chocolate and one of strawberry.

Exercise continues ...

15 Mr Ahmed is choosing his new car.
He can choose a Renault or a Peugeot and he can choose red, blue or black.
If he chooses completely randomly, what is the probability that he will choose

(a) a blue Peugeot?

(b) a black car?

16 Nicola, James and Eleanor are going to a cinema that has two different films showing.
If they each choose at random, what is the probability that they all see the same film?

17 In the game of Monopoly, you throw two dice and your score is the sum of the two numbers.

(a) To buy Park Lane, Hamish needs to score 11.
What is the probability that Hamish can buy Park Lane on his next go?

(b) To get out of jail, Sylvia needs to throw a double.
What is the probability that Sylvia can get out of jail on her next go?

(c) If Sanjay scores 7, he will land on Regent Street.
What is the probability that Sanjay does not land on Regent Street on his next go?

Experimental probability

Sometimes, outcomes are not equally likely. For instance, a dice may be biased, so that it is more likely to fall one way than another. In these situations, you have to rely on experimental or other evidence to **estimate** probabilities.

$$\text{The experimental probability of an event} = \frac{\text{The number of times an event happens}}{\text{The total number of trials}}.$$

Discovery 2.2

You need a large number of trials before you can decide that a dice is biased.

Look at the differences in your results for Discovery 2.1.

How much did they vary?

Were the results for the whole class together closer to the expected result?

Compare your results with the results in Example 2.7 on the next page.

Example 2.7

A biased dice is thrown 1000 times and the results recorded.

Number on dice	1	2	3	4	5	6
Frequency	60	196	84	148	162	350

Use the results to estimate the probability of obtaining a six when the dice is thrown.

Solution

There were 350 sixes out of 1000 throws.

So you estimate the probability of a six as $\frac{350}{1000} = 0.35$.

> **TIP**
>
> Experimental probabilities are usually written as decimals.

Example 2.8

Dan wanted to find the probability of a car driving past his school being red.

He recorded the colour of 200 cars passing the school.

The results are shown in the table.

Colour	Blue	Black	White	Red	Green	Other
Number of cars	58	24	25	38	21	34

Solution

There were 38 red cars out of 200.

So the probability of a car being red is $\frac{38}{200} = 0.19$.

○ Discovery 2.3

What evidence do weather forecasters use to work out the probability of rain?

Exercise 2.3

1 Joe thinks his coin is biased.
   He tosses it 200 times and gets 130 heads and 70 tails.
   (a) What is the experimental probability of getting a head with this coin?
   (b) In 200 tosses, how many heads would you expect if the coin was fair?
   (c) What should Joe do to check if his coin is really biased? *Exercise continues …*

2 Here are the results of 300 throws of a dice.

Number on dice	1	2	3	4	5	6
Frequency	43	39	51	63	49	57

(a) For this dice, calculate the experimental probability of obtaining
   (i) a 6.   (ii) a 2.

(b) For a fair dice, calculate, as a decimal correct to 3 decimal places, the probability of scoring
   (i) a 6.   (ii) a 2.

(c) Do your answers suggest that the dice is fair? Give your reasons.

3 A survey of the type of washing agent used by 500 households had the results shown in the table.

Type of washing agent	Liquid	Powder	Tablet
Number of households	220	175	105

Work out the experimental probability that a household chosen at random will use

(a) liquid.

(b) tablets.

4 Theresa did a survey on the colour of cars going down her road.
These were her results.

Car colour	Black	White	Red	Silver	Blue	Other
Number of cars	16	25	19	28	8	4

Find the experimental probability that the next car will be

(a) silver.

(b) blue.

5 The table shows the results of a survey to find out which supermarket people use most often. Use the table to estimate the probability of a person chosen at random shopping mostly at these supermarkets.

(a) Tesco

(b) Morrison's

Supermarket	Frequency
Tesco	52
Sainsbury's	38
Kwiksave	16
Somerfield	27
Asda	34
Morrison's	18
Other	65
Total	250

Exercise continues …

6 Adam has carried out a survey in his school to find out
   what drinks the students had with their breakfast.
   The results are shown in the table.

   Use this table to estimate the probability that a student
   in Adam's school had these drinks with their breakfast.

   (a) Coffee

   (b) Milk

Drink	Frequency
Tea	28
Coffee	33
Chocolate	15
Milk	24
Fruit juice	31
Other	19
Total	150

7 The table shows a batsman's scores in his last 100
   innings.

   Use the table to estimate the probability that in
   his next innings the batsman will score

   (a) 20 to 29.

   (b) 100 or more.

Runs scored	Frequency
0 to 9	8
10 to 19	12
20 to 29	25
30 to 39	16
40 to 49	14
50 to 99	22
100 or more	3

8 Gail and Saima play chess against each other regularly.
   The results of the last 30 games are shown in the table.

   Use this table to estimate the probability that the next
   game results in these outcomes.

   (a) A win for Gail

   (b) A draw

Gail wins	18
Saima wins	9
Draw	3

9 Ben wants to find out the probability that it will rain on his birthday.
   He finds out that it has rained on two out of his last three birthdays.
   He says that this means that the probability that it will rain on his next birthday is $\frac{2}{3}$.
   Why is he almost certainly wrong?

Exercise continues …

10 The first 100 students to walk into the hall one morning were asked how they had travelled to school that morning.
The table shows the results.

Means of travel	Walk	Car	Bike	Bus
Number of students	37	21	16	26

Use this table to estimate the probability that the next student to walk into the hall had travelled

(a) on foot.　　　　(b) on the bus.　　　　(c) by bike.

● Challenge 2.4

Try this experiment to estimate how many there are in a population.

Put a number (unknown) of discs or counters of one colour in a bag.
Place 10 (or so) discs of a different colour in the bag.
Select a disc at random, note whether it is the first colour or not, and return it to the bag.
Mix the discs well and pick another one out.

How many discs will you need to pick out before you can confidently predict the total number of discs in the bag?

(This method of estimating can be called 'How many fish in the sea?'
It can also be done with cards from a pack of an unknown number of cards.)

What you have learned

- The words *impossible, unlikely, evens, likely* and *certain* can be used to describe how likely an event is to happen
- The probability scale goes from 0 to 1
- Probabilities may be given as fractions or decimals
- $P(event) = \dfrac{\text{The number of ways the event can happen}}{\text{The total number of possible outcomes}}$
- $P(event\ not\ happening) = 1 - P(event)$
- That when listing the outcomes of two events, you need to be systematic to ensure you include all the outcomes
- The experimental probability of an event $= \dfrac{\text{The number of times an event can happen}}{\text{The total number of trials}}$
- For experimental probabilities, you need to carry out a large number of trials to get a good estimate

Mixed exercise 2

1 Choose the best probability word from those below to complete these sentences.

> Impossible   Unlikely   Evens   Likely   Certain

(a) It is .... that I will be age 392 years on my next birthday.

(b) It is .... that a £1 coin will land heads when I toss it.

(c) It is .... that there will be at least one hour of sunshine this week.

2 These six cards are laid face down and mixed up. Then a card is picked.

| 5 | 8 | 4 | 3 | 4 | 4 |

Copy this probability scale.

Impossible   Certain
    0   0.5   1

Put arrows to show the probability of each of the following events happening.
The number on the card is

(a) 4.   (b) less than 9.   (c) an odd number.

3 In her pencil case, Maria has ten pens. She takes one out without looking.

(a) There is an even chance that she takes out a blue pen.
How many blue pens are there?

(b) It is impossible that she takes out a green pen.
How many green pens are there?

4 A fair spinner is labelled 1 to 5.
It is spun once.
Giving your answer as a fraction, find the probability
that it lands on

(a) 2.

(b) an odd number.

(c) a number greater than 3.

5 A letter is chosen at random from the word SATURDAY.
Find the probability that it is

(a) U.   (b) A.   (c) B.

Mixed exercise 2 continues ...

6 (a) What is the probability of getting a 6 with one throw of an ordinary fair dice?

(b) Bethany has a biased dice. The probability of getting a 6 on Bethany's dice is 0.3. What is the probability that Bethany's dice does not land on a 6 when it is thrown?

7 There are 20 marbles in a bag.
The probability of picking out a green marble at random is $\frac{1}{5}$.
How many green marbles are there in the bag?

8 A sweet is taken out of a bag without looking.
The probability that it is red is 0.35.
What is the probability that it is not red?

9 Gary surveyed people to find the activity they were doing at a leisure centre.
The table shows his results.

Calculate an estimate of the probability that the next person to visit the leisure centre is going swimming.

Activity	Frequency
Gym	45
Swimming	72
Ten-pin bowling	40
Ice-skating	16
Other	27

10 Karen has three T-shirts: one red, one white and one yellow.
She has two pairs of jeans: one blue and one black.

(a) Copy and complete this table to show the different colours that she can wear together.

T-shirt	Jeans

(b) One day she picks a T-shirt and a pair of jeans at random.
What is the probability that the T-shirt is yellow and the jeans are blue?

Perimeter, area and volume 1

- Finding the perimeter of simple shapes
- Finding the area of rectangles
- Finding the volume of cuboids

- How to add, subtract and multiply numbers
- How to change between metric units of length
- The meanings of the words *rectangle, square, triangle, equilateral, isosceles, polygon, cube* and *cuboid*

Perimeter

The **perimeter** of a shape is the distance all the way round. It is a length so the units of the answer will be units of length, such as centimetres (cm) or metres (m).

Example 3.1

Find the perimeter of each of these shapes.

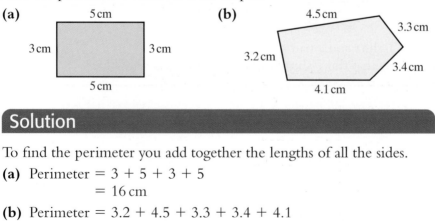

(a)   5 cm

3 cm   3 cm

          5 cm

(b)   4.5 cm

          3.3 cm

3.2 cm

          3.4 cm

          4.1 cm

Solution

To find the perimeter you add together the lengths of all the sides.

(a) Perimeter = 3 + 5 + 3 + 5
            = 16 cm

(b) Perimeter = 3.2 + 4.5 + 3.3 + 3.4 + 4.1
            = 18.5 cm

TIP

Always give the units with your answer. It is often useful to draw a sketch and label each side with its length. Sometimes you need to use your knowledge of shapes to find the lengths.

Check up 3.1

Find the perimeter of each of these shapes.

(a) A square with sides of length 3 cm

(b) An equilateral triangle with sides of length 3 cm

Exercise 3.1

1 Find the perimeter of each of these shapes. All lengths are in centimetres.

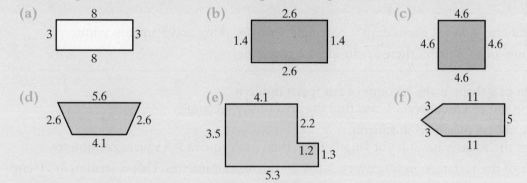

2 A square has sides of length 4.2 m.
What is its perimeter?

3 A rectangle has sides of length 5.3 cm and 4.1 cm.
What is its perimeter?

4 A regular hexagon has sides of length 2.5 m.
What is its perimeter?

5 A five-sided field has sides of length 15.3 m, 24.5 m, 17.3 m, 16 m and 10.2 m.
What is its perimeter?

Challenge 3.1

Find the perimeter of each of these shapes.
All lengths are in centimetres.

(a)   **(b)**

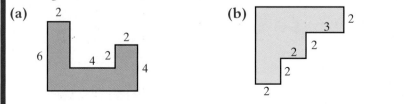

Challenge 3.2

A rectangular sheet of paper measures 20 cm by 15 cm.
A square of side 4 cm is cut out of one corner.
Sketch the remaining piece of paper and find its perimeter.

Area

Discovery 3.1

On centimetre-squared paper draw a rectangle 4 squares long and 5 squares wide.
Count how many squares there are inside the rectangle.

The **area** of a shape is the amount of flat space inside it.
What you did in Discovery 3.1 was find the area of the rectangle.

Areas are always measured in squares.
If each of the squares has sides of length 1 cm, then each square is a square centimetre.
The area of the rectangle in Discovery 3.1 is 20 square centimetres. This is written as 20 cm^2.

Exercise 3.2

1 Find the area of each of these shapes. Give your answers in cm^2.

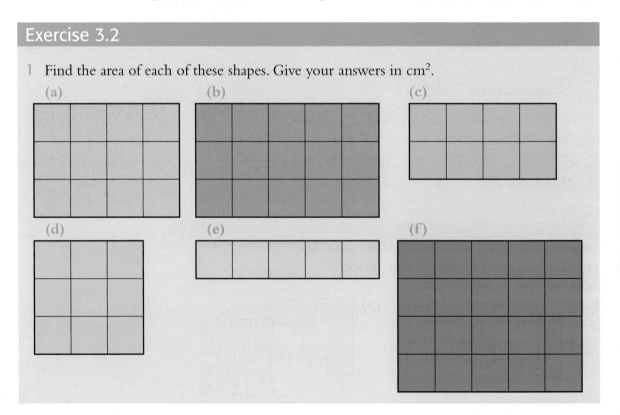

The area of a rectangle

Discovery 3.2

On squared paper, draw four different rectangles that each have an area of 24 squares.

Write down the length and the width of each of your rectangles.

What is the connection between the length and width of a rectangle and its area?

You can work out the area of a rectangle using this formula.

| Area of rectangle = length × width |

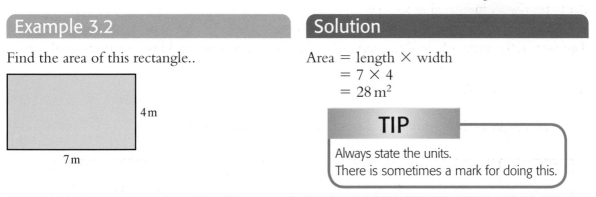

width

length

Areas are measured in square units such as square centimetres (cm^2), square metres (m^2) and square kilometres (km^2).

Example 3.2

Find the area of this rectangle..

4 m

7 m

Solution

Area = length × width
    = 7 × 4
    = 28 m^2

TIP

Always state the units.
There is sometimes a mark for doing this.

Notice in Example 3.2 that the lengths are in metres (m) so the area is in square metres (m^2).

Example 3.3

Find the area of a square with sides of length 4.5 cm.

Solution

It is often useful to draw a sketch and label the length and width.

4.5 cm

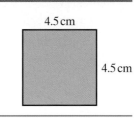

4.5 cm

Area = length × width
    = 4.5 × 4.5
    = 20.25 cm^2

Exercise 3.3

1   Find the area of each of these rectangles.
    Take care to give the correct units in the answer.

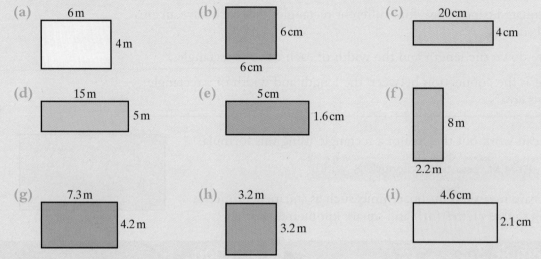

(a)   6 m
      4 m

(b)   6 cm
      6 cm

(c)   20 cm
      4 cm

(d)   15 m
      5 m

(e)   5 cm
      1.6 cm

(f)   8 m
      2.2 m

(g)   7.3 m
      4.2 m

(h)   3.2 m
      3.2 m

(i)   4.6 cm
      2.1 cm

2   A rectangle measures 4.7 cm by 3.6 cm. Find its area.

3   A square has sides of length 2.6 m. Find its area.

4   A rectangle has sides of length 3.62 cm and 4.15 cm. Find its area.

5   A rectangular garden measures 5.6 m by 2.8 m. Find its area.

6   A rectangular pond measures 4.5 m by 8 m. Find the area of the surface of the pond.

7   A patio is a rectangle 4 metres long by 3.5 metres wide.
    (a) Find the area of the patio.
    To pave the patio it costs £24.50 per square metre.
    (b) How much does it cost to pave the patio?

8   A rectangular floor measures 2.5 m by 6 m.
    (a) Find the area of the floor.
    Iain covers the floor with carpet costing £25 per square metre.
    (b) How much does the carpet for the floor cost?

Challenge 3.3

This diagram is the shape of a garden.
Find the area of the garden.

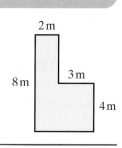

Challenge 3.4

A cuboid is 8 cm long, 5 cm wide and 3 cm high.

(a) Find the total surface area of the cuboid.

(b) What happens to the total surface area of the cuboid if
   (i) all the lengths are doubled?
   (ii) all the lengths are halved?

The volume of a cuboid

Discovery 3.3

The diagram shows a cuboid made of
centimetre cubes.

(a) How many centimetre cubes are
   there in the top layer?

(b) How many centimetre cubes are
   there altogether in the cuboid?

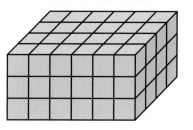

Another cuboid has five layers.
Each layer has four rows of three centimetre cubes.

(c) How many cubes are there in this cuboid?

The number of centimetre cubes in a cuboid is called the **volume** of
the cuboid.

Volumes are always measured in cubes.

If each of the cubes has sides of length 1 cm, then each cube is a
centimetre cube.

The volume of the first cuboid in Discovery 3.3 is 72 centimetre cubes.

This is written as 72 cm^3.

Discovery 3.4

Use multilink cubes to make some cubes and cuboids. For each cube or cuboid
- write down the dimensions (the length, width and height).
- find the volume.

What is the connection between the dimensions of a cuboid and its volume?

You can work out the volume of a cuboid using this formula

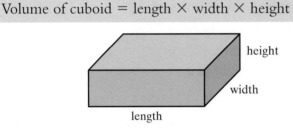

Volume of cuboid = length × width × height

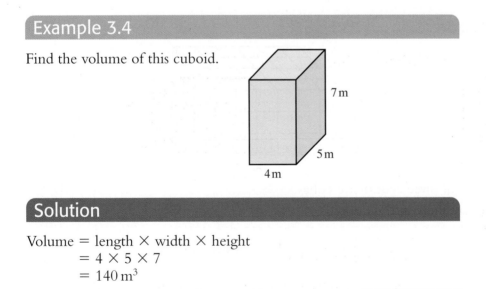

Volumes are measured in cubes so you use cubic units such as centimetre cubes (cm^3), metre cubes (m^3) and kilometre cubes (km^3).

Example 3.4

Find the volume of this cuboid.

7 m

5 m

4 m

Solution

Volume = length × width × height
      = 4 × 5 × 7
      = 140 m^3

Notice in Example 3.4 that the dimensions are in metres (m) so the volume is in metre cubes (m^3).

You must take care with the units. All the dimensions must be in the same units.

Example 3.5

A concrete path is laid. It is a cuboid 20 m long, 1.5 m wide and 10 cm thick. Calculate the volume of concrete used.

Solution

Again, it is usually helpful to draw a sketch.

All units must be the same. Change the thickness (height) from centimetres to metres by dividing by 100.

$10 \div 100 = 0.1$ m

Now work out the volume.

Volume = length $\times$ width $\times$ height
$= 20 \times 1.5 \times 0.1$
$= 3$ m^3

Exercise 3.4

1 Find the volume of each of these cuboids.

(a) 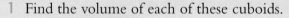 (b)

2 A cuboid has a height of 10 m, a length of 5 m and a width of 3 m. Find its volume.

3 A cube has edges 5 cm long. Find its volume.

4 A shoe box has a base measuring 10 cm by 15 cm and is 30 cm deep. Find its volume.

5 A room is 6 m long, 4 m wide and 3 m high. Work out its volume.

6 Calculate the volume of a cube with edges 3.5 cm long.

Exercise continues ...

7 A cuboid has a height of 6 cm, a length of 12 cm and a width of 3.5 cm.
   Calculate its volume.

8 A filing cabinet has a top measuring 45 cm by 60 cm and is 70 cm high.
   Calculate its volume.

9 The top of a desk is a piece of wood 40 cm wide, 30 cm deep and 1.5 cm thick.
   Find its volume.

10 A piece of glass is 25 cm wide, 60 cm long and 5 mm thick.
    Find the volume of the glass.
    Hint: Be careful with the units.

11 A box of chocolates is a cuboid 15 cm by 12 cm by 8 cm.
    How many of these can be fitted into a carton 1.2 m by 75 cm by 40 cm?

● Challenge 3.5

A cuboid has a volume of 96 cm³.

Using whole numbers only, find the dimensions of as many different cuboids with this volume as you can.

● Challenge 3.6

A cuboid is 12 cm long, 15 cm wide and 30 cm high.

(a) Find the volume of the cuboid.

(b) What would be the volume of the cuboid if all the lengths are reduced to a third of their original size?

What you have learned

- The perimeter of a shape is the distance round the shape
- A perimeter is a length
- Lengths can be measured in units such as centimetres (cm), metres (m) and kilometres (km)
- The area of a rectangle = length × width
- Area is measured in square units such as square centimetres (cm²), square metres (m²) and square kilometres (km²)
- The volume of a cuboid = length × width × height
- Volume is measured in cubic units such as centimetre cubes (cm³), metre cubes (m³) and kilometre cubes (km³)

Mixed exercise 3

1. Find the perimeter of each of these shapes.

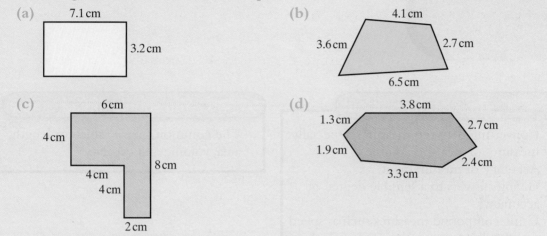

(a) 7.1 cm, 3.2 cm

(b) 4.1 cm, 3.6 cm, 2.7 cm, 6.5 cm

(c) 6 cm, 4 cm, 4 cm, 8 cm, 4 cm, 2 cm

(d) 3.8 cm, 1.3 cm, 2.7 cm, 1.9 cm, 2.4 cm, 3.3 cm

2. A rectangle measures 5.3 cm by 3.4 cm. Find its perimeter and its area.

3. A rectangular field is 12 m long and 15 m wide. Find its perimeter and its area.

4. A square has sides of length 4.6 cm. Find its perimeter and its area.

5. A rectangular patio is 4 m long and 2.5 m wide. Find its perimeter and its area.

6. The diagram shows the shape of a lawn.
   (a) Find the area of part A.
   (b) Find the area of part B.
   (c) Find the total area of the lawn.

   The cost of fertiliser is 75p per square metre.
   (d) How much does enough fertiliser for this lawn cost?

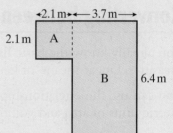

2.1 m ◄─► ◄─ 3.7 m ─►

2.1 m | A

B | 6.4 m

7. A cuboid has a base measuring 4 cm by 5 cm. It is 6 cm high. Find its volume.

8. Find the volume of a cube with edges 2.7 cm long.

9. A packet of paper measures 30 cm by 20 cm by 5 cm. What is its volume?

10. A piece of wood is a 4 m long, 6 m wide and 1.5 cm thick. Find the volume of the wood. Hint: Be careful with the units.

11. A pane of glass is a 15 cm wide, 60 cm long and 5 mm thick. Find the volume of the pane of glass.

12. The top of a desk is a piece of wood 80 cm wide, 50 cm deep and 15 mm thick. Find its volume.

Chapter

4

Measures

This chapter is about

- Converting between measures, especially measures of area and volume
- Accuracy of measurement
- Giving answers to a sensible degree of accuracy
- Using compound measures such as speed and density

You should already know

- The common metric units for length, area, volume and capacity

Converting between measures

You already know the basic **linear** relationships between metric measures. Linear means 'of length'.

You can use these relationships to work out the relationships between metric units of area and volume.

For example:

$1\,cm = 10\,mm$ $1\,m = 100\,cm$

$1\,cm^2 = 1\,cm \times 1\,cm$ $1\,m^2 = 1\,m \times 1\,m$
$1\,cm^2 = 10\,mm \times 10\,mm$ $1\,m^2 = 100\,cm \times 100\,cm$
$1\,cm^2 = 100\,mm^2$ $1\,m^2 = 10\,000\,cm^2$

$1\,cm^3 = 1\,cm \times 1\,cm \times 1\,cm$ $1\,m^3 = 1\,m \times 1\,m \times 1\,m$
$1\,cm^3 = 10\,mm \times 10\,mm \times 10\,mm$ $1\,m^3 = 100\,cm \times 100\,cm \times 100\,cm$
$1\,cm^3 = 1000\,mm^3$ $1\,m^3 = 1\,000\,000\,cm^3$

You can also convert between measures of volume and capacity.

Measures of capacity are used for liquids.

You need to remember that $1000\,cm^3$ is equivalent to 1 litre.

Example 4.1

Change these units.

(a) $5 \, \text{m}^3$ to cm^3 **(b)** $5600 \, \text{cm}^2$ to m^2 **(c)** $330 \, \text{cm}^3$ to litres

Solution

(a) $5 \, \text{m}^3 = 5 \times 1\,000\,000 \, \text{cm}^3$ Convert $1 \, \text{m}^3$ to cm^3 and
         $= 5\,000\,000 \, \text{cm}^3$ multiply by 5.

(b) $5600 \, \text{cm}^2 = 5600 \div 10\,000 \, \text{m}^2$ To convert m^2 to cm^2 you
            $= 0.56 \, \text{m}^2$ multiply, so to convert cm^2 to
                                            m^2 you divide.

TIP

Make sure you have done the right thing by checking your answer makes sense.
If you had multiplied by 10 000 you would have got 56 000 000 m², which is obviously a much larger area than 5600 cm².

(c) $330 \, \text{cm}^3 = 330 \div 1000$ litres   To convert cm^3 to litres you
            $= 0.33$ litres   divide by 1000.

Exercise 4.1

1   Change these units.
    (a) $25 \, \text{m}$ to cm   (b) $42 \, \text{cm}$ to mm   (c) $2.36 \, \text{m}$ to cm   (d) $5.1 \, \text{m}$ to mm

2   Change these units.
    (a) $3 \, \text{m}^2$ to cm^2 (b) $2.3 \, \text{cm}^2$ to mm^2
    (c) $9.52 \, \text{m}^2$ to cm^2 (d) $0.014 \, \text{cm}^2$ to mm^2

3   Change these units.
    (a) $90\,000 \, \text{mm}^2$ to cm^2 (b) $8140 \, \text{mm}^2$ to cm^2
    (c) $7\,200\,000 \, \text{cm}^2$ to m^2 (d) $94\,000 \, \text{cm}^2$ to m^2

4   Change these units.
    (a) $3.2 \, \text{m}^3$ to cm^3 (b) $42 \, \text{cm}^3$ to m^3 (c) $5000 \, \text{cm}^3$ to m^3 (d) $6.42 \, \text{m}^3$ to cm^3

5   Change these units.
    (a) 2.61 litres to cm^3 (b) $9500 \, \text{ml}$ to litres
    (c) 2.4 litres to cm^3 (d) $910 \, \text{cm}^3$ to litres   *Exercise continues ...*

6 What is wrong with this statement?

> The trench I have just dug is 5 m long, 2 m wide and 50 cm deep.
> To fill it in, I would need 500 m³ of concrete.

● Challenge 4.1

Cleopatra is reputed to have had a bath filled with asses' milk.
Today her bath might be filled with cola!

Assuming a can of drink holds 330 millilitres, approximately how
many cans would she need to have a bath in cola?

Accuracy in measurement

All measurements are **approximations**. Measurements are given to
the nearest practical unit.

Measuring a value to the nearest unit means deciding that it is nearer
to one mark on a scale than another; in other words, that the value is
within half a unit of that mark.

Look at this diagram.

Any value within the shaded area is 5 to the nearest unit.

The boundaries for this interval are 4.5 and 5.5. This would be written
as $4.5 \leqslant x < 5.5$.

TIP

Remember that $4.5 \leqslant x < 5.5$ means all values, x, which are greater than or
equal to 4.5 but less than 5.5.
$x < 5.5$ because if $x = 5.5$ it would round up to 6.
Even though x cannot equal 5.5, it can be as close to it as you like. So 5.5 is
the upper bound. Do not use 5.499 or 5.4999, etc.

4.5 is the **lower bound** and 5.5 the **upper bound**.
Any value less than 4.5 is closer to 4 (4 to the nearest unit).
Any value greater than or equal to 5.5 is closer to 6 (6 to the nearest
unit).

Example 4.2

(a) Tom won the 100 m race with a time of 12.2 seconds, to the nearest tenth of a second.
What are the upper and lower bounds for this time?

(b) Copy and complete this statement.

A mass given as 46 kg, to the nearest kilogram, lies between .... kg and .... kg.

Solution

(a) Lower bound = 12.15 seconds, upper bound = 12.25 seconds

(b) A mass given as 46 kg, to the nearest kilogram, lies between 45.5 kg and 46.5 kg.

Exercise 4.2

1 Copy and complete each of these statements.
   (a) A height given as 57 m, to the nearest metre, is between .... m and .... m.
   (b) A volume given as 568 ml, to the nearest millilitre, is between .... ml and .... ml.
   (c) A winning time given as 23.93 seconds, to the nearest hundredth of a second, is between .... seconds and .... seconds.

2 Copy and complete each of these statements.
   (a) A mass given as 634 g, to the nearest gram, is between .... g and .... g.
   (b) A volume given as 234 ml, to the nearest millilitre, is between .... ml and .... ml.
   (c) A height given as 8.3 m, to 1 decimal place, is between .... m and .... m.

3 (a) What is the least and greatest surface area that 3 litres of the paint will cover?
   (b) How many cans of paint are needed to ensure coverage of an area of 100 m²?

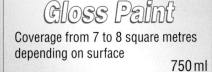

Gloss Paint

Coverage from 7 to 8 square metres depending on surface

750 ml

4 Jessica measures the thickness of a metal sheet with a gauge.
The reading is 4.97 mm accurate to the nearest $\frac{1}{100}$th of a millimetre.
   (a) What is the minimum thickness the sheet could be?
   (b) What is the maximum thickness the sheet could be?

Exercise continues …

5   Gina is fitting a new kitchen.
    She has an oven which is 595 mm wide, to the nearest millimetre.
    Will it definitely fit in a gap which is 60 cm wide, to the nearest centimetre?

6   Two metal blocks are placed together as shown.
    The left-hand block is 6.3 cm long and the right-hand
    block is 8.7 cm.
    Both blocks are 2 cm wide and 2 cm deep.
    All measurements are correct to the nearest millimetre.

    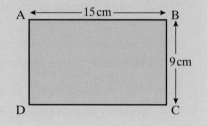

    (a) What is the least and greatest combined length of the two blocks?
    (b) What is the least and greatest depth of the blocks?
    (c) What is the least and greatest width of the blocks?

7   A company manufactures components for the car industry.
    One component consists of a metal block with a hole drilled
    into it. A plastic rod is fixed into the hole.
    The hole is drilled to a depth of 20 mm, to the nearest millimetre.
    The length of the rod is 35 mm, to the nearest millimetre.
    What are the maximum and minimum values of d (the height of
    the rod above the block)?

    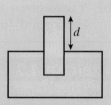

8   The diagram shows a rectangle, ABCD.
    AB = 15 cm and BC = 9 cm.
    All measurements are correct to the nearest centimetre.
    Work out the least and greatest values for the perimeter
    of the rectangle.

Working to a sensible degree of accuracy

Measurements and calculations should not be too accurate for their
purpose.

It is obviously silly to claim that a car journey took 4 hours,
56 minutes and 13 seconds or the distance between two houses is
93 kilometres, 484 metres and 78 centimetres. Answers such as these
would be more sensible rounded to 5 hours and 93 km respectively.

When calculating a measurement, you need to give your answer to a
sensible degree of accuracy.

As a general rule your final answer should not be given to a greater
degree of accuracy than any of the values used in the calculation.

Example 4.3

A table is 1.8 m long and 1.3 m wide. Both measurements are correct to 1 decimal place.

Work out the area of the table.

Give your answer to a sensible degree of accuracy.

Solution

Area = length × width

    = 1.8 × 1.3

    = 2.34   The answer has 2 decimal places.

    = 2.3 m² (to 1 d.p.)   However, the answer should not be more accurate than the original measurements. So you need to round the answer to 1 decimal place.

Exercise 4.3

1 Rewrite each of these statements using sensible values for the measurements.

  (a) It takes 3 minutes and 24.8 seconds to boil an egg.

  (b) It will take me 2 weeks, 5 days, 3 hours and 13 minutes to paint your house.

  (c) Helen's favourite book weighs 2.853 kg.

  (d) The height of the classroom door is 2 metres, 12 centimetres and 54 millimetres.

2 Give your answer to each of these questions to a sensible degree of accuracy.

  (a) Find the length of the side of a square field whose area is 33 m².

  (b) Three friends win £48.32. How much will each receive?

  (c) A strip of card is 2.36 cm long and 0.041 cm wide. Calculate the area of the card.

Problems involving speed, distance and time

This road sign tells drivers that the maximum speed they should be travelling at is 30 mph.

mph stands for 'miles per hour'.

If a car travels 30 miles in 1 hour its average speed is 30 miles per hour or 30 mph.

● Discovery 4.1

(a) Tom walks 6 miles in 2 hours. What is his speed in miles per hour?

(b) Rashid cycles at 10 miles per hour for 3 hours. How far does he cycle?

(c) Mrs Jones drives 150 miles at 50 miles per hour. How long does it take her?

You probably knew the answers to Discovery 4.1 without realising that you were using these three equations.

$$\text{Speed} = \frac{\text{Distance}}{\text{Time}}$$

$$\text{Distance} = \text{Speed} \times \text{Time}$$

$$\text{Time} = \frac{\text{Distance}}{\text{Speed}}$$

You need to learn these.

One way to remember them is using this triangle.

The letters in the triangle go in the order that they occur in the word 'DiSTance'.

To find Speed, cover up the S. You can see $\frac{D}{T}$ so $S = \frac{D}{T}$.

To find Distance, cover up the D. You can see ST so $D = S \times T$.

To find Time, cover up the T. You can see $\frac{D}{S}$ so $T = \frac{D}{S}$.

Average speed

In Discovery 4.1, Mrs Jones travelled 150 miles at 50 mph. It is very unlikely that she travelled at exactly 50 mph all of the time, however.

It is much more likely that some of her journey was faster than 50 mph and some of it slower than this. 50 mph is the **average speed**.

$$\text{Average speed} = \frac{\text{Total distance}}{\text{Total time}}$$

When you use the three equations connecting speed, distance and time you will usually be using average speed.

Units

The unit of speed that has been used so far is miles per hour (mph). That is because the distances were measured in miles and the times were measured in hours.

If the distance is in kilometres (km) and the time in hours (h), the unit for speed is kilometres per hour or km/h.

If the distance is in metres (m) and the time in seconds (s), the unit for speed is metres per second or m/s.

Example 4.4

(a) A train travelled 210 miles in 3 hours.
   What was the average speed?
(b) Rajvee cycled at an average speed of 13 km/h for $2\frac{1}{2}$ hours.
   How far did she cycle?
(c) Jane ran 100 m at an average speed of 8.2 m/s.
   What was her time for the race?

Solution

(a) $\text{Speed} = \dfrac{\text{Distance}}{\text{Time}}$

$= \dfrac{210}{3} = 70 \text{ mph}$

(b) $\text{Distance} = \text{Speed} \times \text{Time}$

$= 13 \times 2\frac{1}{2} = 32.5 \text{ km}$

(c) $\text{Time} = \dfrac{\text{Distance}}{\text{Speed}}$

$= \dfrac{100}{8.2} = 12.2 \text{ seconds (correct to 1 decimal place)}$

Exercise 4.4

1 Alicia cycled for 4 hours at an average speed of 13.5 km/h.
   How far did she cycle?

2 Kieran walked 14 miles in 5 hours.
   What was his average speed? Give the units of your answer.

3 A train travelled 180 miles at an average speed of 60 mph.
   How long did the journey take?

4 Katie cycled 42 km in $3\frac{1}{2}$ hours.
   What was her average speed? Give the units of your answer.

Exercise continues ...

5 Harry swam at 98 metres per minute for 15 minutes.
   How far did he swim?

6 Mollie drove 270 km at an average speed of 60 km/h.
   How long did the journey take?

7 The men's world record for the 100 metres in 2009 was 9.58 seconds.
   What was the average speed for this race?
   Give your answer in metres per second correct to two decimal places.

8 In Europe the speed limits are in kilometres per hour.
   A car travels 13 km on this road in 15 minutes.

   (a) What fraction of an hour is 15 minutes?

   (b) Work out the average speed.
        Has the car broken the speed limit?

9 Jessica swam 180 metres in 2 minutes 30 seconds.

   (a) How many seconds are there in 2 minutes 30 seconds?

   (b) What was Jessica's average speed in metres per second?

10 Matt drove at an average speed of 65 km/h for $2\frac{1}{2}$ hours.
    How far did he drive?

● Challenge 4.2

(a) Tony jogs from his home to the park.
     It is a distance of 2.5 miles and it takes him 20 minutes.
     He then runs back home in 10 minutes.
     What was his average speed for the entire journey?

(b) Samantha walks from her home to the railway station.
     It is a distance of 1.5 miles and it takes her 24 minutes.
     She is just in time for her train.
     Her train journey is 38 miles long and it takes 43 minutes.
     Finally she walks the half mile to her office in 8 minutes.
     What was her average speed for the entire journey?

● Challenge 4.3

Pedro drove 415 miles in 7 hours 51 minutes.

What was his average speed?

● Challenge 4.4

The Moon is about a quarter of a million miles away from Earth.

(a) What could the correct distance be?
Is there a correct distance?

(b) If it were possible, about how long would it take an aeroplane to fly there?

(c) About how long would it take to drive there in a car?

(d) Apollo 11 took about three days to get to the moon.
Estimate its average speed in miles per hour.

You could put this information into a display.

Compound measures

In the previous section you met the formula for calculating speed.

Speed is a **compound measure**.

Some measures are calculated using the same type of measurements. Area, for example, is calculated using length and width, which are both measures of length.

Compound measures are calculated using two different types of measure. Speed is calculated using distance and time.

$$\text{Speed} = \frac{\text{Distance}}{\text{Time}}$$

Compound measures are written with **compound units**.

The units for speed are written in the form distance per time. For example, if the distance is in kilometres and the time is in hours, speed is written as kilometres per hour, or km/h.

Another compound measure is **density**. Density is linked to **mass** and **volume**.

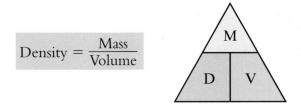

$$\text{Density} = \frac{\text{Mass}}{\text{Volume}}$$

Example 4.5

Gold has a density of 19.3 g/cm³.

Calculate the mass of a gold bar with a volume of 30 cm³.

Solution

$$\text{Density} = \frac{\text{Mass}}{\text{Volume}}$$ First rearrange the formula to make Mass the subject.

$$\begin{aligned}\text{Mass} &= \text{Density} \times \text{Volume} \\ &= 19.3 \times 30 \\ &= 579 \text{ g}\end{aligned}$$

Exercise 4.5

1 The density of aluminium is 2.7 g/cm³.
   What is the volume of a block of aluminium with a mass of 750 g?
   Give your answer to the nearest whole number.

2 Calculate the density of a rock of mass 780 g and volume 84 cm³.
   Give your answer to a suitable degree of accuracy.

3 Calculate the density of a stone of mass 350 g and volume 45 cm³.

4 (a) Calculate the density of a 3 cm³ block of copper of mass 26.7 g.
   (b) What would be the mass of a 17 cm³ block of copper?

5 Gold has a density of 19.3 g/cm³.
   Calculate the mass of a gold bar of volume 1000 cm³.
   Give your answer in kilograms.

6 Air at normal room temperature and pressure has a density of 1.3 kg/m³.
   (a) What mass of air is there in a room which is a cuboid measuring 3 m by 5 m by 3 m?
   (b) What volume of air would have a mass of
      (i)   1 kg?
      (ii) 1 tonne?

7 Calculate the density of a stone of mass 730 g and volume 69 cm³.
   Give your answer correct to 1 decimal place.

8 A town has a population of 74 000 and covers an area of 64 square kilometres.
   Calculate the population density (number of people per square kilometre) of the town.
   Give your answer to the nearest whole number.

> ## What you have learned
>
> - You can convert between measures of length, area and volume
> - $1\,\text{m} = 100\,\text{cm}$; $1\,\text{cm} = 10\,\text{mm}$
> - $1\,\text{m}^2 = 10\,000\,\text{cm}^2$; $1\,\text{cm}^2 = 100\,\text{mm}^2$
> - $1\,\text{m}^3 = 1\,000\,000\,\text{cm}^3$; $1\,\text{cm}^3 = 1000\,\text{mm}^3$
> - You can convert between measures of volume and capacity; $1000\,\text{cm}^3 = 1$ litre
> - All measurements are approximations
> - When calculating a measurement, you need to give your answer to a sensible degree of accuracy; as a general rule your final answer should not be given to a greater degree of accuracy than any of the values used in the calculation
> - $\text{Speed} = \dfrac{\text{Distance}}{\text{Time}}$ and is expressed in units such as m/s
> - Compound measures are calculated from two other measurements; examples include speed and density
> - $\text{Density} = \dfrac{\text{Mass}}{\text{Volume}}$ and is expressed in units such as g/cm^3

Mixed exercise 4

1 Change these units.
   (a) $12\,\text{m}^2$ to cm^2
   (b) $3.71\,\text{cm}^2$ to mm^2
   (c) $0.42\,\text{m}^2$ to cm^2
   (d) $0.05\,\text{cm}^2$ to mm^2

2 Change these units.
   (a) $3\,\text{m}^2$ to mm^2
   (b) $412\,500\,\text{cm}^2$ to m^2
   (c) $9400\,\text{mm}^2$ to cm^2
   (d) $0.06\,\text{m}^2$ to cm^2

3 Change these units.
   (a) 2.13 litres to cm^3
   (b) $5100\,\text{ml}$ to litres
   (c) 421 litres to cm^3
   (d) $91.7\,\text{cm}^3$ to litres

4 Give the lower and upper bounds of each of these measurements.
   (a) 27 cm to the nearest centimetre
   (b) 5.6 cm to the nearest millimetre
   (c) 1.23 m to the nearest centimetre

Mixed exercise 4 continues ...

5 (a) A machine produces pieces of wood.
The length of each piece measures 34 mm, correct to the nearest millimetre.
Between what limits does the actual length lie?

(b) Three of the pieces of wood are put together to make a triangle.
What is the greatest possible perimeter of the triangle?

6 The length of a field is 92.43 m and the width is 58.36 m.
Calculate the area of the field.
Give your answer to a sensible degree of accuracy.

7 A policeman timed a car travelling along a 100 m section of road.
The time taken was 6 seconds.
The length of the road was measured accurate to the nearest 10 cm, and the time was measured accurate to the nearest second.
What is the greatest speed the car could have been travelling at?

8 In a 10 km road race, one runner started at 11:48 and finished at 13:03.

(a) How long did it take this runner to complete the race?

(b) What was his average speed?

9 (a) Eleanor drives to Birmingham on a motorway.
She travels 150 miles in 2 hours 30 minutes.
What is her average speed?

(b) Eleanor drives to Cambridge at an average speed of 57 mph.
The journey takes 3 hours 20 minutes.
How many miles is the journey?

10 The area of the United Kingdom is 243 368 km².
In 1993 the population density was 225.5 people/km².
How many people were there in the UK in 1993?
Give your answer to a sensible degree of accuracy.

5 The area of triangles and parallelograms

▶ This chapter is about

- The area of a triangle
- The area of a parallelogram

▶ You should already know

- The common metric units for length and area
- How to find the area of a rectangle

The area of a triangle

In Chapter 3 you learned

Area of a rectangle = Length × Width

or

Area of a rectangle = $l \times w$.

Look at this diagram.

Area of rectangle ABCD = $l \times w$

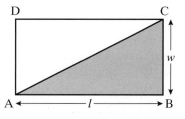

You can see that

Area of triangle ABC = $\frac{1}{2}$ × area of ABCD

$= \frac{1}{2} \times l \times w$.

Now look at a different triangle.

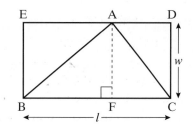

From the diagram you can see that

Area of triangle ABC = $\frac{1}{2}$ × area of BEAF + $\frac{1}{2}$ × area of FADC

$= \frac{1}{2}$ × area of BEDC

$= \frac{1}{2} \times l \times w$.

This shows that the area of any triangle can be found using the formula

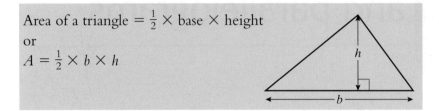

Area of a triangle $= \frac{1}{2} \times$ base $\times$ height

or

$A = \frac{1}{2} \times b \times h$

Note that the height of a triangle, h, is measured at right angles to the base. It is the **perpendicular height** or **altitude** of the triangle.

TIP

You can use any of the sides as the base provided that you use the perpendicular height that goes with it.

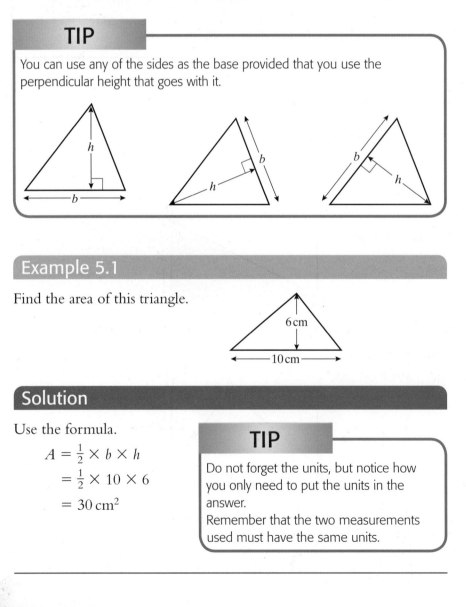

Example 5.1

Find the area of this triangle.

6 cm

10 cm

Solution

Use the formula.

$A = \frac{1}{2} \times b \times h$

$= \frac{1}{2} \times 10 \times 6$

$= 30 \text{ cm}^2$

TIP

Do not forget the units, but notice how you only need to put the units in the answer.

Remember that the two measurements used must have the same units.

Exercise 5.1

1 Find the area of each of these triangles.

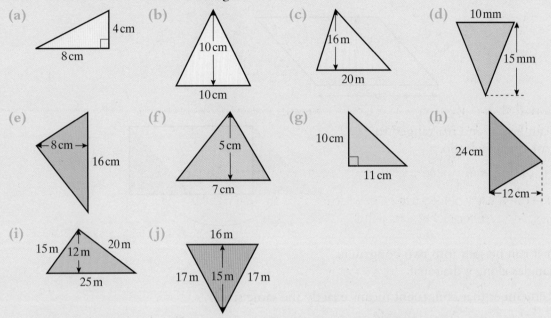

(a) 4 cm 8 cm

(b) 10 cm 10 cm

(c) 16 m 20 m

(d) 10 mm 15 mm

(e) 8 cm 16 cm

(f) 5 cm 7 cm

(g) 10 cm 11 cm

(h) 24 cm 12 cm

(i) 15 m 12 m 20 m 25 m

(j) 16 m 17 m 15 m 17 m

2 A triangle has an area of 12 cm² and a base of 4 cm.
   Find the perpendicular height h.

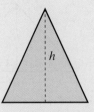

h

3 In triangle ABC, AB = 6 cm, BC = 8 cm and AC = 10 cm. Angle ABC = 90°.
   (a) Find the area of the triangle.
   (b) Find the perpendicular height BD.

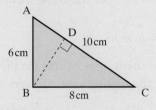

A
D 10 cm
6 cm
B 8 cm C

4 The vertices of a triangle are at A(2, 1), B(5, 1) and C(5, 7).
   Find the area of triangle ABC.

5 The vertices of a triangle are at P(−2, 2), Q(3, 2) and R(5, 6).
   Find the area of triangle PQR.

The area of a parallelogram

There are two ways to find the area of a parallelogram.

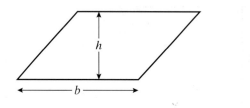

It can be cut and rearranged to
form a rectangle. So,

$$A = b \times h.$$

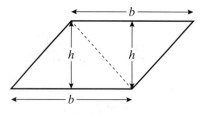

Or it can be split into two congruent
triangles along a diagonal.

(Remember that congruent means exactly the same size.)

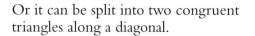

The area of each triangle is

$A = \frac{1}{2} \times b \times h$, so the total area of the parallelogram is

$$A = 2 \times \frac{1}{2} \times b \times h$$
$$= b \times h.$$

Note that the height of the parallelogram is the perpendicular height,
just as in the formula for the area of a triangle.

Area of a parallelogram $= b \times h$ or $A = b \times h$

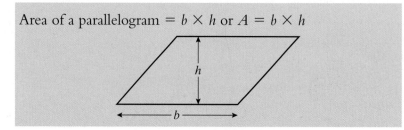

Example 5.2

Find the area of this parallelogram.

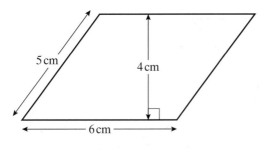

Solution

Use the formula. Make sure you choose the correct measurement for the height.

$$A = b \times h$$
$$= 6 \times 4$$
$$= 24 \, cm^2$$

Exercise 5.2

1 Find the area of each of these parallelograms.

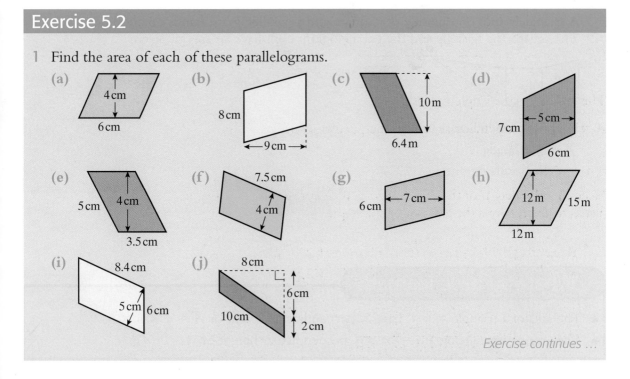

Exercise continues …

2 Find the values of *a*, *b* and *c*.
The lengths are in centimetres.

(a)

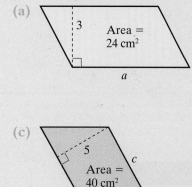

3

Area = 24 cm²

a

(b)

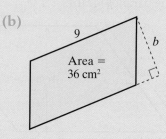

9

Area = 36 cm²

b

(c)

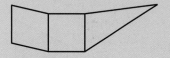

5

c

Area = 40 cm²

3 A triangle has a base of 12 cm and a perpendicular height of 7 cm.
A parallelogram has the same area and a base of 10 cm.
What is the perpendicular height of the parallelogram?

4 A parallelogram has an area of 48 cm² and base of 12 cm.
A triangle has the same perpendicular height as the parallelogram and a base of 10 cm.
What is the area of the triangle?

5 A triangle stands on one side of a square and a parallelogram stands on another.
The square, the triangle and the parallelogram each have an area of 36 cm².

Find the perpendicular height of
(a) the triangle.
(b) the parallelogram.

What you have learned

- The area of a triangle $= \frac{1}{2} \times$ base $\times$ perpendicular height or $A = \frac{1}{2} \times b \times h$
- The area of a parallelogram $=$ base $\times$ perpendicular height or $A = b \times h$

Mixed exercise 5

1 Find the area of each of these triangles.

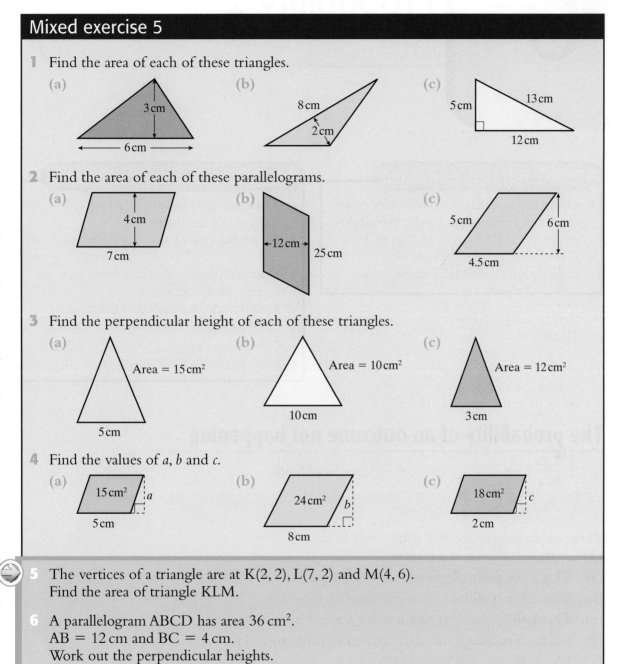

(a)

3 cm

6 cm

(b)

8 cm

2 cm

(c)

13 cm

5 cm

12 cm

2 Find the area of each of these parallelograms.

(a)

4 cm

7 cm

(b)

←12 cm→

25 cm

(c)

5 cm

6 cm

4.5 cm

3 Find the perpendicular height of each of these triangles.

(a)

Area = 15 cm²

5 cm

(b)

Area = 10 cm²

10 cm

(c)

Area = 12 cm²

3 cm

4 Find the values of a, b and c.

(a)

15 cm²

a

5 cm

(b)

24 cm²

b

8 cm

(c)

18 cm²

c

2 cm

5 The vertices of a triangle are at K(2, 2), L(7, 2) and M(4, 6).
Find the area of triangle KLM.

6 A parallelogram ABCD has area 36 cm².
AB = 12 cm and BC = 4 cm.
Work out the perpendicular heights.

Chapter 6

Probability 2

This chapter is about

- Using the fact that the probability of an event not happening and the probability of the event happening add up to 1
- Calculating expected frequency
- Calculating relative frequency

You should already know

- That probabilities are expressed as fractions or decimals
- All probabilities lie on a scale between 0 and 1
- How to find a probability from a set of equally likely outcomes
- How to add decimals and fractions
- How to subtract decimals and fractions from 1
- How to multiply decimals

The probability of an outcome not happening

In Chapter 2 you learned how to calculate probabilities.

Discovery 6.1

There are three pens and five pencils in a box.
One of these is chosen at random.

(a) What is the probability that it is a pen, P(pen)?
(b) What is the probability that it is a pencil, P(pencil)?
(c) What is the probability that it is not a pen, P(not a pen)?
(d) What can you say about your answers to parts (b) and (c)?
(e) What is P(pen) + P(not a pen)?

TIP

P( ) is often used when writing probabilities because it saves time and space.

The probability of an event not happening = 1 − the probability of the event happening.

If p is the probability of an event happening, then this can be written as

P(not happening) = 1 − p.

Example 6.1

(a) The probability that it will rain tomorrow is $\frac{1}{5}$.
What is the probability that it will not rain tomorrow?

(b) The probability that Phil scores a goal in the next match is 0.6.
What is the probability that Phil does not score a goal?

Solution

(a) P(not rain) = 1 − P(rain)

$= 1 - \frac{1}{5}$

$= \frac{4}{5}$

(b) P(not score) = 1 − P(score)

$= 1 - 0.6$

$= 0.4$

Exercise 6.1

1 The probability that Max will get to school late tomorrow is 0.1.
What is the probability that Max will not be late tomorrow?

2 The probability that Charlie has cheese sandwiches for his lunch is $\frac{1}{6}$.
What is the probability that Charlie does not have cheese sandwiches?

3 The probability that Ashley will pass her driving test is 0.85.
What is the probability that Ashley will fail her driving test?

4 The probability that Adam's Mum will cook dinner tonight is $\frac{7}{10}$.
What is the probability that she will not cook dinner?

5 The probability that City will win their next game is 0.43.
What is the probability that City will not win their next game?

6 The probability that Alec will watch TV one night is $\frac{32}{49}$.
What is the probability that he will not watch TV?

Probability involving a given number of different outcomes

Often there are more than two possible outcomes.

If you know the probability of all but one of the outcomes, you can work out the probability of the remaining outcome.

Example 6.2

A bag contains only red, white and blue counters.

The probability of picking a red counter is $\frac{1}{12}$.

The probability of picking a white counter is $\frac{7}{12}$.

What is the probability of picking a blue counter?

Solution

You know that P(not happening) = 1 − P(happening).
So P(not happening) + P (happening) = 1

$$P(\text{not blue}) + P(\text{blue}) = 1$$
$$P(\text{red}) + P(\text{white}) + P(\text{blue}) = 1$$
$$P(\text{blue}) = 1 - [P(\text{red}) + P(\text{white})]$$
$$= 1 - \left(\frac{1}{12} + \frac{7}{12} \right)$$
$$= 1 - \frac{8}{12}$$
$$= \frac{4}{12}$$
$$= \frac{1}{3}$$

There are only red, white and blue counters in the bag, so if a counter is not blue it must be red or white.

When there are a given number of possible outcomes, the sum of the probabilities is equal to 1.

For example, if there are four possible outcomes, A, B, C and D, then

$$P(A) + P(B) + P(C) + P(D) = 1$$

So, for example

$$P(B) = 1 - [P(A) + P(C) + P(D)]$$
or
$$P(B) = 1 - P(A) - P(C) - P(D)$$

Exercise 6.2

1 A shop has black, grey and blue dresses on a rail. Jen picks one at random.
The probability of picking a grey dress is 0.2 and the probability of picking a black dress is 0.1. What is the probability of picking a blue dress?

2 Heather always comes to school by car, bus or bike.
On any day, the probability that Heather will come by car is $\frac{3}{20}$ and the probability that she will come by bus is $\frac{11}{20}$.
What is the probability that Heather will come to school by bike?

3 The probability that the school hockey team will win their next match is 0.4.
The probability that they will lose is 0.25.
What is the probability that they will draw the match?

4 Pat has either boiled eggs, cereal or toast for breakfast.
The probability that she will have toast is $\frac{2}{11}$ and the probability that she will have cereal is $\frac{5}{11}$.
What is the probability that she will have boiled eggs?

5 The table shows the probability of getting some of the scores when a biased six-sided dice is thrown.

Score	1	2	3	4	5	6
Probability	0.27	0.16	0.14		0.22	0.1

What is the probability of getting 4?

6 When it is Jack's birthday, Aunty Chris gives him money or a voucher, or forgets altogether.
The probability that Aunty Chris will give Jack money for his birthday is $\frac{3}{4}$ and the probability that she will give him a voucher is $\frac{1}{5}$.
What is the probability that she forgets?

● Challenge 6.1

The weather forecast says the probability that it will be sunny tomorrow is 0.4.

Terry says this means that the probability that it will rain is 0.6.

Is Terry correct? Why?

Challenge 6.2

A cash bag contains only 5p, 10p and 50p coins.
The total amount of money in the bag is £5.

A coin is chosen from the bag at random.

$P(5p) = \frac{1}{2}$

$P(10p) = \frac{3}{8}$

(a) Work out P(50p).

(b) How many of each kind of coin is there in the bag?

Expected frequency

You can also use probability to predict how often an outcome will occur, or the **expected frequency** of the outcome.

Example 6.3

Each time Ronnie plays a game of snooker, the probability that he will win is $\frac{7}{10}$.

In a season, Ronnie plays 30 games.
How many of the games can he be expected to win?

Solution

The probability $P(win) = \frac{7}{10}$ tells us that Ronnie will win, on average, seven times in every ten games he plays. That is, he will win $\frac{7}{10}$ of the time.

In a season, he can be expected to win $\frac{7}{10}$ of 30 games.

$$\frac{7}{10} \times 30 = \frac{210}{10}$$
$$= 21$$

This is an example of an important result.

> Expected frequency = Probability × Number of trials

Example 6.4

The probability of a child catching measles is 0.2.

Out of the 400 children in a primary school, how many of them might you expect to catch measles?

Solution

The number of 'trials' means the the number of times the probability is tested. Here each of the 400 children has a 0.2 chance of catching measles. The number of trials is the same as the number of children: 400.

Expected frequency = Probability × Number of trials
$$= 0.2 × 400$$
$$= 80 \text{ children}$$

Exercise 6.3

1 The probability that Beverley is late to work is 0.1.
How many times would you expect her to be late in 40 working days?

2 The probability that it will be sunny on any day in April is $\frac{2}{5}$.
On how many of April's 30 days would you expect it to be sunny?

3 The probability that United will win their next match is 0.85.
How many of their next 20 games might you expect them to win?

4 When John is playing darts, the probability that he will hit the bull's eye is $\frac{3}{20}$.
John takes part in a sponsored event and throws 400 darts.
Each dart hitting the bull's eye earns £5 for charity.
How much might he expect to earn for charity?

5 An ordinary six-sided dice is thrown 300 times.
How many times might you expect to score

(a) 5?

(b) an even number?

6 A box contains two yellow balls, three blue balls and five green balls.
A ball is chosen at random and its colour noted.
The ball is then replaced. This is done 250 times.
How many of each colour might you expect to get?

Relative frequency

In Chapter 2 you learned to **estimate** probabilities using **experimental evidence**.

The experimental probability of an event $= \dfrac{\text{The number of times an event happens}}{\text{The total number of trials}}$.

The probability you estimate is known as **relative frequency**.

◉ Discovery 6.2

Copy this table and complete it by following the instructions below.

Number of trials		20	40	60	80	100
Number of heads						
Relative frequency = $\dfrac{\text{Number of heads}}{\text{Number of trials}}$						

- Toss a coin 20 times and record, using tally marks, the number of times it lands on heads.
- Now toss the coin another 20 times and enter the number of heads for all 40 tosses.
- Continue in groups of 20 and record the number of heads for 60, 80 and 100 tosses.
- Calculate the relative frequency of heads for 20, 40, 60, 80 and 100 tosses.
  Give your answers to 2 decimal places.

(a) What do you notice about the values of the relative frequencies?
(b) The probability of getting a head with one toss of a coin is $\frac{1}{2}$ or 0.5. Why is this?
(c) How does your final relative frequency value compare with this value of 0.5?

Relative frequency becomes more accurate the more trials you do.

When using experimental evidence to estimate probability it is advisable to perform at least 100 trials.

Exercise 6.4

1 Ping rolls a dice 500 times and records the number of times each score appears.

Score	1	2	3	4	5	6
Frequency	69	44	85	112	54	136

  (a) Work out the relative frequency of each of the scores.
      Give your answers to 2 decimal places.

  (b) What is the probability of obtaining each score on an ordinary six-sided dice?

  (c) Do you think that Ping's dice is biased? Give a reason for your answer.

2 Rashid notices that 7 out of the 20 cars in the school car park are red.

He says there is a probability of $\frac{7}{20}$ that the next car to come into the car park will be red. Explain what is wrong with this.

3 In a local election, 800 people were asked which party they would vote for.
The results are shown in the table.

Party	Labour	Conservative	Lib. Dem.	Other
Frequency	240	376	139	45

  (a) Work out the relative frequency for each party.
      Give your answers to 2 decimal places.

  (b) Estimate the probability that the next person to be asked will vote Labour.

4 Emma and Rebecca have a coin that they think is biased.
They decide to do an experiment to check.

  (a) Rebecca tosses the coin 20 times and gets a head 10 times.
      She says that the coin is not biased.
      Why do you think she has come to this conclusion?

  (b) Emma tosses the coin 300 times and gets a head 102 times.
      She says that the coin is biased.
      Why do you think she has come to this conclusion?

  (c) Who do you think is correct?
      Give a reason for your answer.

5 Joe made a spinner numbered 1, 2, 3 and 4.
He tested the spinner to see if it was fair.
He spun it 600 times.
The results are shown in the table on the next page.

Exercise continues …

Score	1	2	3	4
Frequency	160	136	158	146

(a) Work out the relative frequency of each of the scores.
Give your answers to 2 decimal places.

(b) Do you think that the spinner is fair?
Give a reason for your answer.

(c) If Joe were to test the spinner again and spin it 900 times, how many times would you expect each of the scores to appear?

6 Samantha carried out a survey into how students travel to school.
She asked 200 students. Here are her results.

Travel by ...	Bus	Car	Bike	Walk
Number of students	49	48	23	80

(a) Explain why it is reasonable for Samantha to use these results to estimate the probabilities of students travelling by the various methods.

(b) Estimate the probability that a randomly selected student will use each of the various methods of getting to school.

● Challenge 6.3

Work in pairs.

Put 10 counters, some red and the rest white, in a bag.

Challenge your partner to work out how many counters there are of each colour.

Hint: You need to devise an experiment with 100 trials.
       At the start of each trial, all 10 counters must be in the bag.

What you have learned

- If three events, A, B and C, cover all possible outcomes then, for example,
  $P(A) = 1 - P(B) - P(C)$
- Expected frequency = Probability × Total number of trials
- Relative frequency $= \dfrac{\text{Number of times an outcome occurs}}{\text{Total number of outcomes}}$
- Relative frequency is a good estimate of probability if there are sufficient trials

Mixed exercise 6

1 The probability that Peter can score 20 with one dart is $\frac{2}{9}$.
What is the probability of him not scoring 20 with one dart?

2 The probability that Carmen will go to the cinema during any week is 0.65.
What is the probability that she will not go to the cinema during one week?

3 Some of the probabilities of the length of time that any car will stay in a car park are shown below.

Time	Up to 30 minutes	30 minutes to 1 hour	1 hour to 2 hours	Over 2 hours
Probability	0.15	0.32	0.4	

What is the probability that a car will stay in the car park for over 2 hours?

4 There are 20 counters in a bag. They are all red, white or blue in colour.
A counter is chosen from the bag at random.
The probability that it is red is $\frac{1}{4}$. The probability that it is white is $\frac{2}{5}$.

(a) What is the probability that it is blue?

(b) How many counters of each colour are there?

5 The probability that Robert goes swimming on any day is 0.4.
There are 30 days in the month of June.
On how many days in June might you expect Robert to go swimming?

6 Holly thinks that a coin may be biased.
To test this, she tosses the coin 30 times. Heads turns up 15 times.
Holly says 'The coin is fair.'

(a) Why does Holly say this?

(b) Is she correct? Give a reason for your answer.

7 In an experiment with a biased dice, the following results were obtained after 400 throws.

Score	1	2	3	4	5	6
Frequency	39	72	57	111	25	96

(a) If the dice was fair, what would you expect the frequency of each score to be?

(b) Use the results to estimate the probability of throwing this dice and getting
  (i)   a 1.
  (ii) an even number.
  (iii) a number greater than 4.

Perimeter, area and volume 2

The circumference of a circle

You should already know that the circumference of a circle is the distance all the way round it; the diameter of a circle is a line all the way across the circle and passing through the centre and the radius is a line from the centre of the circle to the circumference.

◉ Discovery 7.1

Find a number of circular or cylindrical items.
Measure the circumference and diameter of each item and complete a table like this.

Item name	Circumference	Diameter	Circumference ÷ Diameter

What do you notice?

For any circle, $\dfrac{\text{circumference}}{\text{diameter}} \approx 3$.

If it were possible to take very accurate measurements, you would find that $\dfrac{\text{circumference}}{\text{diameter}} = 3.141\,592\ldots$

This number is called **pi** and is represented by the symbol π.

This means that you can write a formula for the circumference of any circle.

> Circumference $= \pi \times$ diameter or $C = \pi d$.

π is a decimal number which does not terminate and does not recur. Scientific calculators store a value for π which you can access using the keys. You can also use an approximation: 3.142 is suitable. In both of these cases you will usually have to round your answer to a suitable degree of accuracy. You can also leave the symbol π in your answer; in this case your answer is exact.

> **TIP**
>
> $\approx$ means 'is roughly equal to'.

Example 7.1

Find the circumference of a circle with a diameter of 45 cm.

Solution

Circumference $= \pi \times$ diameter
$\qquad\qquad\quad = 3.142 \times 45$
$\qquad\qquad\quad = 141.39 = 141.4\,\text{cm}$ (1 d.p.)

You could do this calculation on your calculator, using the $\boxed{\pi}$ key.

Input $\boxed{\pi}\,\boxed{\times}\,\boxed{4}\,\boxed{5}\,\boxed{=}$.

The answer on your display will be 141.371 67.

> **TIP**
>
> 3.142 is an approximation for π so your answer is not exact and should be rounded. You will often be told to what accuracy to give your answer. Here the answer is given correct to 1 decimal place.

If you are working without a calculator or perhaps if you will be using your answer in further calculations, you can leave π in your answer. The circumference of the circle in Example 7.1 is 45π cm.

Exercise 7.1

1 Find the circumferences of the circles with these diameters.

   (a) 12 cm   (b) 25 cm   (c) 90 cm   (d) 37 mm

   (e) 66 mm   (f) 27 cm   (g) 52 cm   (h) 4.7 cm

   (i) 9.2 cm   (j) 7.3 m   (k) 2.9 m   (l) 1.23 m

Exercise continues …

2 Find the circumferences of the circles with these radii.

   (a) 8 cm   (b) 30 cm   (c) 65 cm

   (d) 59 mm   (e) 0.7 m   (f) 1.35 m

3 Find the circumferences of the circles with these radii. Leave your answers in terms of π.

   (a) 4 m   (b) 13 cm   (c) 3.5 mm

4 Mr Jones has a circular pond of radius 3 m.
He wants to surround the pond with a gravel
path 1 metre wide.
Edging strip is needed on each side of the path.
What length of edging strip is needed altogether?

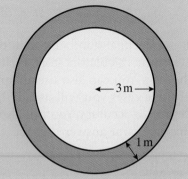

5 Jane is riding her bike.
The diameter of the wheels is 70 cm.
How far has she cycled when the wheels have rotated 100 full turns?
Give your answer in metres.

The area of a circle

The area of a circle is the space it covers.

● Discovery 7.2

Take a disc of paper and cut it into
12 narrow sectors, all the same size.
Arrange them, reversing every other
piece, like this.
This is nearly a rectangle. If you had
cut the disc into 100 sectors it would
be more accurate.

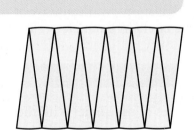

(a) What are the dimensions of the rectangle?

(b) What is its area?

The height of the rectangle in Discovery 7.2 is the radius of the circle, r.

The width is half the circumference of the circle, $\frac{1}{2}\pi d$ or πr.

This gives a formula to calculate the area of a circle.

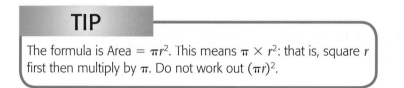

Area $= \pi r^2$ where r is the radius of the circle.

> **TIP**
>
> The formula is Area $= \pi r^2$. This means $\pi \times r^2$: that is, square r first then multiply by π. Do not work out $(\pi r)^2$.

Example 7.2

Find the area of a circle with a radius of 23 cm.

Solution

Area $= \pi r^2$
$= 3.142 \times 23^2$
$= 1662.118$
$= 1662 \text{ cm}^2$ (to the nearest whole number)

You could do this calculation on your calculator, using the $\boxed{\pi}$ key.

Input $\boxed{\pi}\,\boxed{\times}\,\boxed{2}\,\boxed{3}\,\boxed{x^2}\,\boxed{=}$.

The answer on your display will be 1661.9025.

As before, you can also give your answer in terms of π.

Example 7.3

Find the area of a circle of radius 5 cm, leaving π in your answer.

Solution

Area $= \pi r^2$
$= \pi \times 5^2$
$= \pi \times 25$
$= 25\pi \text{ cm}^2$

Example 7.4

A circular pond of radius 3 m is surrounded
by a path 2 m wide.
Find the area of the path.
Give your answer as a multiple of π.

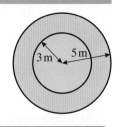

Solution

Area of path = area of large circle − area of small circle
$$= \pi \times 5^2 - \pi \times 3^2$$
$$= 25\pi - 9\pi$$
$$= 16\pi \text{ m}^2$$

You learned about **collecting
like terms** in Unit A Chapter 15.
You can treat π in the same way.

Exercise 7.2

1 Find the areas of the circles with these radii.

(a) 14 cm	(b) 28 cm	(c) 80 cm	(d) 35 mm
(e) 62 mm	(f) 43 cm	(g) 55 cm	(h) 4.9 cm
(i) 9.7 cm	(j) 3.4 m	(k) 2.6 m	(l) 1.25 m

2 Find the areas of the circles with these diameters.

(a) 16 cm	(b) 24 cm	(c) 70 cm	(d) 36 mm
(e) 82 mm	(f) 48 cm	(g) 54 cm	(h) 4.4 cm
(i) 9.8 cm	(j) 3.8 m	(k) 2.8 m	(l) 2.34 m

3 Find the areas of the circles with these diameters. Leave your answers in terms of π.

(a) 8 cm   (b) 6 m   (c) 9 mm   (d) 15 cm

● Challenge 7.1

Mr Jones has a circular pond of radius 3 m.
He wants to surround the pond with a gravel path 1 metre wide.
Mr Jones uses this formula to calculate how much gravel he needs
for his path.

Tonnes of gravel needed = area of path in square metres $\times$ 0.034

Calculate how much gravel he needs.

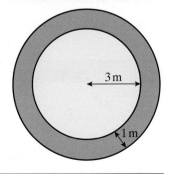

The area of complex shapes

In Chapter 3 you learned that the formula for the area of a rectangle is

$$\text{Area} = \text{length} \times \text{width} \quad \text{or} \quad A = l \times w.$$

In Chapter 5 you learned that the formula for the area of a triangle is

$$\text{Area} = \tfrac{1}{2} \times \text{base} \times \text{height} \quad \text{or} \quad A = \tfrac{1}{2} \times b \times h.$$

You can use these formulae to find the area of more complex shapes, which can be broken down into rectangles and right-angled triangles.

Example 7.5

Find the area of this shape.

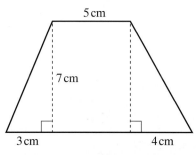

Solution

Work out the area of the rectangle and each of the triangles separately and then add them together to find the area of the whole shape.

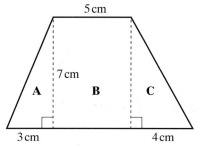

Area of shape = area of triangle A + area of rectangle B + area of triangle C

$$= \frac{3 \times 7}{2} \qquad + \qquad 5 \times 7 \qquad + \qquad \frac{4 \times 7}{2}$$

$$= 10.5 \qquad\quad + \qquad\quad 35 \qquad\quad + \qquad\quad 14$$

$$= 59.5 \ \text{cm}^2$$

Exercise 7.3

Find the area of each of these shapes.
Break them down into rectangles and
right-angled triangles first.

> **TIP**
>
> In cases like the shape in question 2 it is easier
> to find the area of the rectangle and subtract
> the area of the triangle taken out of it.

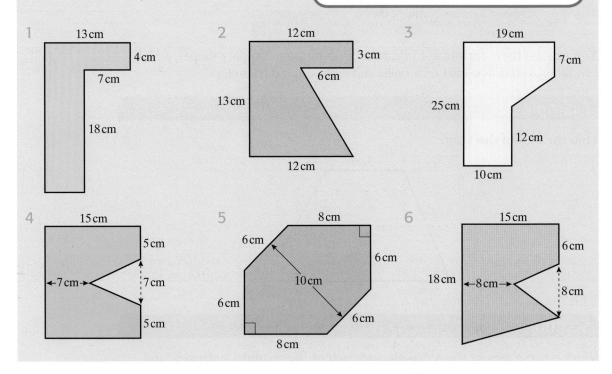

Challenge 7.2

A door is in the shape of a rectangle with
a semicircle on top.

What is its area?

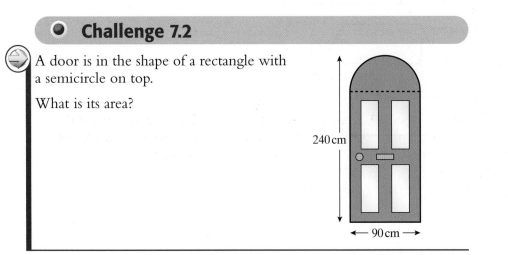

The volume of complex shapes

You learned in Chapter 3 that the formula for the volume of a cuboid is

Volume = length × width × height   or $V = l \times w \times h$.

It is possible to find the volume of shapes made from cuboids by breaking them down into smaller parts.

Example 7.6

Find the volume of this shape.

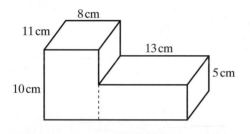

Solution

This shape can be broken down into two cuboids, A and B.

Work out the volumes of these two cuboids and add them together to find the volume of the whole shape.

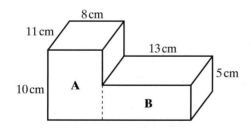

Volume of shape = volume of cuboid A + volume of cuboid B   The width of cuboid B is
$=$ 8 × 11 × 10   +   13 × 11 × 5   the same as the width of
$=$ 880   +   715   cuboid A.
$=$ 1595 cm³

Exercise 7.4

Find the volume of each of these shapes.

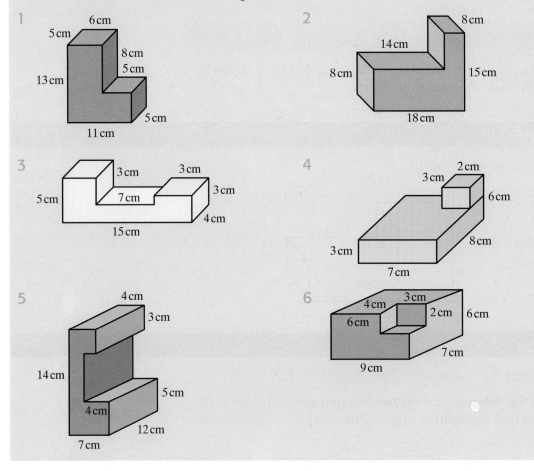

1.
6 cm
5 cm
8 cm
5 cm
13 cm
11 cm
5 cm

2.
8 cm
14 cm
8 cm
15 cm
18 cm

3.
3 cm
3 cm
3 cm
5 cm
7 cm
4 cm
15 cm

4.
2 cm
3 cm
6 cm
8 cm
3 cm
7 cm

5.
4 cm
3 cm
14 cm
5 cm
4 cm
12 cm
7 cm

6.
4 cm
3 cm
2 cm
6 cm
6 cm
7 cm
9 cm

The volume of a prism

A **prism** is a three-dimensional object that is the same 'shape'
throughout. The correct definition is that the object has a **uniform
cross-section**.

In this diagram the shaded area is the cross-section.

Looking at the shape from point F you see the cross-section
as an L-shape. If you were to cut through the shape along
the dotted line you would still see the same cross-section.

You could cut the shape into slices, each 1 cm thick.
The volume of each slice, in centimetres cubed, would be the
area of the cross-section × 1.

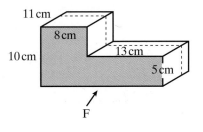

11 cm
8 cm
10 cm
13 cm
5 cm
F

As the shape is 11 cm thick, you would have 11 identical slices.

So the volume of the whole shape would be the area of the cross-section × 11.

This tells you that the formula for the volume of a prism is

> Volume = area of cross-section × length.

The area of the cross-section (shaded) = (10 × 8) + (13 × 5)
$$= 80 + 65$$
$$= 145 \text{ cm}^2$$

Volume = 145 × 11
$$= 1595 \text{ cm}^3$$

This is the same answer as in Example 7.6, when the volume of this shape was found by breaking it down into cuboids.

The formula works for any prism.

Example 7.7

This prism has a cross-section of area 374 cm², and is 26 cm long. Find its volume.

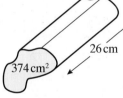

Solution

Volume = area of cross-section × length
$$= 374 \times 26$$
$$= 9724 \text{ cm}^3$$

Exercise 7.5

Find the volume of each of these shapes.

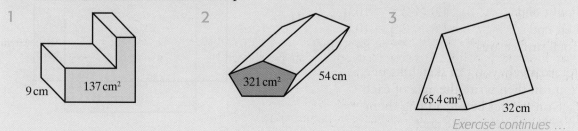

1 9 cm 137 cm²

2 321 cm² 54 cm

3 65.4 cm² 32 cm

Exercise continues …

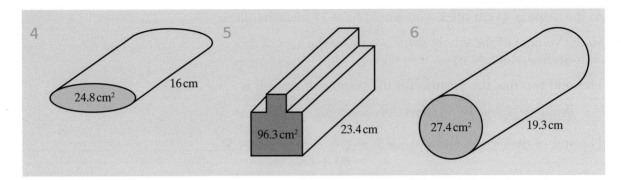

The surface area of three-dimensional shapes

The surface area of a solid is the total area of all its faces. To find
the surface area of a shape, you find the area of each of its faces
and add them together. The faces can be rectangles, triangles or
other shapes. You might find it helpful to visualize or draw the
net of the solid.

Example 7.8

Here is a cardboard box with no lid.

Calculate its surface area.

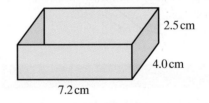

Solution

To find the surface area, you can make a list of the size of the faces and then add up their areas.

Base	$7.2 \times 4.0 = 28.8$	
Front	$7.2 \times 2.5 = 18.0$	
Back	$7.2 \times 2.5 = 18.0$	
Right end	$4.0 \times 2.5 = 10.0$	
Left end	$4.0 \times 2.5 = 10.0$	
Total surface area	$= 84.8 \text{ cm}^2$	

Alternatively, you can sketch its net.
You can then write the area of each
face on the net before adding them
to get the total surface area.

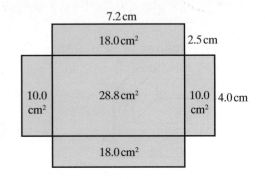

Exercise 7.6

1 Calculate the surface area of each of these cuboids.

(a)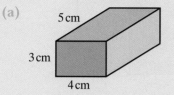
5 cm
3 cm
4 cm

(b)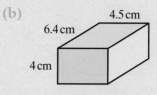
4.5 cm
6.4 cm
4 cm

2 The diagram shows the net of a shoe box.
Calculate its surface area.

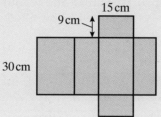
15 cm
9 cm
30 cm

3 A freezer is in the shape of a cuboid.
It is 0.6 m wide, 0.6 m deep and 1.4 m high.
Calculate its surface area.

4 This is a sketch of a net for a triangular prism.
Calculate its surface area.

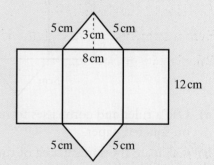
5 cm 3 cm 5 cm
8 cm
12 cm
5 cm 5 cm

5 This is a sketch of the net for a square-based pyramid.
Calculate the surface area of the pyramid.

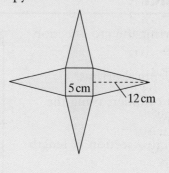
5 cm
12 cm

Exercise continues ...

6 These two cuboids have the same volume.

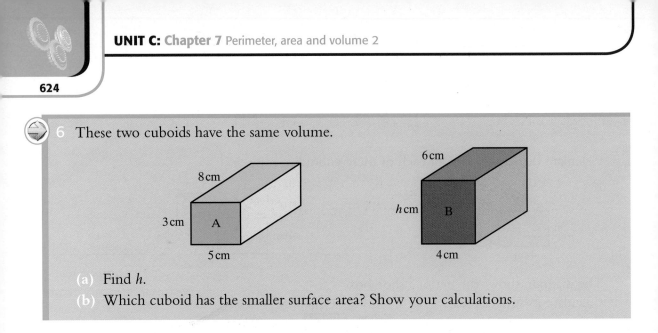

(a) Find h.

(b) Which cuboid has the smaller surface area? Show your calculations.

Challenge 7.3

This is a sketch of the net for a triangular prism.

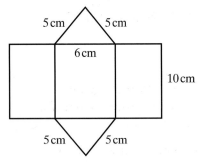

(a) Use a ruler and compasses to make an accurate drawing of the net on squared paper.

(b) Calculate the surface area of the prism.

The volume of a cylinder

A **cylinder** is a special kind of prism: the cross-section is always a circle.

Cylinder **A** and cylinder **B** are identical prisms.

You can find the volume of both cylinders using the formula for the volume of a prism.

Volume of cylinder **A** = area of cross-section × length
$$= 77 \times 18$$
$$= 1386 \text{ cm}^3$$

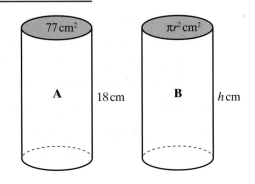

Volume of cylinder **B** = area of cross-section (area of circle)
$$\times \text{ length (height)}$$
$$= \pi r^2 \times h \, \text{cm}^3$$

This gives you the formula for the volume of any cylinder:

Volume = $\pi r^2 h$ where r is the radius of the circle and h is the height of the cylinder.

Example 7.9

Find the volume of a cylinder with radius 13 cm and height 50 cm.

Solution

Volume = $\pi r^2 h$
$$= 3.142 \times 13^2 \times 50$$
$$= 26\,549.9$$
$$= 26\,550 \, \text{cm}^3 \text{ (to the nearest whole number)}$$

You could do this calculation on your calculator, using the $\boxed{\pi}$ key.

Input $\boxed{\pi}$ $\boxed{\times}$ $\boxed{1}$ $\boxed{3}$ $\boxed{x^2}$ $\boxed{\times}$ $\boxed{5}$ $\boxed{0}$ $\boxed{=}$.

The answer on your display will be 26 546.458.

Rounded to the nearest whole number, this is 26 546 cm³. This is different from the answer you get using 3.142 as an approximation for π because your calculator uses a more accurate value for π.

You could also calculate the exact answer by giving your answer in terms of π.

Volume = $\pi \times 13^2 \times 50$
$$= 8450\pi \, \text{cm}^3$$

Exercise 7.7

1 Find the volume of the cylinders with these dimensions.
   (a) Radius 8 cm and height 35 cm
   (b) Radius 14 cm and height 42 cm
   (c) Radius 20 cm and height 90 cm
   (d) Radius 12 mm and height 55 mm
   (e) Radius 25 mm and height 6 mm
   (f) Radius 0.7 mm and height 75 mm
   (g) Radius 3 m and height 25 m
   (h) Radius 5.8 m and height 3.5 m

2 A cylindrical glass has a radius of 4 cm and a height of 9 cm.
   Find the volume of the glass in terms of π.

Challenge 7.4

This block of wood is a cuboid.

A hole with a diameter of 20 mm has been drilled through it.

Calculate the remaining volume.

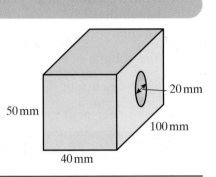

The surface area of a cylinder

You can probably think of lots of examples of cylinders. Some of them, like the inner tube of a roll of kitchen paper, have no ends: these are called **open cylinders**. Others, like a can of beans, do have ends: these are called **closed cylinders**.

Curved surface area

If you took an open cylinder, cut straight down its length and opened it out, you would get a rectangle. The **curved surface area** of the cylinder has become a flat shape.

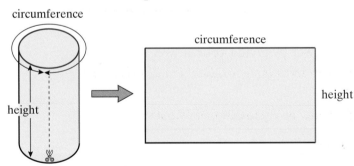

The area of the rectangle is circumference × height.

You know that the formula for the circumference of a circle is

Circumference = π × diameter   or $C = \pi d$.

So the formula for the area of the curved surface of any cylinder is

Curved surface area = π × diameter × height   or πdh.

This formula is usually written in terms of the radius. You know that the radius is half the length of the diameter, or $d = 2r$.

So you can also write the formula for the area of the curved surface area of any cylinder as

Curved surface area = 2 × π × radius × height   or $2\pi rh$.

Example 7.10

Find the curved surface area of a cylinder with radius 4 cm and height 0.7 cm.

Solution

Curved surface area = $2\pi rh$
$$= 2 \times 3.142 \times 4 \times 0.7$$
$$= 17.5952$$
$$= 17.6 \, cm^2 \text{ (correct to 1 decimal place)}$$

You could do this calculation on your calculator, using the $\boxed{\pi}$ key.

Input $\boxed{2}$ $\boxed{\times}$ $\boxed{\pi}$ $\boxed{\times}$ $\boxed{4}$ $\boxed{\times}$ $\boxed{0}$ $\boxed{.}$ $\boxed{7}$ $\boxed{=}$.

The answer on your display will be 17.592 919.

You can also calculate an exact answer by giving your answer in terms of π.

Curved surface area = $2\pi rh$
$$= 2 \times \pi \times 4 \times 0.7$$
$$= 5.6\pi \, cm^2$$

Total surface area

The total surface area of a closed cylinder is made of the curved surface area and the area of the two circular ends.

So the formula for the total surface area of a (closed) cylinder is

Total surface area = $2\pi rh + 2\pi r^2$.

Example 7.11

Find the total surface area of a cylinder with radius 13 cm and height 1.5 cm.

Solution

Total surface area = $2\pi rh + 2\pi r^2$
$$= (2 \times 3.142 \times 13 \times 1.5) + (2 \times 3.142 \times 13^2)$$
$$= 122.538 + 1061.996$$
$$= 1184.534$$
$$= 1185 \, cm^2 \text{ (to the nearest whole number)}$$

You could do this calculation on your calculator, using the $\boxed{\pi}$ key.

Input $\boxed{(}\,\boxed{2}\,\boxed{\times}\,\boxed{\pi}\,\boxed{\times}\,\boxed{1}\,\boxed{3}\,\boxed{\times}\,\boxed{1}\,\boxed{.}\,\boxed{5}\,\boxed{)}\,\boxed{+}$
$\boxed{(}\,\boxed{2}\,\boxed{\times}\,\boxed{\pi}\,\boxed{\times}\,\boxed{1}\,\boxed{3}\,\boxed{x^2}\,\boxed{)}\,\boxed{=}$.

The answer on your display will be 1184.3804.

Rounded to the nearest whole number, this is 1184 cm².
This is different from the answer you get using 3.142 as an
approximation for π because your calculator uses a more accurate
value for π.

Again, you can also calculate an exact answer by giving your answer in
terms of π.

$$\begin{aligned}
\text{Total surface area} &= 2\pi rh + 2\pi r^2 \\
&= 2 \times \pi \times 13 \times 1.5 + 2 \times \pi \times 13^2 \\
&= 39\pi + 338\pi \\
&= 377\pi \text{ cm}^2
\end{aligned}$$

Exercise 7.8

1 Find the curved surface areas of the cylinders with these dimensions.
   (a) Radius 12 cm and height 24 cm   (b) Radius 11 cm and height 33 cm
   (c) Radius 30 cm and height 15 cm   (d) Radius 18 mm and height 35 mm
   (e) Radius 15 mm and height 4 mm   (f) Radius 1.3 mm and height 57 mm
   (g) Radius 2.1 m and height 10 m   (h) Radius 3.5 m and height 3.5 m

2 Find the total surface areas of the cylinders with these dimensions.
   (a) Radius 14 cm and height 10 cm   (b) Radius 21 cm and height 32 cm
   (c) Radius 35 cm and height 12 cm   (d) Radius 18 mm and height 9 mm
   (e) Radius 25 mm and height 6 mm   (f) Radius 3.5 mm and height 50 mm
   (g) Radius 1.8 m and height 15 m   (h) Radius 2.5 m and height 1.3 m

3 (a) Find the curved surface area of a cylinder with a radius of 7 cm and a height of
        10 cm.
        Give your answer in terms of π.
   (b) Find the total surface area of a cylinder with a radius of 5 cm and a height of 3 cm.
        Give your answer in terms of π.

Exercise continues …

4 The roller on a machine has diameter 60 cm and width 1 m.
   It is used to roll a rectangle of tarmac measuring 70 m by 3 m.
   How many revolutions will the roller make to cover the tarmac once?

What you have learned

- The circumference of a circle $= \pi d$ or $2\pi r$
- The area of a circle $= \pi r^2$
- You can leave your answer to a calculation involving π in terms of π; the answer is then exact
- The area of a complex shape can be found by breaking the shape down into simpler shapes such as rectangles and right-angled triangles
- The volume of a complex shape can be found by breaking the shape down into simpler shapes such as cuboids
- A prism is a three-dimensional object that has a uniform cross-section
- The volume of a prism = area of cross-section × length
- The surface area of a three-dimensional shape is the total area of all its faces; you can look at its net to help you
- A cylinder is a special kind of prism with a circular cross-section
- The volume of a cylinder $= \pi r^2 h$
- The curved surface area of a cylinder $= 2\pi r h$ or $\pi d h$
- The total surface area of a (closed) cylinder $= 2\pi r h + 2\pi r^2$

Mixed exercise 7

1 Find the circumferences of the circles with these diameters.
   (a) 14.2 cm   (b) 29.7 cm   (c) 65 cm   (d) 32.1 mm

2 Find the areas of the circles with these dimensions.
   (a) Radius 6.36 cm   (b) Radius 2.79 m
   (c) Diameter 9.4 mm   (d) Diameter 12.6 cm

3 Andrew's house boat has a circular glass window.
   Its diameter is 40 cm.
   Find the area of the glass in terms of π.

Mixed exercise 7 continues ...

4 Work out the area of each of these shapes.

(a)

8.3 cm

8.7 cm

3.9 cm

15.4 cm

(b)

3.5 cm   2.5 cm

4 cm

8 cm

10.5 cm

(c)

6 cm

15 cm

9 cm

7 cm

11 cm

(d)

17.9 cm

6 cm

11.7 cm

6.5 cm

6 cm   12.7 cm

5 Find the volume of each of these shapes.

(a)

2 cm

7 cm

12 cm

4 cm

6 cm

(b)

0.5 cm   1.5 cm

2.5 cm

3 cm

5 cm

4 cm

6 Find the volume of each of these prisms.

(a)

37.4 cm

543 cm²

(b)

93.4 mm²

16.9 mm

(c)

74.9 cm²

75 cm

(d)

0.58 m²

1.53 m

Mixed exercise 7 continues …

7 Calculate the surface area of each of these cuboids.

(a)

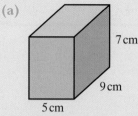

7 cm
9 cm
5 cm

(b)

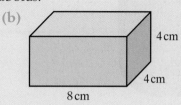

4 cm
4 cm
8 cm

8 Alex has persuaded his mum to let him paint his room black.
It is a cuboid measuring 4.2 m by 3.2 m and is 2.6 m high.

(a) Calculate the total area of the four walls and the ceiling.

(b) 1 litre of paint covers 13 m^2.
The total area of the windows and doors is 5 m^2.
How much paint does he need?

9 Find the volumes of the cylinders with these dimensions.

(a) Radius 6 mm and height 23 mm   (b) Radius 17 mm and height 3.6 mm

(c) Radius 22 cm and height 70 cm   (d) Radius 12 cm and height 0.4 cm

(e) Radius 35 m and height 6 m   (f) Radius 1.8 m and height 2.7 m

10 Find the curved surface areas of the cylinders with these dimensions.

(a) Radius 9.6 m and height 27.5 m   (b) Radius 23.6 cm and height 16.4 cm

(c) Radius 1.7 cm and height 1.5 cm   (d) Radius 16.7 mm and height 6.4 mm

11 Find the total surface areas of the cylinders with these dimensions.

(a) Radius 23 mm and height 13 mm   (b) Radius 3.6 m and height 1.4 m

(c) Radius 2.65 cm and height 7.8 cm   (d) Radius 4.7 cm and height 13.8 cm

12 A cylinder has a radius of 3 cm and a height of 5 cm.
By leaving your answer in terms of π, find the exact value of

(a) the volume.

(b) the curved surface area.

(c) the total surface area.

Chapter 8

Using a calculator

▶ This chapter is about

- The order in which your calculator does calculations
- Fractions on your calculator
- Rounding to a given number of significant figures

▶ You should already know

- How to round to the nearest whole number, 10, 100, …
- How to round to a given number of decimal places
- How to round to 1 significant figure
- How to work with fractions without a calculator
- How to find the circumference and area of a circle

Order of operations

Your calculator always follows the correct order of operations. This means that it does brackets first, then powers (such as squares), then multiplication and division and lastly addition and subtraction.

If you want to change the normal order of doing things you need to give your calculator different instructions.

Sometimes the simplest way of doing this is to press the ⊜ key in the middle of a calculation. This is shown in the following example.

Example 8.1

Work out $\dfrac{5.9 + 3.4}{3.1}$.

Solution

You need to work out the addition first.

Press ⑤ ⦁ ⑨ ⊕ ③ ⦁ ④ ⊜. You should see 9.3.

Now press ⊘ ③ ⦁ ① ⊜. The answer is 3.

Using brackets

Sometimes you need other ways of changing the order of operations.

For example, in the calculation $\dfrac{5.52 + 3.45}{2.3 + 1.6}$, you need to add

5.52 + 3.45, then add 2.3 + 1.6 before doing the division.

One way to do this is to write down the answers to the two addition sums and then do the division

$$\frac{5.52 + 3.45}{2.3 + 1.6} = \frac{8.97}{3.9} = 2.3$$

A more efficient way to do it is to use brackets.

You do the calculation as $(5.52 + 3.45) \div (2.3 + 1.6)$.

This is the sequence of keys to press.

(5 . 5 2 + 3 . 4 5) ÷ (2 . 3 + 1 . 6) =

Check up 8.1

Enter the sequence above in your calculator and check that you get 2.3.

Example 8.2

Use your calculator to work out these calculations without writing down the answers to middle stages.

(a) $\sqrt{5.2 + 2.7}$

(b) $\dfrac{5.2}{3.7 \times 2.8}$

Solution

(a) You need to work out 5.2 + 2.7 before finding the square root.
You use a bracket so that the addition is done first.
$\sqrt{(5.2 + 2.7)} = 2.811$ correct to 3 decimal places.

(b) You need to work out 3.7 × 2.8 before doing the division.
You use a bracket so that the multiplication is done first.
$5.2 \div (3.7 \times 2.8) = 0.502$ correct to 3 decimal places.

Exercise 8.1

Work these out on your calculator without writing down the answers to middle stages. If the answers are not exact, give them correct to 2 decimal places.

1 $\dfrac{5.2 + 10.3}{3.1}$

2 $\dfrac{127 - 31}{25}$

3 $\dfrac{9.3 + 12.3}{8.2 - 3.4}$

4 $\sqrt{15.7 - 3.8}$

5 $6.2 + \dfrac{7.2}{2.4}$

6 $(6.2 + 1.7)^2$

7 $\dfrac{5.3}{2.6 \times 1.7}$

8 $\dfrac{2.6^2}{1.7 + 0.82}$

9 $2.8 \times (5.2 - 3.6)$

10 $\dfrac{6.2 \times 3.8}{22.7 - 13.8}$

11 $\dfrac{5.3}{\sqrt{6.2 + 2.7}}$

12 $\dfrac{5 + \sqrt{25 + 12}}{6}$

Fractions on your calculator

You learned how to work with fractions without a calculator in Unit B. Not all calculators are the same and how they display fractions varies a lot. They might not look exactly like they do here. You need to know how to use *your* calculator.

When a calculator is allowed you can use the fraction key.

The fraction key looks like this $\boxed{a^b\!/_c}$.

To enter a fraction such as $\frac{2}{5}$ into your calculator you press

$\boxed{2}\ \boxed{a^b\!/_c}\ \boxed{5}\ \boxed{=}$.

Your display will look like this. $\boxed{2\,\lrcorner\,5}$

This is the calculator's way of showing the fraction $\frac{2}{5}$.

● Discovery 8.1

Some calculators may have the ⌐ symbol a different way round.

Check now what you see when you press $\boxed{2}\ \boxed{a^b\!/_c}\ \boxed{5}\ \boxed{=}$.

To do a calculation like $\frac{2}{5} + \frac{1}{2}$, the sequence of keys is

$\boxed{2}\ \boxed{a^b\!/_c}\ \boxed{5}\ \boxed{+}\ \boxed{1}\ \boxed{a^b\!/_c}\ \boxed{2}\ \boxed{=}$.

This is what you should see on your display. $\boxed{9\,\lrcorner\,10}$

You must, of course, write this down as $\frac{9}{10}$ for your answer.

Example 8.3

Use your calculator to work out $\frac{3}{4} + \frac{5}{6}$.

Solution

This is the sequence of keys to press.

$\boxed{3}\;\boxed{a^b\!/_c}\;\boxed{4}\;\boxed{+}\;\boxed{5}\;\boxed{a^b\!/_c}\;\boxed{6}\;\boxed{=}$.

The display on your calculator should look like this. $\boxed{1\lrcorner 7\lrcorner 12}$

This is the calculator's way of showing the mixed number $1\frac{7}{12}$.

So the answer is $1\frac{7}{12}$.

To enter a mixed number such as $2\frac{3}{5}$ into your calculator you press

$\boxed{2}\;\boxed{a^b\!/_c}\;\boxed{3}\;\boxed{a^b\!/_c}\;\boxed{5}\;\boxed{=}$.

Your display will look like this. $\boxed{2\lrcorner 3\lrcorner 5}$

Example 8.4

Use your calculator to work out these.

(a) $2\frac{3}{5} - 1\frac{1}{4}$ **(b)** $2\frac{2}{3} \times 3\frac{3}{4}$

Solution

(a) This is the sequence of buttons to press.

$\boxed{2}\;\boxed{a^b\!/_c}\;\boxed{3}\;\boxed{a^b\!/_c}\;\boxed{5}\;\boxed{-}\;\boxed{1}\;\boxed{a^b\!/_c}\;\boxed{1}\;\boxed{a^b\!/_c}\;\boxed{4}\;\boxed{=}$

The display on your calculator should look like this. $\boxed{1\lrcorner 7\lrcorner 20}$

So the answer is $1\frac{7}{20}$.

(b) This is the sequence of buttons to press.

$\boxed{2}\;\boxed{a^b\!/_c}\;\boxed{2}\;\boxed{a^b\!/_c}\;\boxed{3}\;\boxed{\times}\;\boxed{3}\;\boxed{a^b\!/_c}\;\boxed{3}\;\boxed{a^b\!/_c}\;\boxed{4}\;\boxed{=}$

The answer is 10.

Cancelling fractions

You can write a fraction in its **lowest terms** by dividing the numerator and the denominator by the same number.

For example $\frac{8}{12} = \frac{2}{3}$ (by dividing both the numerator and the denominator by 4).

You can also do this on a calculator.

When you press $\boxed{8}$ $\boxed{a^b/_c}$ $\boxed{1}$ $\boxed{2}$, you should see $\boxed{8 \lrcorner 12}$.

When you press $\boxed{=}$, the display changes to $\boxed{2 \lrcorner 3}$, meaning $\frac{2}{3}$.

When you do calculations with fractions on your calculator, it will automatically give the answer as a fraction in its lowest terms.

If you do a calculation which is a mixture of fractions and decimals, your calculator will give the answer as a decimal.

Example 8.5

Use your calculator to work out $2\frac{3}{4} \times 1.5$.

Solution

This is the sequence of keys to press.

$\boxed{2}$ $\boxed{a^b/_c}$ $\boxed{3}$ $\boxed{a^b/_c}$ $\boxed{4}$ $\boxed{\times}$ $\boxed{1}$ $\boxed{.}$ $\boxed{5}$ $\boxed{=}$

The answer is 4.125.

Improper fractions

If you enter an improper fraction into your calculator and press the $\boxed{=}$ key, the calculator will automatically change it to a mixed number.

Example 8.6

Use your calculator to change $\frac{187}{25}$ to a mixed number.

Solution

This is the sequence of keys to press.

$\boxed{1}$ $\boxed{8}$ $\boxed{7}$ $\boxed{a^b/_c}$ $\boxed{2}$ $\boxed{5}$ $\boxed{=}$

The display on your calculator should look like this.

$\boxed{7 \lrcorner 12 \lrcorner 25}$

So the answer is $7\frac{12}{25}$.

Exercise 8.2

1   Work out these.

(a) $\frac{2}{7} + \frac{1}{3}$

(b) $\frac{3}{4} - \frac{2}{5}$

(c) $\frac{5}{8} \times \frac{4}{11}$

(d) $\frac{11}{12} \div \frac{5}{8}$

(e) $2\frac{3}{7} + 3\frac{1}{2}$

(f) $5\frac{2}{3} - 3\frac{3}{4}$

(g) $4\frac{2}{7} \times 3$

(h) $5\frac{7}{8} \div 1\frac{5}{6}$

Exercise continues …

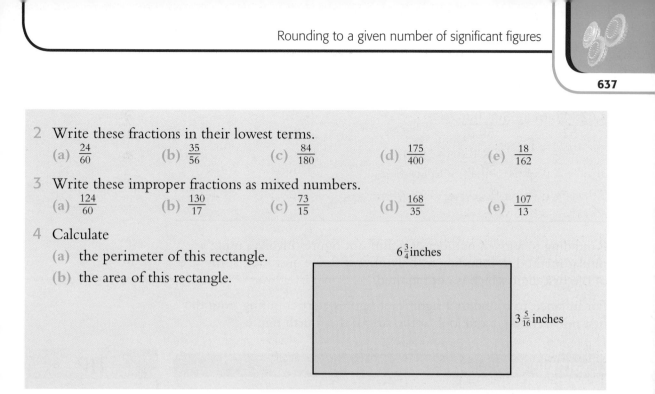

2   Write these fractions in their lowest terms.

(a) $\frac{24}{60}$　　　(b) $\frac{35}{56}$　　　(c) $\frac{84}{180}$　　　(d) $\frac{175}{400}$　　　(e) $\frac{18}{162}$

3   Write these improper fractions as mixed numbers.

(a) $\frac{124}{60}$　　　(b) $\frac{130}{17}$　　　(c) $\frac{73}{15}$　　　(d) $\frac{168}{35}$　　　(e) $\frac{107}{13}$

4   Calculate

(a) the perimeter of this rectangle.

(b) the area of this rectangle.

$6\frac{3}{4}$ inches

$3\frac{5}{16}$ inches

Rounding to a given number of significant figures

You should already know how to round to 1 significant figure. You often do this to estimate to answer to a problem. The next example reminds you of this.

Example 8.7

Estimate the answer to this calculation.

$$\frac{4.62 \times 0.61}{52}$$

Solution

Round each number in the calculation to 1 significant figure.

$4.62 = 5$ to 1 s.f.	The second non-zero digit is 6 so round 4 up to 5.
$0.61 = 0.6$ to 1 s.f.	The second non-zero digit is 1 so the 6 stays as it is.
	Looking at the place value, the 6 is 0.6, which stays as it is.
$52 = 50$ to 1 s.f.	The second non-zero digit is 2 so the 5 stays as it is.
	Looking at the place value, the 5 is 50, which stays as it is.

$$\frac{4.62 \times 0.61}{52} \approx \frac{5 \times 0.6}{50}$$

$$= \frac{{}^{1}\cancel{5} \times 0.6}{\cancel{50}_{10}}$$

$$= \frac{0.6}{10} = 0.06$$

Rounding to a given number of significant figures involves using a similar method to rounding to 1 significant figure: just look at the size of the first digit which is not required.

For instance, to round to 3 significant figures, start counting from the first non-zero digit and look at the size of the fourth figure.

Example 8.8

(a) Round 52 617 to 2 significant figures.

(b) Round 0.072 618 to 3 significant figures.

(c) Round 17 082 to 3 significant figures.

TIP

Always state the accuracy of your answers, when you have rounded them.

Solution

(a) 52|617 = 53 000 to 2 s.f.   To round to 2 significant figures, look at the third figure.
It is 6, so the second figure changes from 2 to 3.
Remember to add zeros for placeholders.

(b) 0.072 6|18 = 0.0726 to 3 s.f.   The first significant figure is 7.
To round to 3 significant figures, look at the fourth significant figure.
It is 1, so the third figure is unchanged.

(c) 17 0|82 = 17 100 to 3 s.f.   The 0 in the middle here is a significant figure.
To round to 3 significant figures, look at the fourth figure.
It is 8, so the third figure changes from 0 to 1.
Remember to add zeros for placeholders.

You use rounding to 2 or 3 significant figures when you do calculations that involve a combination of big and small numbers or that result in very big numbers or very small numbers.

Example 8.9

A red blood cell is approximately flat and circular.

It has a diameter of about 0.007 mm.

Without using your calculator, work out the area of one side of a red blood cell.

Use $\pi = 3$ in your calculation and give your answer to 2 significant figures.

Solution

$$\text{Area of a circle} = \pi r^2$$
$$= 3 \times 0.007^2$$

$0.007 \times 0.007 = 0.000\,049$

$0.000\,049 \times 3 = 0.000\,147$

The area of one side of a red blood cell is $0.000\,15\,\text{mm}^2$ (to 2 s.f.).

Example 8.10

Jupiter is the largest planet in our solar system.

It has a radius of 71 500 km (to 3 s.f.).

Use your calculator to find its circumference.

Solution

$$\text{Circumference of a circle} = 2\pi r$$
$$= 2 \times \pi \times 71\,500$$
$$= 449\,247.7495$$

This is the answer your calculator gives but you were told that the radius was correct to 3 significant figures so you shouldn't give your answer more accurately than that.

The circumference of Jupiter is 449 000 km (to 3 s.f.).

Exercise 8.3

1 Round each of these numbers to 1 significant figure.
   (a) 14.9   (b) 167   (c) 21.2   (d) 794
   (e) 6027   (f) 0.013   (g) 0.58   (h) 0.037
   (i) 1.0042   (j) 20 053   (k) 0.069   (l) 1942

2 Round each of these numbers to 2 significant figures.
   (a) 17.6   (b) 184.2   (c) 5672   (d) 97 520
   (e) 50.43   (f) 0.172   (g) 0.0387   (h) 0.006 12
   (i) 0.0307   (j) 0.994

3 Round each of these numbers to 3 significant figures.
   (a) 8.261   (b) 69.77   (c) 16 285   (d) 207.51
   (e) 12 524   (f) 7.103   (g) 50.87   (h) 0.4162
   (i) 0.038 62   (j) 3.141 59

4 At the local derby match in Sheffield, the attendance was 15 870 men, 10 740 women
   and 8475 juniors.
   The *Sheffield Star* reported that there were 35 000 people at the match.
   Explain this.

5 Over the three days of the Totley Show there were 4700 visitors.
   On average, how many visitors were there each day correct to 2 significant figures?

6 The river Nile is 4169 miles long.
   1 mile is approximately 1.6 kilometres.
   How long is the Nile, correct to the nearest kilometre.

7 Six friends have a meal together and the bill comes to £28.71.
   They divide the bill equally.
   How much does each pay?

8 At the garage, Pauline bought 16.5 litres of petrol at 103.9 pence per litre.
   She paid with a £20 note.
   How much change did she get?

9 The circular pond in Denby Park has a radius of 5.8 m.
   Find the surface area of the pond correct to 3 significant figures.

Exercise continues . . .

10 Work out these.
    Give your answers to 3 significant figures.

(a) $\sqrt{84}$

(b) $\dfrac{1083}{8.2}$

(c) $43.7 \times 18.9 \times 29.3$

(d) $\dfrac{2.46}{18.5}$

(e) $\dfrac{29}{41.6}$

(f) $\dfrac{283 \times 97}{724}$

(g) $\dfrac{614 \times 0.83}{3.7 \times 2.18}$

(h) $\dfrac{6.72}{0.051 \times 39.7}$

(i) $\sqrt{39 \times 80}$

Challenge 8.1

Write a number which will round to 500 to 1 significant figure.
Write a number which will round to 500 to 2 significant figures.
Write a number which will round to 500 to 3 significant figures.

Compare your results with your classmates.
What do you notice?

What you have learned

- How to change the order in which your calculator does calculations using the ⊟ key and brackets
- How to work with fractions and mixed numbers on your calculator using the $\boxed{a^b_c}$ key
- To round to a given number of significant figures, you look at the size of the first digit which is not required; if it is 5 or more, you round up; if it is less than 5 you round down; if necessary, you use zeros as placeholders to show the value of the number
- You often need to round the answer to a calculation to a sensible degree of accuracy; rounding to a given number of significant figures is often used when the number is very big or very small

Mixed exercise 8

1 Work out these calculations without writing down the answers to any middle stages.

(a) $\dfrac{7.83 - 3.24}{1.53}$

(b) $\dfrac{22.61}{1.7 \times 3.8}$

2 Work out $\sqrt{5.6^2 - 4 \times 1.3 \times 5}$.
Give your answer correct to 2 decimal places

3 Use your calculator to work out these.

(a) $\frac{2}{11} + \frac{5}{6}$ (b) $\frac{7}{8} - \frac{3}{5}$ (c) $2\frac{2}{7} \times 1\frac{3}{8}$ (d) $8\frac{2}{5} \div 2\frac{7}{10}$

4 Round each of these numbers to 1 significant figure.

(a) 6.0   (b) 985   (c) 0.32   (d) 0.57   (e) 45 218

5 Round each of these numbers to 2 significant figures.

(a) 9.16   (b) 4.72   (c) 0.0137   (d) 164 600   (e) 507

6 Round each of these numbers to 3 significant figures.

(a) 1482   (b) 10.16   (c) 0.021 85   (d) 20 952   (e) 0.005 619

7 In 2009 Dougal Jerram attempted to abseil down into the crater of the volcano Erta Ale in Ethiopia.
The crater contains the planet's oldest lava lake and the team used 3-D imaging to map the inside of the volcano.
The crater is roughly circular and has a diameter of approximately 130 m.
Find its area.
Give your answer to 2 significant figures.

8 Work out these.
Give your answers to 3 significant figures.

(a) $\sqrt{107}$ (b) $\sqrt{0.00547}$ (c) $\dfrac{983}{5.2}$ (d) $72.7 \times 19.6 \times 3.3$

(e) $\dfrac{59}{1.96}$ (f) $\dfrac{586 \times 97}{187}$ (g) $\dfrac{318 \times 0.72}{5.1 \times 2.09}$ (h) $\dfrac{512 + 93.2}{8.02 - 2.57}$

Trial and improvement

▶ **This chapter is about**

- Solving equations by trial and improvement

▶ **You should already know**

- How to substitute numbers into formulae

Solving equations by trial and improvement

Sometimes you will need to solve an equation by **trial and improvement**. This means that you substitute different values into the equation until you find a solution.

It is important that you work systematically and do not just choose the numbers you try at random.

First you need to find two numbers between which the solution lies.

Next you try the number halfway between these two numbers.

You continue this process until you find the answer to the required degree of accuracy.

Example 9.1

Find a solution of the equation $x^3 - x = 40$.

Give your answer correct to 1 decimal place.

Solution

$x^3 - x = 40$

Try $x = 3$	$3^3 - 3 = 24$	Too small.　Try a larger number.
Try $x = 4$	$4^3 - 4 = 60$	Too large. The solution must lie between 3 and 4.
Try $x = 3.5$	$3.5^3 - 3.5 = 39.375$	Too small.　Try a larger number.
Try $x = 3.6$	$3.6^3 - 3.6 = 43.056$	Too large. The solution must lie between 3.5 and 3.6.

Try $x = 3.55$ $3.55^3 - 3.55 = 41.188 \ldots$ Too large. The solution must lie between 3.5 and 3.55.

Any number between 3.5 and 3.55 rounds to 3.5 so the answer is $x = 3.5$, correct to 1 decimal place.

Example 9.2

(a) Show that $x^3 - 3x = 6$ has a solution between $x = 2$ and $x = 3$.

(b) Find the solution correct to 1 decimal place.

Solution

(a) $x^3 - 3x = 6$

When $x = 2$ $2^3 - 3 \times 2 = 2$ Too small.
When $x = 3$ $3^3 - 3 \times 3 = 18$ Too large.
6 is between 2 and 18. Therefore there is a solution of $x^3 - 3x = 6$ between $x = 2$ and $x = 3$.

(b) Try $x = 2.5$ $2.5^3 - 3 \times 2.5 = 8.125$ Too large.   Try a smaller number.
Try $x = 2.3$ $2.3^3 - 3 \times 2.3 = 5.267$ Too small.   Try a larger number.
Try $x = 2.4$ $2.4^3 - 3 \times 2.4 = 6.624$ Too large.   Try a smaller number.
Try $x = 2.35$ $2.35^3 - 3 \times 2.35 = 5.927 \ldots$ Too small. The solution must lie between 2.35 and 2.4.

Any number between 2.35 and 2.4 rounds to 2.4 so the answer is $x = 2.4$, correct to 1 decimal place.

Exercise 9.1

1   Find a solution, between $x = 1$ and $x = 2$, to the equation $x^3 = 5$.
Give your answer correct to 1 decimal place.

2   (a) Show that a solution to the equation $x^3 - 5x = 8$ lies between $x = 2$ and $x = 3$.
(b) Find the solution correct to 1 decimal place.

3   (a) Show that a solution to the equation $x^3 - x = 90$ lies between $x = 4$ and $x = 5$.
(b) Find the solution correct to 1 decimal place.

4   Find a solution to the equation $x^3 + x = 15$.
Give your answer correct to 1 decimal place.

5   Find a solution to the equation $x^3 + x^2 = 100$.
Give your answer correct to 2 decimal places.

Exercise continues …

6 Which whole number, when cubed, gives a value closest to 10 000?

7 Use trial and improvement to find which number, when squared, gives 1000. Give your answer correct to 1 decimal place.

8 A number, added to the square of this number, gives 10.
   (a) Write this as a formula.
   (b) Find the number correct to 1 decimal place.

9 The product of two whole numbers, the difference of which is 4, is 621.
   (a) Write this as formula in terms of x.
   (b) Use trial and improvement to find the two numbers.

10 Use trial and improvement to find which number, when squared, gives 61. Give your answer correct to 1 decimal place.

What you have learned

- To find the solution to an equation by trial and improvement you first need to find two numbers between which the solution lies; you then try the number halfway between these two numbers and continue the process until you find the answer to the required degree of accuracy

Mixed exercise 9

1 (a) Show that a solution to the equation $x^3 + 4x = 12$ lies between $x = 1$ and $x = 2$.
  (b) Find the solution correct to 1 decimal place.

2 (a) Show that a solution to the equation $x^3 - x^2 = 28$ lies between $x = 3$ and $x = 4$.
  (b) Find the solution correct to 1 decimal place.

3 A number, added to the cube of this number, gives 100.
  (a) Write this as a formula.
  (b) Find the number correct to 1 decimal place.

Chapter 10

Enlargement

▶ **This chapter is about**

- The scale factor of enlargement
- Similar triangles
- The area and volume of similar shapes

▶ **You should already know**

- How to recognise and describe an enlargement using a scale factor and a centre of enlargement
- How to work with ratios
- How to find the area of a triangle

The scale factor of enlargement

The object and image of an enlargement are **similar**. The angles of the object and the image are the same but all the sides of the object have been enlarged by the same scale factor in the image.

You should already know how to work out the scale factor of an enlargement. It is part of the description of an enlargement.

Example 10.1

Describe the transformation that maps triangle A on to triangle B.

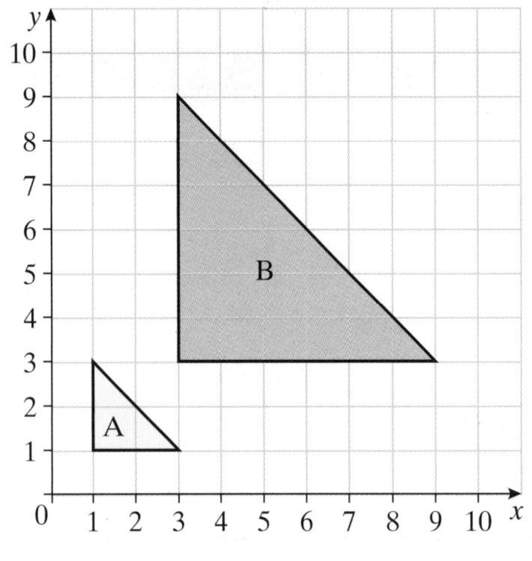

Solution

The transformation is an enlargement because the two triangles are similar.

Each side has been enlarged by the same scale factor.

The length of the base of triangle A is 2 units.
The length of the base of triangle B is 6 units.

$$\text{Scale factor} = \frac{\text{Length of image}}{\text{Length of object}}$$

$$= \frac{6}{2} = 3$$

So the transformation that maps triangle A on to triangle B is an enlargement, scale factor 3, centre the origin.

> **TIP**
>
> You can use any pair of sides to work out the scale factor.
> On a grid, horizontal or vertical sides are the easiest.

When you know two triangles are similar and are given some lengths, you can use the scale factor of enlargement to work out the length of corresponding sides in the triangles.

Example 10.2

Triangles ABC and DEF are similar.

Calculate the length of EF.

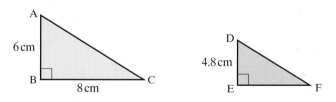

Solution

Length of AB = 6 cm
Length of DE = 4.8 cm

$$\text{Scale factor} = \frac{\text{Length of image}}{\text{Length of object}}$$

$$= \frac{4.8}{6} = 0.8$$

Length of EF = Length of BC × Scale factor
$$= 8 \times 0.8$$
$$= 6.4\,\text{cm}$$

Sometimes you have to work out lengths in both triangles.

You might find this triangle helpful.

You use it in a similar way to the triangle for the formula linking distance, time and speed that you met in Chapter 4.

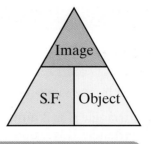

Example 10.3

Triangle ABC is an enlargement of triangle XBY.
Calculate the lengths of AC and XB.

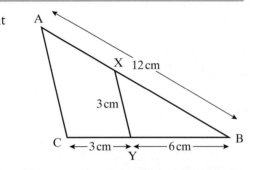

Solution

It is helpful to draw the two triangles separately.

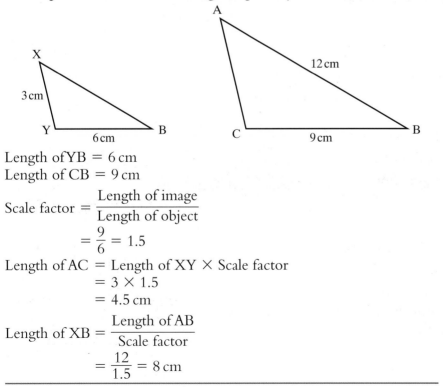

Length of YB = 6 cm
Length of CB = 9 cm

$$\text{Scale factor} = \frac{\text{Length of image}}{\text{Length of object}}$$

$$= \frac{9}{6} = 1.5$$

Length of AC = Length of XY × Scale factor
$$= 3 \times 1.5$$
$$= 4.5 \text{ cm}$$

$$\text{Length of XB} = \frac{\text{Length of AB}}{\text{Scale factor}}$$

$$= \frac{12}{1.5} = 8 \text{ cm}$$

TIP

Make sure you use a pair of **corresponding** sides.

Exercise 10.1

1   Triangles ABC and XYZ are similar.
    Calculate the lengths of XY and YZ.

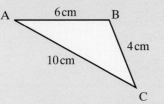

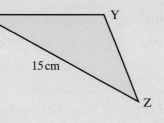

2   Triangles DEF and UVW are similar.
    Calculate the lengths of UV and UW.

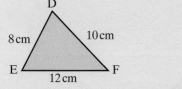

3   Triangles ABC and PQR are similar.
    Calculate the lengths of PQ and QR.

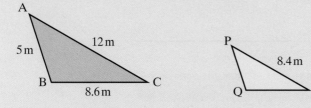

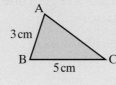

4   Triangles ABC and PQR are similar.
    (a) Calculate the length of PQ.
    (b) Calculate the length of AC.

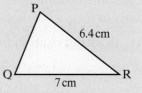

5   Triangles ABC and AXY are similar.
    (a   Calculate the length of XY.
    (b) Calculate the length of AC.
    (c) Work out the perimeter of triangles ABC and AXY.
         What do you notice?

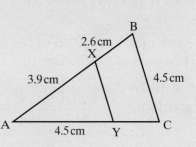

The area and volume of similar shapes

◉ Discovery 10.1

Here are two similar triangles.

The sides of the large triangle are three times the length of those of the small triangle.

The scale factor is 3.

It is a linear scale factor because it describes the length of the lines.

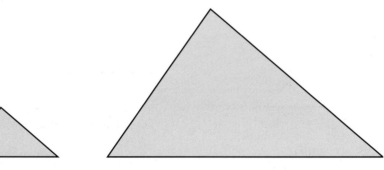

Tessellate the small triangle into the large one.

How many will fit?

◉ Discovery 10.2

Use some centimetre cubes to make a larger cube with each side three times as long. (The linear scale factor is 3.)

Are the small cube and the large cube similar?

What is the volume scale factor?

What is the connection between the volume scale factor and the linear scale factor?

The results of Discoveries 10.1 and 10.2 can be generalised.

For similar shapes, the area scale factor is the square of the linear scale factor.

For similar solids, the volume scale factor is the cube of the linear scale factor.

Exercise 10.2

1 Two lemonade bottles are similar.
One has height of 30 cm and the other has a height of 15 cm.

(a) The circumference of the larger bottle is 25 cm.
What is the circumference of the smaller bottle?

(b) The area of the label on the smaller bottle is 20 cm².
Explain why the area of the label on the larger bottle is not 40 cm².

(c) The smaller bottle holds 187.5 millilitres.
How much does the larger bottle hold?

What you have learned

- The object and image of an enlargement are similar
- Similar shapes have the same angles but all the sides of the object have been enlarged by the same scale factor
- Scale factor = $\dfrac{\text{Length of image}}{\text{Length of object}}$
- When you know two triangles are similar and are given some lengths, you can use the scale factor of enlargement to work out the lengths of corresponding sides in the triangles
- The area and volume scale factors are not the same as the linear scale factor

Mixed exercise 10

1 Triangles ABC and DEF are similar.
Calculate the lengths of DE and EF.

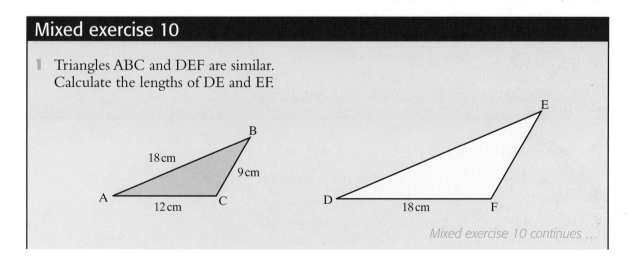

Mixed exercise 10 continues ...

2 Triangles ABC and PQR are similar.
Calculate the lengths of PR and QR.

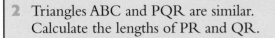

3 Triangle ABC is an enlargement of triangle XBY.
Calculate the lengths of AC and XB.

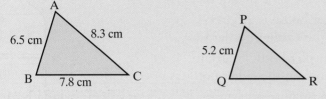

4 Triangles PQR and XYZ are similar.

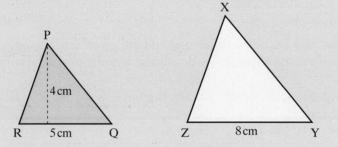

(a) What is the linear scale factor of the enlargement?

(b) Find the height of triangle XYZ.

(c) Calculate the area of triangle PQR.

(d) Calculate the area of triangle XYZ.

(e) What do you notice about the ratio of the areas?

Graphs

▶ This chapter is about

- Distance–time graphs
- Drawing and interpreting graphs of real-life situations
- Drawing graphs of quadratic functions
- Solving equations using quadratic graphs
- Using coordinates in three dimensions

▶ You should already know

- How to plot and interpret simple straight-line graphs involving conversions, distance–time and other real-life situations
- How to use the relationship between distance, speed and time
- How to plot and read points in all four quadrants in two dimensions
- How to substitute numbers into equations
- How to square numbers
- How to draw straight-line graphs from their equations
- How to rearrange equations
- How to add, subtract and multiply negative numbers

Distance–time graphs

In Unit B Chapter 10 you learned about travel graphs. These can also be called distance–time graphs.

Check up 11.1

James walked to the bus stop and waited for the bus.
When the bus arrived he got on the bus and it took him to school without stopping.

Which of these distance–time graphs best shows James' journey to school? Explain your answer.

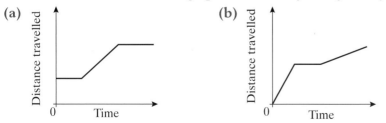

(a)   Distance travelled / Time   **(b)**   Distance travelled / Time

(c)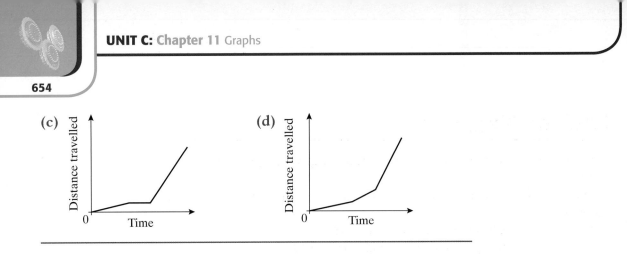

(d)

In Chapter 4 you met this formula for calculating speed.

$$\text{Speed} = \frac{\text{Distance}}{\text{Time}}$$

You can work out speed from a distance–time graph.

⦿ Discovery 11.1

James walked to the bus stop at 4 km/h. This took him 15 minutes.

He waited 5 minutes at the bus stop.

The bus journey was 12 km and took 20 minutes. The bus went at a constant speed.

(a) Copy these axes and draw an accurate graph of James' journey.

(b) What was the speed of the bus in km/h?

(c) After 30 minutes, how far was James away from home?

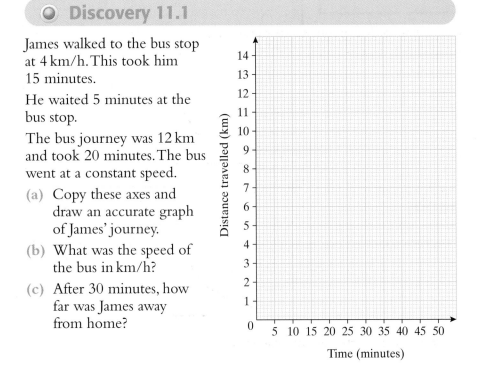

When a graph illustrates real quantities, the gradient represents the **rate of change**.

When the graph shows distance (vertical) against time (horizontal), the rate of change is equal to the **speed**.

TIP

Speed is also known as **velocity**.

Other real-life graphs

When you are asked to answer questions about a given graph, you should:
- look carefully at the labels on the axes to see what the graph represents.
- check what the units are on each of the axes.
- look to see whether the lines are straight or curved.

If the graph is straight, the rate of change is constant.
The steeper the line, the higher the rate of change.

A horizontal line represents a part of the graph where there is no change in the quantity on the vertical axis.

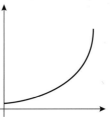

If the graph is curved as in this diagram, the rate of change is increasing.

If the graph is curved as in this diagram, the rate of change is decreasing.

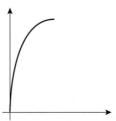

Example 11.1

The graph shows the cost of printing tickets.

(a) Find the total cost of printing 250 tickets.

(b) The cost consists of a fixed charge and an additional charge for each ticket printed.
   (i) What is the fixed charge?
   (ii) Find the additional charge for each ticket printed.
   (iii) Find the total cost of printing 800 tickets.

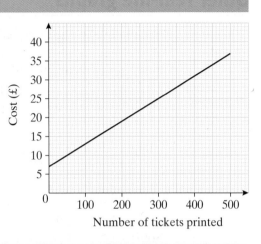

Solution

(a) Draw a line from 250 on the Number of tickets printed axis, to meet the straight line.
Then draw a horizontal line and read off the value where it meets the Cost axis.

The total cost of printing 250 tickets is £22.

(b) (i) Read from the graph the cost of zero tickets (where the graph cuts the Cost axis).

The fixed charge is £7.

(ii) 250 tickets cost £22.
Fixed charge is £7.
So the additional charge for 250 tickets is $22 - 7 = £15$.

The additional charge per ticket is $\frac{15}{250} = £0.06$ or 6p

(iii) Cost in $£ = 7 +$ number of tickets $\times 0.06$

$$Cost\ of\ 800\ tickets = 7 + 800 \times 0.06$$
$$= 7 + 48$$
$$= £55$$

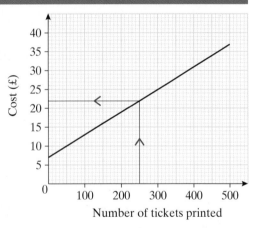

TIP

Work in either pounds or pence.

If you work in pounds, you won't have to convert your final answer back from pence.

Exercise 11.1

1 Jane and Halima live in the same block of flats and go to the same school.
The graphs represent their journeys home from school.
  (a) Describe Halima's journey home.
  (b) After how many minutes did Halima overtake Jane?
  (c) Calculate Jane's speed in
      (i) kilometres per minute.
      (ii) kilometres per hour.
  (d) Calculate Halima's fastest speed in
      (i) kilometres per minute.
      (ii) kilometres per hour.

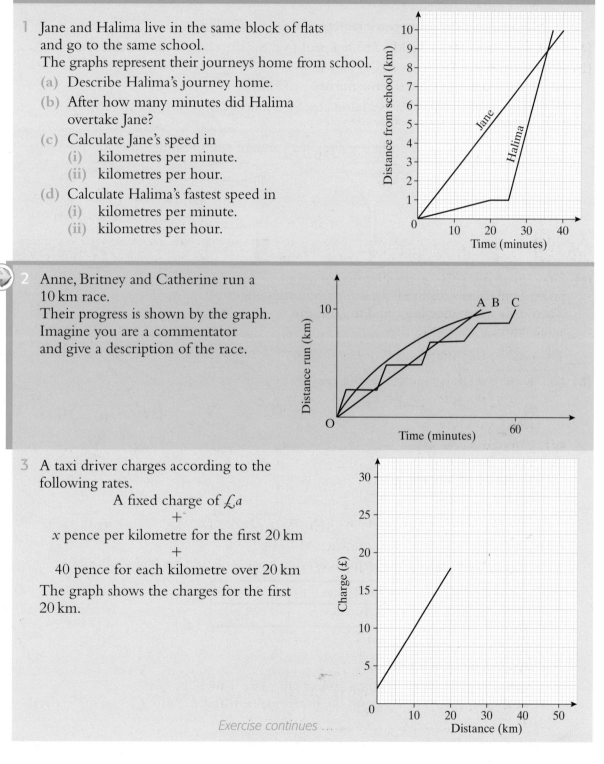

2 Anne, Britney and Catherine run a 10 km race.
Their progress is shown by the graph.
Imagine you are a commentator and give a description of the race.

3 A taxi driver charges according to the following rates.
  A fixed charge of £a
  +
  x pence per kilometre for the first 20 km
  +
  40 pence for each kilometre over 20 km
The graph shows the charges for the first 20 km.

Exercise continues …

(a) What is the fixed charge, £*a*?

(b) Calculate *x*, the charge per kilometre for the first 20 km.

(c) Copy the graph and add a line segment to show the charges for distances from 20 km to 50 km.

(d) What is the total charge for a journey of 35 km?

(e) What is the average cost per kilometre for a journey of 35 km?

4 Water is poured into each of these glasses at a constant rate until they are full.

These graphs show depth of water (*d*) against time (*t*).
Choose the most suitable graph for each glass.

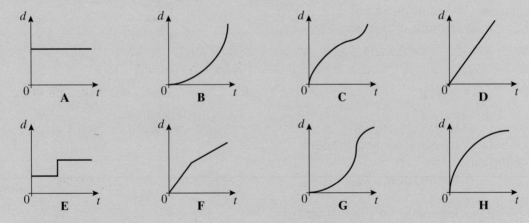

5 An office supplies firm advertises the following price structure for boxes of paper.

Number of boxes	1 to 4	5 to 9	10 or more
Price per box	£6.65	£5.50	£4.65

(a) How much do 9 boxes cost?

(b) How much do 10 boxes cost?

(c) Draw a graph to show the total cost of orders for 1 to 12 boxes.
Use a scale of 1 cm to 1 box on the horizontal axis and 2 cm to £10 on the vertical axis.

Exercise continues …

6 The table shows the cost of sending first class mail in 2006.
The graph shows the information in the first two rows in the table.

Maximum weight	First class cost
60 g	30p
100 g	46p
150 g	64p
200 g	79p
250 g	94p
300 g	£1.07

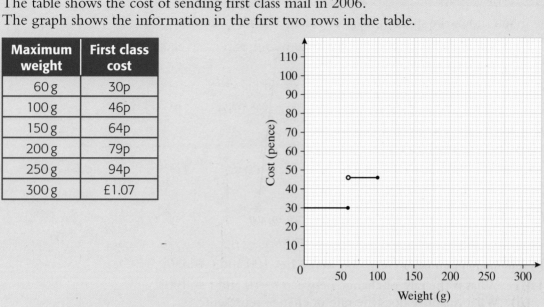

(a) What was the cost of sending a letter that weighs
(i)   60 g?
(ii) 60.1 g?

(b) (i)   What is the meaning of the dot at the right of the lower line?
(ii) What is the meaning of the circle at the left of the second line?

(c) Copy the graph and add lines to show the cost of sending first class mail weighing up to 300 g.

(d) Osman posted one letter weighing 95 g and another weighing 153 g. What was the total cost?

7 A water company makes the following charges for customers with a water meter.

Basic charge	£20.00
Charge per cubic metre for the first 100 cubic metres used	£1.10
Charge per cubic metre for water used over 100 cubic metres	£0.80

(a) Draw a graph to show the charge for up to 150 cubic metres.
Use a scale of 1 cm to 10 cubic metres on the horizontal axis and 1 cm to £10 on the vertical axis.

(b) Customers can choose instead to pay a fixed amount of £120.
For what amounts of water is it cheaper to have a water meter?

Challenge 11.1

The graph shows the speed (v m/s) of a train at time t seconds.

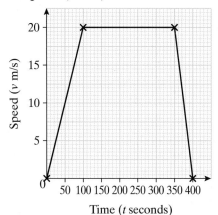

(a) What is happening between the times $t = 100$ and $t = 350$?

(b) (i) What is the rate of change between $t = 0$ and $t = 100$?
    (ii) What quantity does the rate of change represent?
    (iii) What are the units of the rate of change?

(c) (i) What is the rate of change between $t = 350$ and $t = 400$?
    (ii) What quantity does the rate of change represent?

Quadratic graphs

A **quadratic function** is a function where the highest power of x is 2.

So the function will have an x^2 term.
It may also have an x term and a numerical term.
It will *not* have a term with any other power of x.

The function $y = x^2 + 2x - 3$ is a typical quadratic function.

Check up 11.2

State whether or not each of these functions is quadratic.

(a) $y = x^2$
(b) $y = x^2 + 5x - 4$
(c) $y = \dfrac{5}{x}$
(d) $y = x^2 - 3x$
(e) $y = x^2 - 3$
(f) $y = x^3 + 5x^2 - 2$
(g) $y = x(x - 2)$

As with all graphs of functions of the form '$y =$ ', to plot the graph you must first choose some values of x. Quadratic functions do not produce straight-line graphs so it is best to complete a table of values.

The simplest quadratic function is $y = x^2$.

x	-3	-2	-1	0	1	2	3
$y = x^2$	9	4	1	0	1	4	9

TIP

Remember that the square of a negative number is positive.

The points can then be plotted and joined up with a smooth curve.

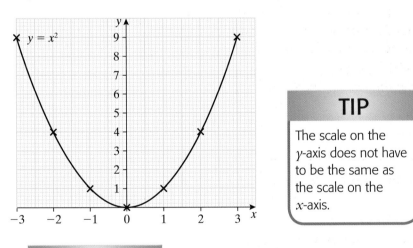

TIP

The scale on the y-axis does not have to be the same as the scale on the x-axis.

TIP

Turn your paper round and draw the curve from the inside. The sweep of your hand will give a smoother curve.

Draw the curve without taking your pencil away from the paper.

Look ahead to the next point as you draw the curve.

You can read off from your graph the value of y for any value of x or the value of x for any value of y.

For some quadratic graphs you may need extra rows in your table to get the final y values, as is shown in the next example.

Example 11.2

(a) Complete the table of values for $y = x^2 - 2x$.

(b) Plot the graph of $y = x^2 - 2x$.

(c) Use your graph to
    (i)   find the value of y when $x = 2.6$.
    (ii) solve $x^2 - 2x = 5$.

Solution

(a)

x	−2	−1	0	1	2	3	4
x^2	4	1	0	1	4	9	16
$-2x$	4	2	0	−2	−4	−6	−8
$y = x^2 - 2x$	8	3	0	−1	0	3	8

TIP

The second and third rows are included in the table only to make the calculation of the y values easier: for this graph, add the numbers in the second and third rows to find the y values.

The values you plot are the x values (first row) and y values (last row).

(b) $y = x^2 - 2x$

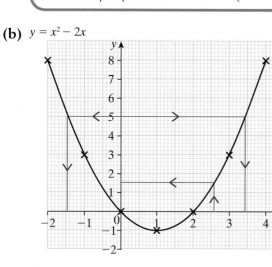

(c) (i) $y = 1.5$ Read up from $x = 2.6$.
    (ii) $x = -1.4$ or $x = 3.4$ $x^2 - 2x = 5$ means that $y = 5$.
                                        Reading across from 5, you will see that there are two possible answers.

Example 11.3

(a) Complete the table of values for $y = x^2 + 3x - 2$.

(b) Plot the graph of $y = x^2 + 3x - 2$.

(c) Use your graph to
      (i)   find the value of y when $x = -4.3$.
      (ii) solve $x^2 + 3x - 2 = 0$.

Solution

(a)

x	−5	−4	−3	−2	−1	0	1	2
x^2	25	16	9	4	1	0	1	4
$3x$	−15	−12	−9	−6	−3	0	3	6
−2	−2	−2	−2	−2	−2	−2	−2	−2
$y = x^2 + 3x - 2$	8	2	−2	−4	−4	−2	2	8

(b)

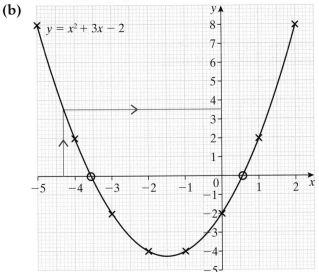

TIP

The lowest y values in the table are both −4, but the actual curve goes below −4. In such situations, it is often useful to work out the coordinates of the lowest (or highest) point of the curve.

Because the curve is symmetrical, the lowest point of $y = x^2 + 3x - 2$ must lie halfway between $x = -2$ and $x = -1$, that is, at $x = -1.5$.

When $x = -1.5$, $y = (-1.5)^2 + 3 \times -1.5 - 2 = 2.25 - 4.5 - 2 = -4.25$.

(c) (i) $y = 3.5$ Read up from $x = -4.3$.
(ii) $x = -3.6$ or $x = 0.6$ $x^2 + 3x - 2 = 0$ means that $y = 0$.
Reading off the graph when
$y = 0$, you will see that there are two
possible answers.

All quadratic graphs are the same basic shape. This shape is called a
parabola.

The three you have seen so far were ∪-shaped. In these graphs, the x^2
term was positive.

If the x^2 term is negative, the parabola is the other way up (∩).

> **TIP**
>
> If your graph is
> not shaped like a
> parabola, go back
> and check your table.

Example 11.4

(a) Complete the table of values for $y = 5 - x^2$.
(b) Plot the graph of $y = 5 - x^2$.
(c) Use your graph to solve
     (i) $5 - x^2 = 0$.   **(ii)** $5 - x^2 = 3$.

Solution

(a)

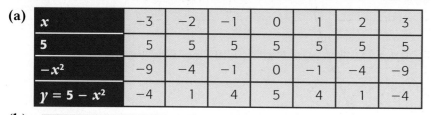

x	-3	-2	-1	0	1	2	3
5	5	5	5	5	5	5	5
$-x^2$	-9	-4	-1	0	-1	-4	-9
$y = 5 - x^2$	-4	1	4	5	4	1	-4

(b)

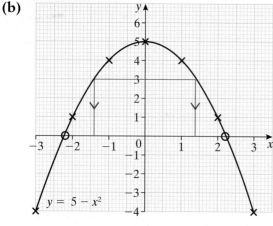

(c) (i) $x = -2.2$ or $x = 2.2$ Read off at $y = 0$.
      (ii) $x = -1.4$ or $x = 1.4$ Read off at $y = 3$.

Exercise 11.2

1 (a) Copy and complete the table of values for $y = x^2 - 2$.

x	-3	-2	-1	0	1	2	3
x^2	9					4	
-2	-2					-2	
$y = x^2 - 2$	7					2	

   (b) Plot the graph of $y = x^2 - 2$.
        Use a scale of 2 cm to 1 unit on the x-axis and 1 cm to 1 unit on the y-axis.

   (c) Use your graph to
        (i)   find the value of y when $x = 2.3$.
        (ii) solve $x^2 - 2 = 4$.

2 (a) Copy and complete the table of values for $y = x^2 - 4x$.

x	-1	0	1	2	3	4	5
x^2					9		
$-4x$					-12		
$y = x^2 - 4x$					-3		

   (b) Plot the graph of $y = x^2 - 4x$.
        Use a scale of 2 cm to 1 unit on the x-axis and 1 cm to 1 unit on the y-axis.

   (c) Use your graph to
        (i)   find the value of y when $x = 4.2$.
        (ii) solve $x^2 - 4x = -2$.

3 (a) Copy and complete the table of values for $y = x^2 + x - 3$.

x	-4	-3	-2	-1	0	1	2	3
x^2			4					
x			-2					
-3			-3					
$y = x^2 + x - 3$			-1					

   (b) Plot the graph of $y = x^2 + x - 3$.
        Use a scale of 2 cm to 1 unit on the x-axis and 1 cm to 1 unit on the y-axis.

   (c) Use your graph to
        (i)   find the value of y when $x = 0.7$.
        (ii) solve $x^2 + x - 3 = 0$.

Exercise continues …

4 (a) Make a table of values for $y = x^2 - 3x + 4$. Choose values of x from -2 to 5.

  (b) Plot the graph of $y = x^2 - 3x + 4$.
     Use a scale of 2 cm to 1 unit on the x-axis and 1 cm to 1 unit on the y-axis.

  (c) Use your graph to
     (i)   find the minimum value of y.
     (ii) solve $x^2 - 3x + 4 = 10$.

5 (a) Copy and complete the table of values for $y = 3x - x^2$.

x	-2	-1	0	1	2	3	4	5
$3x$				3			12	
$-x^2$				-1			-16	
$y = 3x - x^2$				2			-4	

  (b) Plot the graph of $y = 3x - x^2$.
     Use a scale of 2 cm to 1 unit on the x-axis and 1 cm to 1 unit on the y-axis.

  (c) Use your graph to
     (i)   find the maximum value of y.
     (ii) solve $3x - x^2 = -2$.

6 (a) Make a table of values for $y = x^2 - x - 5$. Choose values of x from -3 to 4.

  (b) Plot the graph of $y = x^2 - x - 5$.
     Use a scale of 2 cm to 1 unit on the x-axis and 1 cm to 1 unit on the y-axis.

  (c) Use your graph to solve
     (i) $x^2 - x - 5 = 0$.   (ii) $x^2 - x - 5 = 3$.

7 (a) Make a table of values for $y = 2x^2 - 5$. Choose values of x from -3 to 3.

  (b) Plot the graph of $y = 2x^2 - 5$.
     Use a scale of 2 cm to 1 unit on the x-axis and 1 cm to 1 unit on the y-axis.

  (c) Use your graph to solve
     (i) $2x^2 - 5 = 0$.   (ii) $2x^2 - 5 = 10$.

8 The total surface area (A cm^2) of this cube is given by $A = 6x^2$.

  (a) Make a table of values for $A = 6x^2$. Choose values of
     x from 0 to 5.

  (b) Plot the graph of $A = 6x^2$.
     Use a scale of 2 cm to 1 unit on the x-axis and 1 cm to
     10 units on the A-axis.

  (c) Use your graph to find the side of a cube with surface area
     (i)   20 cm^2.   (ii) 80 cm^2.

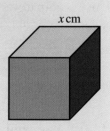

x cm

Three-dimensional coordinates

You already know how to describe a point using coordinates in two dimensions.
You use x- and y-coordinates.

If you are working in three dimensions you need a third coordinate.
This is known as the z-coordinate.

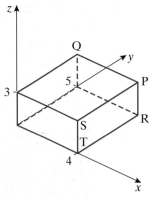

> ## TIP
>
> Notice that the x- and y-axes lie flat with the z-axis vertical.

The coordinates of point P are $(4, 5, 3)$.
That is, 4 in the x direction, 5 in the y direction and 3 in the z direction.

> ## TIP
>
> As with two-dimensional coordinates, three-dimensional coordinates are written in alphabetical order: x then y then z.

Check up 11.3

What are the coordinates of the points Q, R, S and T in the diagram above?

Example 11.5

A is the point $(4, 2, 3)$ and B is the point $(2, 6, 9)$.
What are the coordinates of the midpoint of AB?

Solution

The coordinates of the midpoint of a line are the means of the coordinates of the two endpoints.

In three dimensions:

$$\text{Midpoint of line with coordinates } (a, b, c) \text{ and } (d, e, f) = \left(\frac{a+d}{2}, \frac{b+e}{2}, \frac{c+f}{2} \right).$$

For A$(4, 2, 3)$ and B$(2, 6, 9)$ the coordinates of the midpoint of the

line AB are $\left(\dfrac{4+2}{2}, \dfrac{2+6}{2}, \dfrac{3+9}{2} \right) = (3, 4, 6)$.

Exercise 11.3

1 The diagram shows the outline of a cuboid.
 The coordinates of point A are $(5, 0, 0)$.
 The coordinates of point B are $(0, 3, 0)$.
 The coordinates of point C are $(0, 0, 2)$.
 Write down the coordinates of

 (a) point D.

 (b) point E.

 (c) point F.

 (d) point G.

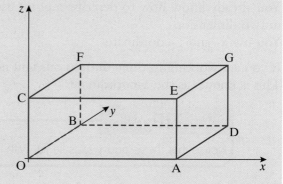

2 VOABC is a square-based pyramid.
 A is the point $(6, 0, 0)$.
 N is the centre of the base.
 The perpendicular height VN of the pyramid is 5 units.
 Write down the coordinates of

 (a) point C.

 (b) point B.

 (c) point N.

 (d) point V.

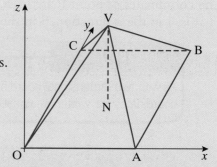

3 The diagram shows the outline of a cuboid.
 The coordinates of point A are $(8, 0, 0)$.
 The coordinates of point B are $(0, 6, 0)$.
 The coordinates of point C are $(0, 0, 4)$.
 L is the midpoint of AD.
 M is the midpoint of EG.
 N is the midpoint of FG.
 Write down the coordinates of

 (a) point D.

 (b) point L.

 (c) point M.

 (d) point N.

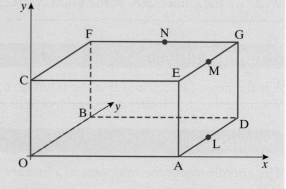

What you have learned

- When you are asked to answer questions about a graph, you must look carefully at the labels and units on the axes and see whether the line is straight or curved

- A straight line represents a constant rate of change, and the steeper the line, the greater the rate of change
- A horizontal line means that there is no change in the quantity on the y-axis
- A convex curve (viewed from below) represents an increasing rate of change
- A concave curve (viewed from below) represents a decreasing rate of change
- The rate of change on a distance–time graph is the speed
- On a cost graph, the value where the graph cuts the cost axis is the fixed charge
- A quadratic function has x^2 as the highest power of x. It may also have an x term and a numerical term. It will not have a term with any other power of x
- The shape of all quadratic graphs is a parabola. If the x^2 term is positive the curve is U-shaped. If the x^2 term is negative the curve is ∩-shaped
- In three dimensions a point has three coordinates, the third being the z-coordinate

Mixed exercise 11

1 The graph shows an energy supplier's quarterly charges for up to 500 kWh of electricity.

(a) What is the cost if 350 kWh are used?
The cost is made up of a fixed charge plus an amount per kWh of electricity used.

(b) (i) What is the fixed charge?
(ii) Calculate the cost per kWh in pence.

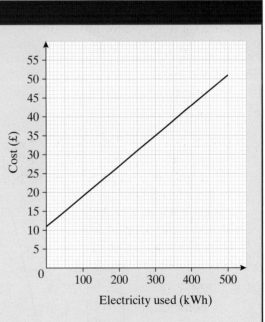

2 The same energy supplier makes a fixed charge for gas of £15 per quarter.
In addition to this there is a charge of 2p per kWh.

(a) Draw a graph to show the quarterly bill for up to 1500 kWh of gas used.
Use a scale of 1 cm to 100 kWh on the horizontal axis and 2 cm to £10 on the vertical axis.

(b) Look at this graph and the graph in question 1.
Which is cheaper: 400 kWh of electricity or 400 kWh of gas? By how much?

Mixed exercise 11 continues …

3 This table shows the cost of sending letters by second class post in 2006.

Maximum weight	Second class cost
60 g	21p
100 g	35p
150 g	47p
200 g	58p
250 g	71p
300 g	83p

(a) Look back at the graph for first class post in Exercise 11.1, question **6**. Draw a similar graph for second class post.
Use a scale of 2 cm to 50 g on the horizontal axis and 1 cm to 10p on the vertical axis.

(b) Wasim posted second class letters weighing 45 g, 60 g, 170 g, 200 g and 240 g. What was the total cost?

4 Water is poured into these vases at a constant rate.
Sketch the graphs of depth of water (vertical) against time (horizontal).

(a)

(b)

5 The graph shows Kerry's Saturday morning shopping trip.

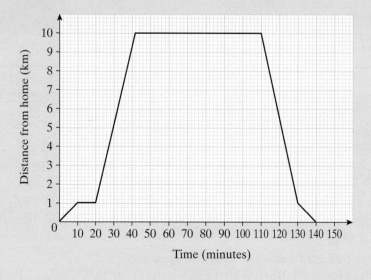

Mixed exercise 11 continues …

(a) What happened between 10 and 20 minutes after Kerry left home?

(b) How long did she spend at the shops?

(c) She caught a bus home. What was the speed of the bus?

(d) How far from Kerry's home is
    (i)   the bus stop?
    (ii)   the shopping centre?

6 Which of these functions are quadratic?
For each of the functions that is quadratic, state whether the graph is ∪-shaped or
∩-shaped.

(a) $y = x^2 + 3x$ (b) $y = x^3 + 5x^2 + 3$ (c) $y = 5 + 3x - x^2$

(d) $y = (x + 1)(x - 3)$ (e) $y = \dfrac{4}{x^2}$ (f) $y = x^2(x + 1)$

(g) $y = x(5 - 2x)$

7 (a) Copy and complete the table of values for $y = x^2 + 3x$.

x	−5	−4	−3	−2	−1	0	1	2
x^2	25			4				4
$3x$	−15			−6				6
$y = x^2 + 3x$	10			−2				10

(b) Plot the graph of $y = x^2 + 3x$.
Use a scale of 2 cm to 1 unit on the x-axis and 1 cm to 1 unit on the y-axis.

(c) Use your graph to
    (i)   find the minimum value of y.
    (ii)   solve $x^2 + 3x = 3$.

8 (a) Copy and complete the table of values for $y = (x + 3)(x - 2)$.

x	−4	−3	−2	−1	0	1	2	3
$(x + 3)$			1		3			6
$(x - 2)$			−4		−2			1
$y = (x + 3)(x - 2)$			−4		−6			6

(b) Plot the graph of $y = (x + 3)(x - 2)$.
Use a scale of 2 cm to 1 unit on the x-axis and 1 cm to 1 unit on the y-axis.

(c) Use your graph to
    (i)   find the minimum value of y.
    (ii)   solve $(x + 3)(x - 2) = -2$.

Mixed exercise 11 continues …

9 (a) Make a table of values for $y = x^2 - 2x - 1$. Choose values of x from -2 to 4.

(b) Plot the graph of $y = x^2 - 2x - 1$.
Use a scale of 2 cm to 1 unit on the x-axis and 1 cm to 1 unit on the y-axis.

(c) Use your graph to solve
(i) $x^2 - 2x - 1 = 0$.
(ii) $x^2 - 2x - 1 = 4$.

10 (a) Make a table of values for $y = 5x - x^2$. Choose values of x from -1 to 6.

(b) Plot the graph of $y = 5x - x^2$.
Use a scale of 2 cm to 1 unit on the x-axis and 1 cm to 1 unit on the y-axis.

(c) Use your graph to
(i) solve $5x - x^2 = 3$.
(ii) find the maximum value of y.

11 The diagram shows the outline of a house.

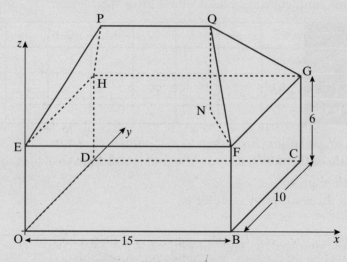

All the measurements are in metres. All the walls are vertical.
E, F, G, H and N are in the same horizontal plane.

(a) Using the axes on the diagram, write down the coordinates of these points.
(i) B   (ii) H   (iii) G

The coordinates of Q are $(11, 5, 9)$. N is vertically below Q.

(b) Write down the coordinates of N.

Percentages

▶ **This chapter is about**

- Repeated percentage change
- Solving problems

▶ **You should already know**

- How to find a percentage of a quantity
- How to find a percentage increase or decrease

Percentage increase and decrease

In Unit B Chapter 12, you learned two ways to find a percentage of a quantity. You can convert the percentage to a fraction or you can convert it to a decimal. It is usually easiest to use decimals when you are using your calculator.

Percentage increase

To find a percentage increase you find the increase and then add it to the original amount. The next example reminds you about this.

Example 12.1

Increase £240 by 23%.

Solution

Increase = £240 × 0.23   23% = 0.23 as a decimal.
       = £55.20

New amount = £240 + 55.20   Add the increase to the original
           = £295.20   amount.

There is a quicker way to do the same calculation.

To increase a quantity by 23% you need to find the original quantity plus 23%.

This means that to increase £240 by 23% you need to find

$$\begin{array}{r} 100\% \text{ of } £240 \\ + \quad 23\% \text{ of } £240 \\ \hline = 123\% \text{ of } £240 \end{array}$$

The decimal equivalent of 123% is 1.23.

The calculation can therefore be done in one stage

$$240 \times 1.23 = £295.20$$

The number that you multiply the original quantity by (here 1.23) is called the **multiplier**.

Example 12.2

Amir's salary is £17 000 per year.
He receives a 3% increase.
Find his new salary.

Solution

Amir's new salary is 103% of his original salary so the multiplier is 1.03.

£17 000 × 1.03 = £17 510

This method is much quicker when repeated calculations are needed, such as calculating compound interest.

When you invest money in a savings account you often receive compound interest. This is a percentage of the total amount in the account. It is different from simple interest, when interest is paid only on the original amount invested.

Example 12.3

Jane invests £1500 for the full 5 years.

What will her investment be worth at the end of the 5 years?

> Invest now and receive a guaranteed 6% compound interest over 5 years.

Solution

At the end of year 1 the investment will be worth $£1500 \times 1.06 = £1590.00$

At the end of year 2 the investment will be worth $£1590 \times 1.06 = £1685.40$
This is the same as $£1500 \times 1.06 \times 1.06 = £1685.40$
or $£1500 \times 1.06^2 = £1685.40$

At the end of year 3 the investment will be worth $£1685.40 \times 1.06 = £1786.524$
This is the same as $£1500 \times 1.06 \times 1.06 \times 1.06 = £1786.524$
or $£1500 \times 1.06^3 = £1786.524$

At the end of year 4 the investment will be worth $£1786.524 \times 1.06 = £1893.7154$
This is the same as $£1500 \times 1.06 \times 1.06 \times 1.06 \times 1.06 = £1893.7154$
or $£1500 \times 1.06^4 = £1893.7154$

At the end of year 5 the investment will be worth $£1893.7154 \times 1.06 = £2007.34$
This is the same as $£1500 \times 1.06 \times 1.06 \times 1.06 \times 1.06 \times 1.06 = £2007.34$
or $£1500 \times 1.06^5 = £2007.34$
(to the nearest penny)

Notice that at the end of year n you multiply £1500 by 1.06^n.

Use the powers key on your calculator.

$\boxed{1}\,\boxed{5}\,\boxed{0}\,\boxed{0}\,\boxed{\times}\,\boxed{1}\,\boxed{.}\,\boxed{0}\,\boxed{6}\,\boxed{\wedge}\,\boxed{5}\,\boxed{=}$

The result is £2007.34 when rounded, the same as above.

Percentage decrease

Percentage decrease can be done in a similar way.

Example 12.4

Kieran buys a DVD recorder in the sale.
The original price was £225.
Calculate the sale price.

> **SALE!**
> **15%** off everything

Solution

A percentage decrease of 15% is the same as $100\% - 15\% = 85\%$.
So the multiplier is 0.85.
$£225 \times 0.85 = £191.25$

Again, this method is very useful for repeated calculations.

Example 12.5

The value of a car decreases by 12% every year.
Zara's car cost £9000 when new.
Calculate its value 4 years later. Give your answer to the nearest pound.

Solution

100% − 12% = 88%

Value after 4 years = £9000 × 0.88⁴   At the end of year 4 you
$$= £5397.26$$ multiply £9000 by 0.88⁴.
$$= £5397 \text{ to the nearest pound}$$

> **TIP**
>
> Percentage decrease in monetary value is often called 'depreciation'.

Exercise 12.1

1 Write down the multiplier that will increase an amount by
   (a) 13%.   (b) 20%.   (c) 68%.   (d) 8%.
   (e) 2%.   (f) 17.5%.   (g) 100%.   (h) 150%.

2 Write down the multiplier that will decrease an amount by
   (a) 14%.   (b) 20%.   (c) 45%.   (d) 7%.
   (e) 3%.   (f) 23%.   (g) 86%.   (h) 16.5%.

3 Sanjay earns £4.60 per hour from his Saturday job.
   If he receives a 4% increase, how much will he earn?
   Give your answer to the nearest penny.

4 In a sale all items were reduced by 30%. Abi bought a pair of shoes.
   The original price was £42. What was the sale price?

5 Mark invested £2400 at 5% compound interest.
   What was the investment worth at the end of 4 years?
   Give your answer to the nearest pound.

6 This painting was worth £15 000 in 2004.
   The painting increased in value by 15% every year for 6 years.
   How much was it worth at the end of the 6 years?
   Give your answer to the nearest pound.

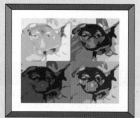

Exercise continues …

7 The value of a car decreased by 9% per year.
When it was new it was worth £14 000.
What was its value after 5 years?
Give your answer to the nearest pound.

8 House prices rose by 12% in 2003, 11% in 2004 and 7% in 2005.
At the start of 2003 the price of a house was £120 000.
What was the price at the end of 2005?
Give your answer to the nearest pound.

9 The value of an investment rose by 8% in 2007 and fell by 8% in 2008.
If the value of the investment was £3000 at the start of 2007, what was the value at the end of 2008?

Challenge 12.1

In 2009, the government introduced a car scrappage scheme.

When you trade in a car that is over 10 years old, you receive a £2000 reduction in the price of a new car.

At Selman motors you can either have the car scrappage amount or a reduction of 16%.

What is the original price of the car when the reduced price is the same in both cases?

Solving problems

When solving a problem, break it down into steps.

Read the question carefully and then ask yourself these questions.
• What am I asked to find?
• What information have I been given?
• What methods can I apply?

If you can't see how to find what you need straight away, ask yourself what you can find with the information you are given. Then, knowing that information, ask yourself what you can find next that is relevant.

Many of the complex problems that you meet in everyday life concern money. For example, people have to pay **income tax**. This is calculated as a percentage of what you earn.

Everyone is entitled to a personal allowance (income that is not taxed). For the tax year 2009–2010 this was £6475.

Income tax in excess of the personal allowance is known as taxable income and is taxed at different rates. For the tax year 2009–2010 the tax rates were as follows.

Tax bands	Taxable income (£)
Basic rate 20%	0–37 400
Higher rate 40%	Over 37 400

Example 12.6

In the tax year 2009–10 Stacey earned £28 500.

Calculate how much tax she had to pay.

Solution

First subtract the personal allowance from Stacey's total income to find her taxable income.

$$\text{Taxable Income} = £28\,500 - £6475$$
$$= £22\,025$$

Stacey's taxable income is less than £37 400 so she pays all her tax at the basic rate.

$$\text{Income tax at basic rate} = 20\% \text{ of } £22\,025$$
$$= 0.2 \times 22\,025$$
$$= £4405$$

Index numbers

The **Retail Price Index** (RPI) is used by the government to help keep track of how much certain basic items cost. It helps to show how much your money is worth year on year.

The system started in the 1940s and the base price was reset to 100 in January 1987. You can think of this base RPI number as being 100% of the price at the time.

In December 2009 the RPI for all items was 218. This tells you that there has been an increase of 118% in the price of these items since January 1987.

However, the RPI for all items excluding housing costs was 201. This tells you that there has been a smaller increase, of 101%, if housing costs are not included.

The RPI is often referred to in the media, when monthly figures are published: people need price increases to be kept small, otherwise they will in effect be poorer unless their income increases.

As well as the RPI, there are other index numbers in use, such as the **Average Earnings Index**. This has a base set to 100 in the year 2000, so the current index values show comparisons with earnings in the year 2000.

You can find more information about these and other index numbers on the government's statistics website, www.statistics.gov.uk.

> **TIP**
>
> This may seem complicated, but index numbers are really just percentages.

Example 12.7

In December 2008, the RPI for all items excluding housing costs was 192.4.

In December 2009, this same RPI was 201.

Calculate the percentage increase during that 12-month period.

Solution

Increase in RPI during the year $= 201 - 192.4$ First work out the
$= 8.6$ increase in the RPI.

Percentage increase $= \dfrac{\text{Increase}}{\text{Original}} \times 100$

$= \dfrac{8.6}{192.4} \times 100$

$= 4.47\%$ (to 2 decimal places)

Exercise 12.2

1 Pierre bought 680 g of cheese at £7.25 a kilogram.
   He also bought some peppers at 69p each.
   The total cost was £8.38.
   How many peppers did he buy?

2 Two families share the cost of a meal in the ratio $3:2$.
   They spend £38.40 on food and £13.80 on drinks.
   How much do the families each pay for the meal?

Exercise continues …

3 A recipe for four people uses 200 ml of milk.
Janna makes this recipe for six people. She uses milk from a full 1 litre carton.
How much milk is left after she has made the recipe?

4 At the beginning of a journey, the mileometer in Steve's car read 18 174.
At the end of the journey it read 18 309. His journey took 2 hours 30 minutes.
Calculate his average speed.

6 Mr Brown's electricity bill showed that he had used 2316 units of electricity at 7.3p per unit.
He also pays a standing charge of £12.95. VAT on the total bill was at the rate of 5%.
Calculate the total bill including VAT.

6 In June 2008 the Retail Price Index (RPI) excluding housing was 193.2.
During the next 12 months it increased by 1.76%.
What was this RPI in June 2009?

7 In 2003 the Average Earning Index (AEI) was 111.9. In 2008 it was 136.1.
Calculate the percentage increase in earnings over this five-year period.

What you have learned

- To increase a quantity by a percentage you can use a multiplier; for example to increase by 12%, you multiply by 1.12; to increase by 7.5% you multiply by 1.075
- To decrease a quantity by a percentage you can use a multiplier; for example to decrease by 15%, you multiply by 0.85; to decrease by 8% you multiply by 0.92
- To perform repeated percentage change you use the multiplier to the power n where n is the number of time periods; for example to find the amount of money in a bank account after it has been invested at 6% compound interest for 5 years you multiply by 1.06^5
- To solve a problem, break it down into steps
- Consider what you have been asked to find, what information you have been given and what methods you can apply
- If you can't immediately see how to find what you need, consider what you can find with the information and then look at the problem again

Mixed exercise 12

1 Birkdale Coaches claim that over 85% of the seats in their coaches are fitted with safety belts. If this is true of a 53-seater coach, what is the smallest number of seats that must have seat belts fitted

Mixed exercise 12 continues …

2 Last year Vicky paid tax at 20% on £37 400 and 30% on £5000.
What was the total amount she paid in tax?

3 Bertram Cars offers to sell a car listed at £12 000 at a 7.5% discount.
Vulcan Cars offer to reduce the price of the same car by £850.
Which garage offers the better deal?
How much cheaper is that garage?

4 Sam invested £3500 at 6% compound interest.
What was the investment worth at the end of 7 years?
Give your answer to the nearest pound.

5 In a sale, prices were reduced by 10% every day.
A pair of jeans originally cost £45.
Nicola bought a pair of jeans on the fourth day of the sale.
How much did she pay for them?
Give your answer to the nearest penny.

6 The insurance premium for Della's car was £360.
The firm reduced it by 12% for each year she had no claim.
What was the cost after six years with no claims?
Give the answer to the nearest pound.

7 Mordovia has high inflation.
In 2009 it was 15% a month for the first six months and 12.5% for the next six months.
A car cost 78 000 scuds (their unit of currency) in January 2009.
How much did it cost

(a) after six months?

(b) in January 2010?

Give your answers to the nearest whole number.

8 Ken bought 400 g of meat at £6.95 a kilogram.
He also bought some melons at £1.40 each. He paid £6.98.
How many melons did he buy?

9 Mrs Singh's electricity bill showed that she had used 1054 units of electricity at 7.5p per unit.
She also had to pay a standing charge of £13.25.
VAT on the total bill was at the rate of 5%.
Calculate the total bill including VAT.

10 In 2006 the Average Earnings Index (AEI) was 126.4.
During the next 12 months it increased by 4.03%.
What was this AEI in 2007?